T0077257

THE WALTZ OF REASON

Also by Karl Sigmund

Exact Thinking in Demented Times: The Vienna Circle and the Epic Quest for the Foundations of Science

Games of Life: Explorations in Ecology, Evolution and Behavior

The Calculus of Selfishness

THE WALTZ OF REASON

The Entanglement of Mathematics and Philosophy

KARL SIGMUND

BASIC BOOKS
New York

Cover design by Ann Kirchner

Cover image © CTK / Alamy Stock Photo

Basic Books
Hachette Book Group
1290 Avenue of the Americas, New York, NY 10104
www.basicbooks.com

Printed in the United States of America

First Edition: December 2023

Published by Basic Books, an imprint of Hachette Book Group, Inc. The Basic Books name and logo is a registered trademark of the Hachette Book Group.

The Hachette Speakers Bureau provides a wide range of authors for speaking events. To find out more, go to hachettespeakersbureau.com or email HachetteSpeakers@hbgusa.com.

Basic books may be purchased in bulk for business, educational, or promotional use. For more information, please contact your local bookseller or the Hachette Book Group Special Markets Department at special.markets@hbgusa.com.

The publisher is not responsible for websites (or their content) that are not owned by the publisher.

Library of Congress Cataloging-in-Publication Data has been applied for.

ISBNs: 9781541602694 (hardcover), 9781541602700 (ebook)

LSC-C

Printing 1, 2023

Contents

Introduction

Mathematics and philosophy have dated for ages. This book describes in a not-too-serious vein some of their most memorable encounters.

The two fields seem to be made for each other. On the one hand, mathematics is a useful tool for both theoretical and practical philosophy. For instance, the theory of knowledge deals with topics such as space and chance that are at the core of geometry and probability theory; ethics deals with notions such as fairness and the social contract, on which game theory sheds a light; and so on. On the other hand, mathematics itself is among the most puzzling and rewarding sources of philosophical problems. It is obviously not an empirical science, so why is it so practical? Is it invented or discovered? And what makes mathematical knowledge seem so secure?

Such questions will lead to entertaining, informative, and occasionally rambling excursions into the history of human thought, involving some remarkably original characters, all of them celebrities from the Dead Thinkers' Society. Philosophical and mathematical ways of thinking are utterly distinct, but have often progressed on twin tracks. There were times when philosophers and mathematicians could not be told apart. Those times are receding, but the fact remains that the two fields have wonderful ways of stimulating and often surprising each other. It seems that the two siblings from Greece are fated to remain eternally entangled, in a complex and occasionally dizzying waltz—and not without, sometimes, stepping on each other's toes. Let us follow their evolutions.

There are many illustrations and very few formulas in the book. It is meant as a sightseer's guide and no more. The text will not say too much on the philosophy of mathematics itself, an arduous field, but it will make up for this by describing applications of mathematics to all kinds of philosophical questions, from morals to logic, and emphasizing the historical quirks of an age-old quest. That quest is guaranteed to remain open-ended, especially as the current explosion of artificial intelligence is likely to sweep some card-houses of reason from the table, and deal us a new hand.

The first part of the book has to do with space, number, algorithm, axiom, and proof. It traces the evolution of the self-image of mathematics, a short coming-of-age version of the long official history that leads from Euclid to Turing, or more precisely, from Thales to Hales (no pun intended!). Thales was the semi-legendary figure from Miletus, on the shore of the Ionian Sea, who may (or may not) have first conceived, some hundred generations ago, the idea of a mathematical proof conveying insight and certainty. Thomas Hales is the contemporary US mathematician who has become famous for a proof that was so complex that nobody could be completely sure of its validity. Thereupon, Hales removed all doubts by convincing a computer.

The development of mathematics between these two milestones has involved drastic changes in perspective, which greatly exercised the minds of mathematicians and philosophers alike. On the menu were the role of spatial intuition; the fate of the parallel axiom; the divorce of mathematical and physical space; the nature and purpose of number; infinity, surrounded by its halo of taboos and scandals; and the incestuous relations of mathematics with logic. All these issues underwent centuries of development, punctuated by major revolutions, and all led to strange encounters (almost encounters of the third kind) between mathematicians and philosophers.

The second part of this book deals with chance and the continuum. The latter provided thinkers such as Zeno of Elea with riddles to confuse the best minds and filled ancient mathematics with unease. It was only in the aftermath of the Renaissance, during the heyday of the alchemists, that some intellectual adventurers began to develop an infinitesimal calculus. They aimed to reach for the limit by a daredevil "fast forward"—named "passage to the limit"—and to divide the finite into infinitely many infinitely small parts. At almost the same time, probability was tamed. This was the age when mathematicians, with the confidence of sleepwalkers, went beyond common sense. Nobody understood properly how chance fits with causality, nor how an infinitesimal could be smaller than anything and yet not zero. All that seemed to matter was that the stuff worked. In due time, it emerged that the calculus of chances and the computation of volumes could be handled by the same analytical tools. Mathematicians got used to defying reason. They also began to vex philosophers.

The third part of the book turns to practical philosophy: morality and economics, politics and law. Plato had once proposed that ideal rulers should start out by doing mathematics for ten years. Mercifully, this suggestion fell flat. But 2000 years later, some mathematicians did indeed turn to reflect on what was good and desirable. This started harmlessly enough, with the investigation of voting schemes, at a time when democracy was the pipe dream of radicals and the only republic was the "republic of scholars." A little later, the Benthamite notion of a "felicity calculus" attracted ridicule, but that of "utility" took hold of economics. Eventually, utility turned on itself and cast a shadow on our optimistic self-image as "rational beings."

Toward the middle of last century, under the innocuous heading of "theory of games," mathematics began to investigate conflicts of interest. Such conflicts are the *raison d'être* for all morals and laws. Today, it seems hard to understand how philosophers could ever have cogitated on selfishness, cooperation, or the social contract without resorting to the Prisoner's Dilemma or the Stag Hunt game. Similarly, the notion of fairness or the evolution of ownership norms are by now fully established as mathematical issues (which does not necessarily mean that we understand them any better).

The fourth and last part of this book tries to approach mathematics from the outside, as if landing on an unknown shore and fraternizing with the native tribe. The first chapter takes a look at the language of mathematics, or more precisely, at its writing. This quasi-graphological approach reveals that the tribe is currently undergoing a change, one that seems to accelerate at a dizzying pace. This upheaval is caused by digitalization, of course (a word that refers to the fingers used for counting). The computer, that mathematical brainchild, is radically transforming mathematics in more ways than can be reckoned yet. The next chapter is a tip of the hat to the philosophy of mathematics, a venerable discipline that seems, today, to deal almost as much with itself as with mathematics. And the last chapter turns to what may appear, to many eyes, to be the greatest riddle: namely, why does mathematics provide so much delight for some (but only some) of us?

As for myself, I have loved mathematics since as far as I can think back. I well remember the cozy evening at home when little me carefully measured the angles of a triangle, added them up, and discovered that my dad had been right! While growing up in Vienna, I soon crossed traces of Ludwig Wittgenstein, Kurt Gödel, and the Vienna Circle, and I could not help wondering about their disparate views. There were other formative experiences. A large part of my professional life (maybe the best) was spent teaching mathematics to undergraduates and watching how, within a few months, they acquired a specific mindset, as budding mathematicians. It felt like witnessing an initiation rite. My scientific research dealt first with dynamical systems on the borderline of deterministic and probabilistic models. Later, I turned to evolutionary game theory. The former field offered food for thought on theoretical philosophy, the latter on practical philosophy. Despite this heady diet, I am an expert on neither, alas! In fact, I have not touched one topic, in this book, on which there are not many better experts. My excuse is that I have merely intended to cover a vast field lightly, in a series of leisurely strolls, sometimes returning to the same spot from another side, sometimes relaxing to enjoy the view.

This being said, I must confess that I have occasionally experienced, while writing this book, a sort of *The Old Man and the Sea* feeling: namely, that I had hooked a fish who is far, far too big for me and drags me and my skiff, hell knows where. . . .

I can only take it philosophically.

PART I

1

Geometry

Memories of a Nameless

The Art of Unforgetting

The opening scene is set in Athens, in the villa of a shady politico named Anytus. Young Meno, an up-and-coming military leader, is on a visit. Socrates also happens to be around. Meno is quick to seize the occasion and asks him whether virtue can be taught—a bait that never failed to hook Socrates. The gambit secured Meno's passport to eternity. He died soon after, in the Persian Wars, under dubious circumstances, but a Platonic dialogue was named after him.

Figure 1.1. Socrates (469–399 BCE).

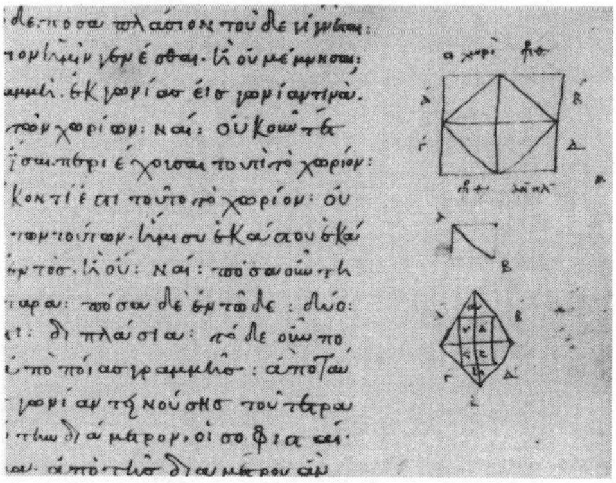

Figure 1.2. A page of *Meno*.

Meno is one of the earliest works of Plato. This is where, for the first time, the philosopher presents one of his favorite ideas: There is a knowledge that we learn by remembering. Our immortal soul has known it all along. We only need to dig it up.

The Greeks have a word for this recovery of buried knowledge: anamnesis. The notion may strike us as hopelessly outdated, a leftover from an age of superstition. However, it led to one of the most venerable, and indeed thrilling, encounters between philosophy and mathematics. This came about when Socrates proposed to defend his curious idea by performing, for Meno's benefit, an experiment on total recall.

One of the slaves hanging around is asked to approach, a boy whose utter lack of education is beyond any doubt. By skillful questioning, Socrates leads him to the discovery of a geometric theorem that the boy had certainly never been told before. Hence, concludes Socrates, he must have known it all along. The boy had merely been unaware of it. With gentle probing, his submerged knowledge has come to light. In modern parlance, part of the subconscious had become conscious, like in a session on the couch of Dr. Freud. Socrates himself likened his role to that of a midwife—he merely had helped the boy *unforget* what he had forgotten.

The whole episode took merely a quarter of an hour. With that, the boy returned to his lowly sphere of ignorance. He neither learned what all the questioning had been about, nor that it had brought him his fifteen minutes of fame—or rather (since he had never been asked for his name), fifteen minutes of immortality. Socrates and Meno returned to their discussion of what virtue is about.

Tellingly, Socrates had chosen geometry for his experiment, not any other science such as physics or geography. With his well-honed dialectical skills, he could probably have led the slave boy to also remember that Crete is an island or that everything is made from water, fire, air, and earth. Socrates preferred to focus on a geometric theorem, because nothing can pass more plausibly for an eternal truth.

The boy had been asked to construct, for a given square, a square of twice its area. We know, from school, that the length of its side must be $\sqrt{2}$ times the length of the side of the original square. It must therefore be as long as its diagonal. But talking of square roots was out of bounds. Actually, Socrates did not even mention a square. He spoke of a quadrilateral whose sides have equal length (Figure 1.3). This is not enough to define a square; but Socrates added—no doubt using a figure such as Figure 1.4—that its diagonals are of equal length. This guarantees that the quadrilateral with equal sides is indeed a square.

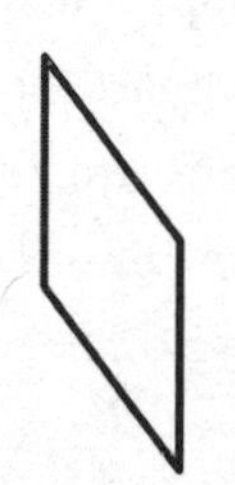

Figure 1.3. A rhombus is a quadrilateral whose sides have equal length.

Figure 1.4. A square is a rhombus whose diagonals have equal length.

Socrates could have demanded instead that all angles are equal. This condition would also guarantee that the quadrilateral with equal sides is a square. But he preferred to introduce diagonals right at the start of the Q and A session, in a casual way—a neat trick, as these diagonals will eventually turn out to yield the solution (Figure 1.5). Toward this end, Socrates allowed the boy to follow his own way, gently correcting his mistakes. Double the side length? No, this would not do; it leads to a square having *four* times the area of the original square. Multiply the side length by one and a half? No, still too large. And so the dialogue winds on, until, in the end, Socrates has coached the correct answer from the boy.

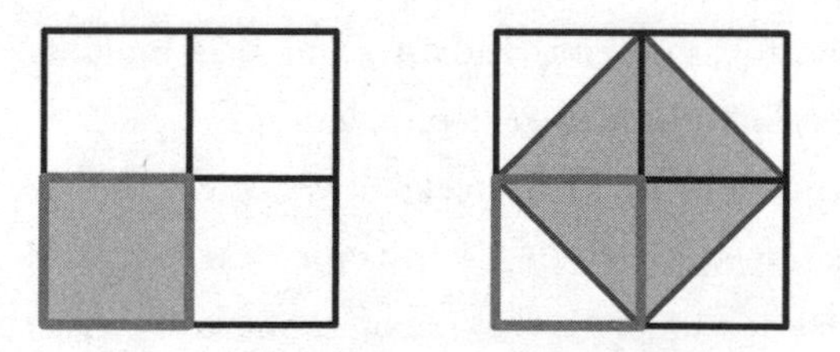

Figure 1.5. The gray square on the right has twice the area of the gray square on the left.

We shall see in the next chapter that the corresponding problem in three dimensions is unsolvable. It is the Delian problem, which owes its name to the tiny island of Delos in the Aegean Sea. Some thirty years before Socrates had his chat with Meno, a plague had raged through Greece. Pericles had died from it. As usual in a pandemic, experts knew exactly what to do. They said, “Go ask an oracle.” Lo and indeed, the oracle offered advice: double the size of the cube-shaped altar that stands in the temple of Apollo, and you will thereby appease the gods. Today, the plague is gone, the altar nowhere to be found. Apollo has quit. But the Delian problem remains unsolved. In fact, we know—we have *proof*—that it will remain unsolved for all time. How wise of Socrates not to have asked the boy to recall the solution.

Three Angles for Euclid

Geometry was the first branch of mathematics to really flourish. It may have done so because of its obvious use to architects, sailors, and field surveyors. More likely, it flourished because it is beautiful. Even the simplest geometric figures, such as the triangle, are fascinating. In music, the triangle is a marginal instrument hidden somewhere in the back of the orchestra. In mathematics, triangles shine in the front row—the very first objects to fascinate early Greek thinkers, such as Thales of Miletus or Pythagoras of Samos. Triangles are also the first mathematical figures likely to raise a child’s interest.

One of the oldest geometric theorems is that of Pythagoras (Figure 1.6). If a, b, and c are the lengths of the sides of a right triangle (c being that of the hypotenuse, the side opposite the right angle), then $a^2 + b^2 = c^2$. This fact had been known to Egyptians, Indians, and Babylonians, but Pythagoras was (possibly? probably?) the first to offer a proof.

What is a proof? For the old Greeks, it was an argument to make everyone see why the statement is true—just as the slave boy suddenly came to see why the diagonal of a square is the side of a square of twice the size.

There exist many proofs of the Pythagorean theorem. The most common proof is based on Figure 1.7 (which in the case $a = b$ is more or less straight from *Meno*). The large square, with side length $a + b$, is divided into five pieces: the four right triangles and the square formed by their hypotenuses. Each triangle has area $\frac{ab}{2}$, which altogether yields $2ab$. We obtain $c^2 = (a + b)^2 - 2ab$ by simply removing the right triangles from the large square.

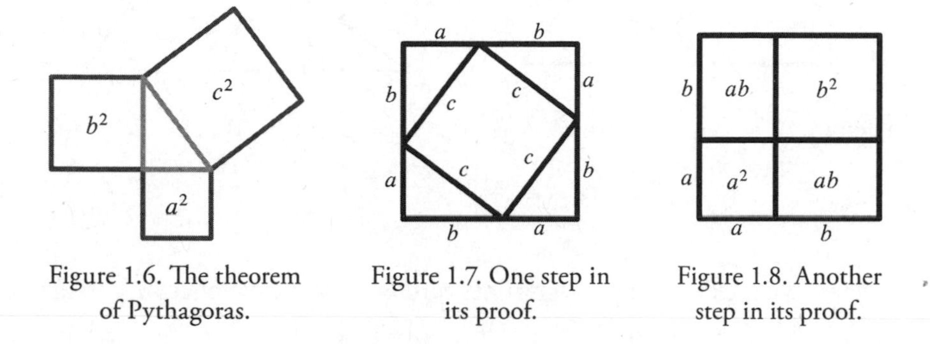

Figure 1.6. The theorem of Pythagoras.

Figure 1.7. One step in its proof.

Figure 1.8. Another step in its proof.

Another decomposition of the same "large" square with sides of length $a + b$ (see Figure 1.8) yields the equation $(a + b)^2 = a^2 + 2ab + b^2$. Substituting this into the previous equation, we obtain $c^2 = a^2 + b^2$, as had to be proved.

At school we learn a few things about triangles (less than our elders did, by the way). For example, the perpendicular bisectors of the three sides of a triangle intersect in one point. This is immediately obvious from Figure 1.9. Indeed, let P be the point where the perpendicular bisectors of the sides AB and AC intersect. Being on the perpendicular bisector of AB, P is equidistant from A and B. Similarly, P is equidistant from A and C. Hence, P is equidistant from B and C, and thus on the perpendicular bisector of BC. We thereby also see that P is the center of the circle through A, B, and C.

Here is a very similar theorem: the three altitudes of a triangle intersect in one point (see Figure 1.10). But now the proof is a bit more demanding. Can you find it yourself? Do you remember having seen it in school (or in some earlier life, as Plato might expect)?

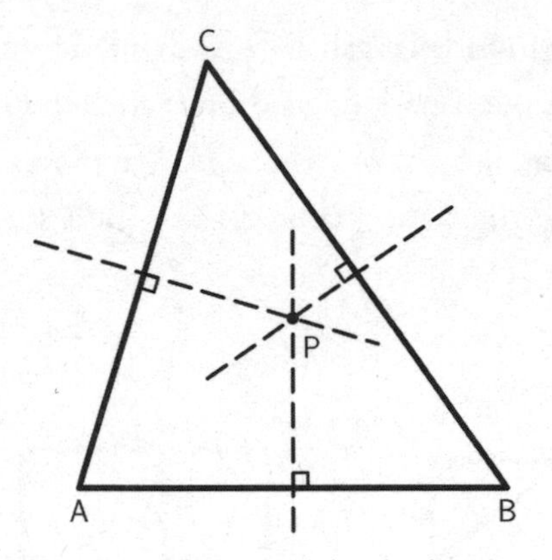

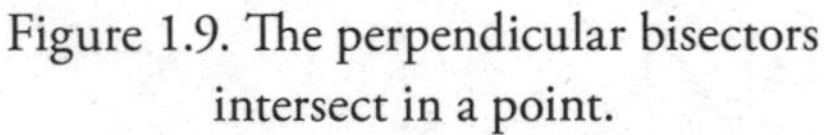

Figure 1.9. The perpendicular bisectors intersect in a point.

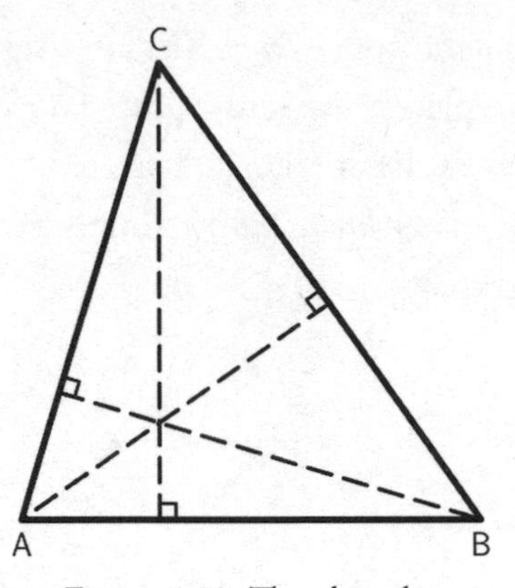

Figure 1.10. The altitudes intersect in a point.

Here is a trick that helps. Draw lines through A, B, and C that are parallel to the three opposite sides of the triangle (see Figure 1.11). This yields a large triangle. It is easy to see, using parallelograms, that the points A, B, and C are the midpoints of the sides of the new triangle (Figure 1.12). The altitudes of triangle ABC are perpendicular to these sides. Hence, they are the perpendicular bisectors of the new triangle. As such, they intersect in one point, as was to be proved.

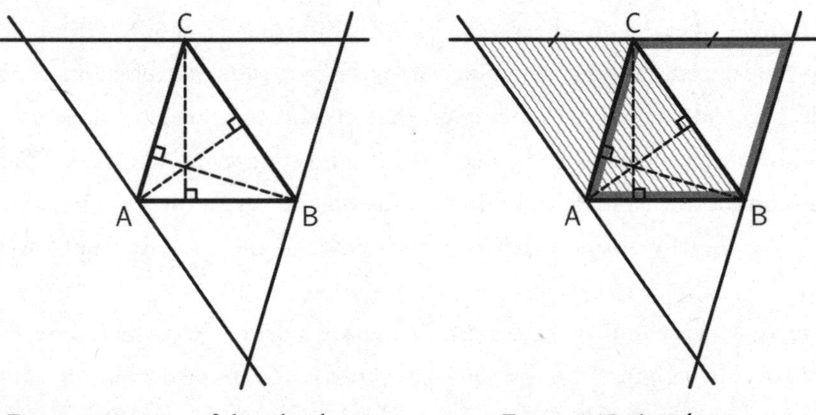

Figure 1.11. A proof that the three altitudes intersect in a point.

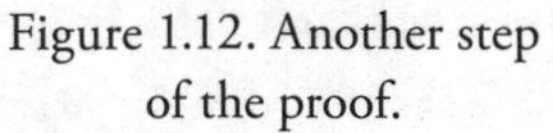

Figure 1.12. Another step of the proof.

Many theorems about triangles are 2000 years old, but many more are much newer. In fact, the Greek geometers would tear their hair out if they were told how much they had missed.

For example, the triangle whose vertices A′, B′, and C′ are the feet of the three altitudes of an acute triangle ABC (i.e., the so-called pedal triangle A′B′C′, see Figure 1.13) has the smallest perimeter of all triangles inscribed in ABC (i.e., having their three vertices on the three sides opposite A, B, and C). An elastic ribbon stretched across the three sides of triangle ABC would, by contracting, end up in A′B′C′. The old Greeks had no elastic ribbons, but the theorem would have thrilled them. It was discovered by Leonhard Euler in the eighteenth century.

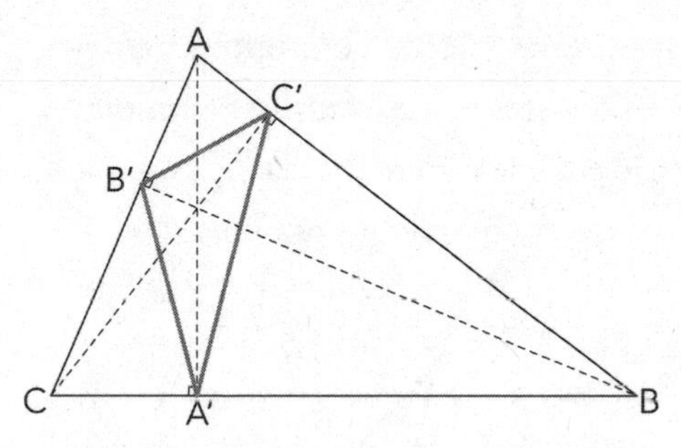

Figure 1.13. A theorem of Euler.

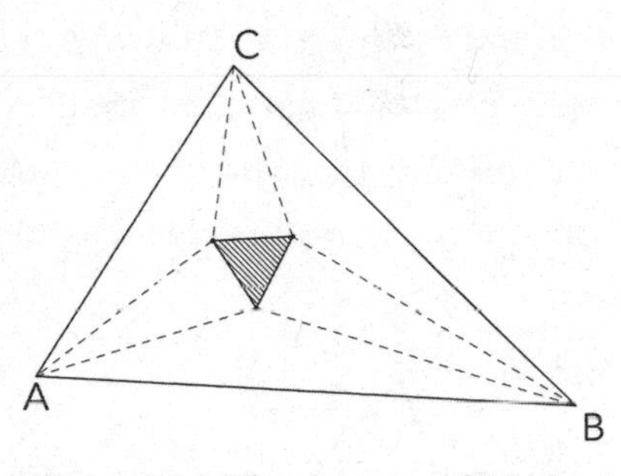

Figure 1.14. The theorem of Morley.

Here is another example: It is well known that the bisectors of the angles of a triangle meet in a point. Let us, instead, look at the trisectors (the lines dividing angles into three equal parts). It turns out that, for any triangle, the three points of intersection of adjacent trisectors of different angles form an equilateral triangle (Figure 1.14). It is an almost magical result. The proof is tricky, to say the least. In the words of H. S. M. Coxeter, the foremost geometer of the last century, "Much trouble is experienced if we try a direct approach." Yet ten years ago, unfazed by this warning, the British mathematician John H. Conway found a wonderfully clever, one-page proof.

Flawed Evidence

Proofs in Euclidean geometry can be extremely sophisticated, full of ingenious and surprising moves. But each step seems immediately evident. This appeal to evidence was what convinced Plato and Pythagoras that the truth of geometric arguments could not be questioned. To *see* the square over the diagonal is to *understand* that its area is twice that of the original square. To *see* that the attitudes of one triangle are the perpendicular bisectors of another is to *understand* that they intersect in a point.

In many languages, "I see" means "I understand." So does the expression "This is clear to me." The Latin root of the word *evidence* is *videre*, meaning "to see."

However, to see is to believe, they say, and to believe means to believe that you know. This can easily lead to fatal mistakes. Sense perceptions may be sense deceptions, as said Descartes. The old Greeks were familiar with optical illusions. They also knew of geometric fallacies. Here is a classic example: the "proof" that all triangles ABC are isosceles (an obvious nonsense).

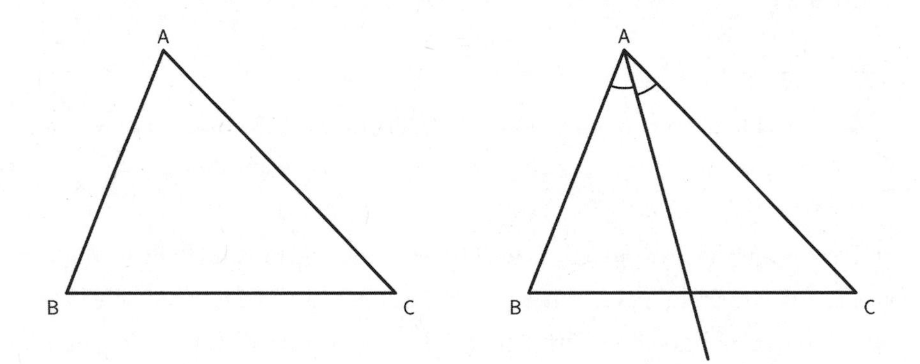

Figure 1.15. A triangle (clearly non-isosceles) and the bisector of angle A.

Indeed, let us consider the bisector of angle A (Figure 1.15). If it is perpendicular to the side BC, then the triangle is obviously isosceles. In this case, we are done. Thus, suppose that it is *not* perpendicular to BC. Then, it is not parallel to the perpendicular bisector of BC. Hence, it intersects that line in a point P (see Figure 1.16).

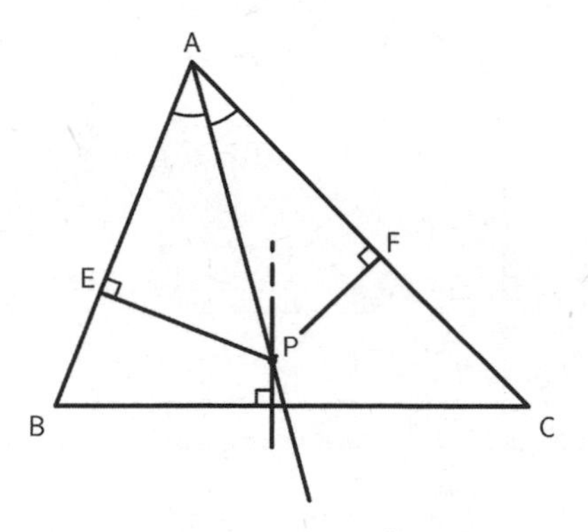

Figure 1.16. One step to fallacy.

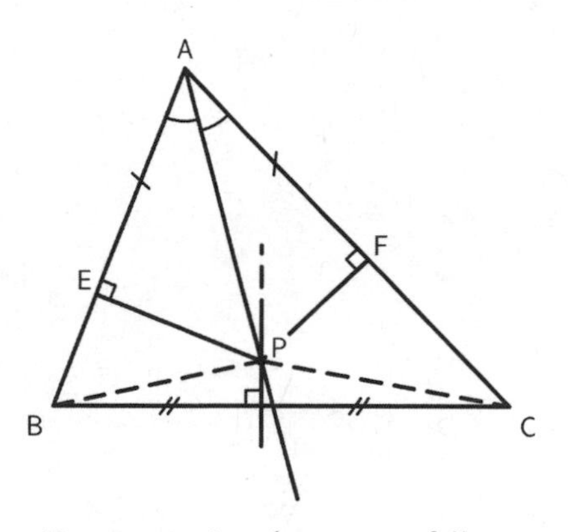

Figure 1.17. Another step to fallacy.

Let us draw from P the perpendiculars to the sides AB and AC of the triangle, and denote their feet by E and F, respectively. The two right triangles APE and APF have equal angles and a common side. Hence, they must be congruent. In particular, AE and AF have equal length. The two right triangles BEP and CFP are also congruent (see Figure 1.17). Indeed, PE and PF are of equal length because P is on the bisector of angle A; and PB and PC are of equal length because P is on the perpendicular bisector of BC. But if AE and AF have the same length, and EB and FC have the same length, then we only have to add equals to equals to obtain that AB and AC are of the same length. This means that the triangle ABC is isosceles.

Here is another, more modern geometric fallacy: an example that by decomposing and rearranging polygons, we can reduce their area (see Figure 1.18). This is preposterous, of course; but what we seem to "see," right in front of our eyes, is that a triangle is dissected into four parts that, when rearranged, look just as before, except for a tooth gap—a small indentation on the bottom side. This must be an illusion. It cannot be! We look again: the four pieces are two triangular pieces and two *polyominos*—polygons made up of squares of the same size (one made of seven, the other of eight). They are merely shifted around, and do not change their area. Yet in the rearrangement, one square is lost. What kind of legerdemain is at work?

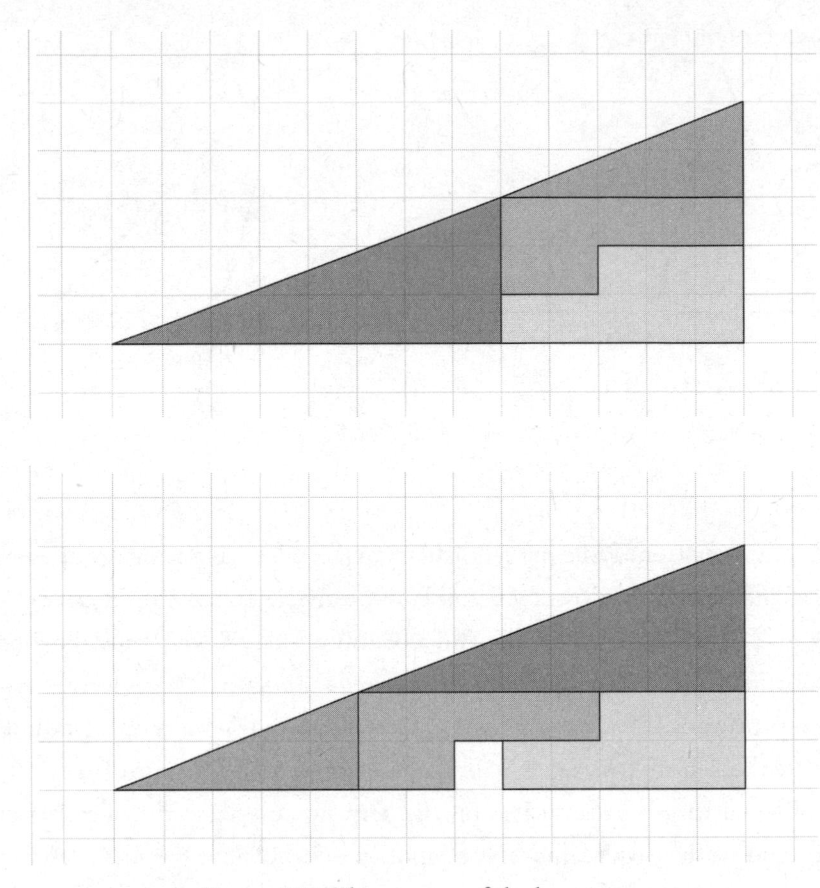

Figure 1.18. The mystery of the lost square.

In both examples the outcome is so obviously wrong that we start immediately to look for some mistake in the argument. It will not take long to discover the fallacies. (And they are pointed out in the references.) But what if a geometric theorem does *not* immediately strike us as wrong? What if *no* mental alarm bells start ringing?

As a matter of fact, wrong proofs abound in mathematics. Usually, their life span is short. Colleagues of the mathematician who erred will very willingly point out the mistake. But what if no colleague is ever interested in the result? How can one make sure that the theorem is valid? Evidence can mislead.

Euclid Writes a Bestseller

To avoid the pitfalls of intuition, Euclid decided to use it as little as possible. His idea was to start with a few statements placed up front, in plain sight; to accept them "on evidence" as given; and then *never* to recur to evidence again. Everything else should be deduced from the little that is given. That deduction must follow from strictly logical arguments, without any further appeal to intuition.

Not much is known about Euclid (except that he should never be confused with another Euclid, who hailed from Megara). Our Euclid apparently lived around 300 BCE in Alexandria. He may have frequented Plato's academy in his youth. Or maybe not. The one thing that really counts is that he wrote a book, *The Elements*, which summed up the mathematical knowledge of his time in masterful clarity. With this book, mathematics ceased being a ragtag bundle of results and became an organized whole. If there ever was a game changer in mathematics, it was Euclid.

Euclid's book set the standard for the next few thousand years. It reigned uncontested. For centuries, only the Bible sold more copies. More important than its success as a best- and long-seller was that it shaped mathematics by laying down the ground rules. A theory has to start with primitive concepts that remain undefined and primitive statements (the axioms) that remain unproved. All other concepts have to be defined in terms of the primitive concepts, and all other statements proved from the primitive statements. No theorem can hold without a proof. The proofs have to be chains of logical deductions without recourse to intuitive evidence. Any appeal to visualization was considered off-limits. In the nineteenth century, a German geometer insisted on holding his classes in a completely darkened lecture room. Arguing "from figures" was taboo.

It turned out that Euclid did not completely fulfill his own standards. Again and again, he unwittingly used intuition. This cannot surprise anyone. Human error is hard to avoid.

To keep things in perspective, we should bear in mind that the old Greeks were wrong on almost any count in anything they said. They believed that the Sun and planets wheeled around the Earth; they attributed health to the balance of blood, bile, and phlegm; they vastly overestimated the number of gods; and so on. When seen against this background, the achievements of Greek geometers are simply stupendous, and their mistakes no more than tiny imperfections.

The primitive concepts of Euclidean geometry were point, line (meaning a straight line), and distance. Here are the axioms in slightly updated form:

1. Any two points lie on a line.
2. Any line can be extended.
3. There exists a circle with any center and any radius.
4. All right angles are equal to one another.
5. Through any point not on a given line, there is a unique parallel to that line.

A right angle is an angle equal to its supplement, and the circles introduce the notion of distance. The last axiom is a story in itself, and will be dealt with in the next section.

During the nineteenth century, it became obvious that Euclid's theorems were all correct, but some of their proofs incomplete. Occasionally, arguments of Euclid used "evidence" that was not warranted by the axioms: for instance, he assumed tacitly that if A, B, and C are three points on a line, with B between A and C, then C is not between A and B; or that a line intersecting a side of a triangle has to intersect another side, too. Unconsciously, or at least unwittingly, the spatial reasoning used in real life made things look "obvious" that had not been properly proved. Such glitches occurred from the first theorem onward.

In 1899, the famous mathematician David Hilbert wrote a little book on *The Foundations of Geometry*, where he used twenty-three axioms to derive Euclidean geometry in an indisputable way. (Later, this number was slightly reduced.)

The notion of "betweenness," that blind spot of Greek geometers, was given a starring role on Hilbert's list of primitives. In newer versions of geometry, the notion even replaced "line," which joined the rank of concepts that could be defined. Indeed, we can start with any two points A and B, and then define a *segment* as the set of all points between A and B; next, we define two *rays*, one as the set of all points C such that B is between A and C, and the other as the set of all points C such that A is between C and B. The line *g* through A and B is then defined as the set consisting of the two points A and B, the two rays, and the segment. Next, we need (as an axiom!) that there is a point that is not on that line. We thus obtain the plane through that point and the line *g*. Next, we define angles, and triangles.

Now we are well-equipped to show, for instance, that if any points A, B, . . . , X are *not* all on one line, then there exists a line containing *only two* of these points. This is Sylvester's theorem—another gem the Greeks had overlooked.

Hilbert's book did much more than merely repair a few holes in Euclid's *Elements*. It got rid of "evidence" and "intuition" altogether, evincing them out of their last retreat, the primitive concepts and axioms. Hilbert did not even *try* to give any meaning to the primitive concepts, besides that they obey the axioms.

Euclid's statement "A point is that which has no part" finds no place in Hilbert's book. The question "What is a point?" is as meaningless for modern geometry as "What is a rook?" is for chess. All that chess players, including chess computers, need to know is how a "rook" can move, which pieces it can capture, and so on. A rook may look like a little tower and be made of ebony, but this is irrelevant. It is the same with geometric concepts. What mathematicians *think* when they think of a point is their private affair. It may help a child to be shown a grain of sand or a dot on a blackboard, or better still, a star in the sky, to get the point (if I may say so when I mean to visualize "a point"). But the logic of a proof must proceed without any visualization—as in the geometry lectures of a German professor that were held in a pitch-dark auditorium.

Figure 1.19. David Hilbert (1862–1943) refined the rules of the game.

Parallel Actions

The fifth axiom of Euclid, the parallel axiom, has a special status. Euclid actually used a different statement, which is equivalent. Geometers immediately sensed that this fifth axiom was less evident than the others.

Two lines are said to be parallel if they do not have a unique point in common (which means that the lines either coincide or have no point in common). The parallel axiom may seem evident, but only at first sight. If we are standing on a straight railway track, we see that the rails seem to intersect in a point on the horizon. We know that they don't, but cannot advance any "evidence": We don't see that they don't meet.

The old Greeks had no railways, but a fine sense for geometry, and knew that the parallel axiom is far from obvious. Indeed, lines extend to infinity (or, to speak more carefully, they have no end). So, how can one say that two lines will never, ever intersect? We cannot survey them entirely. Who can say what happens beyond our visual radius, beyond the limits of perception?

For thousands of years, the consensus was that the parallel axiom is true, no person of sound mind can seriously doubt it, yet it is not so evidently evident as one would like an axiom to be. Compared to the other axioms, it falls short. This is why generation after generation of geometers tried to get rid, not of that statement as such, but of its status as an axiom. They attempted to downgrade the axiom to the rating of a theorem. For this purpose, they

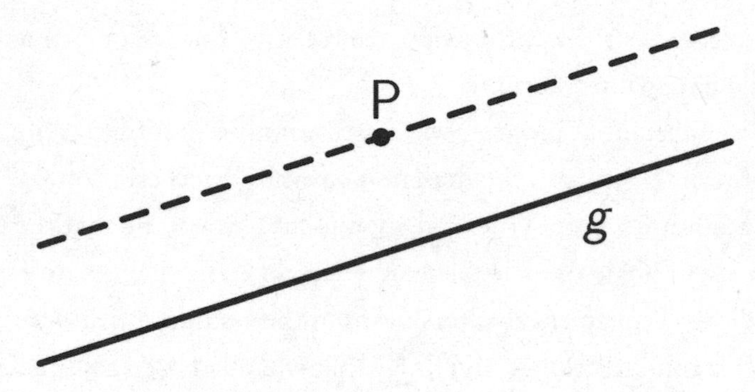

Figure 1.20. The parallel axiom: through any point P not on the line g, there is a unique parallel to g.

Figure 1.21. A parallel illusion.

had to derive it as a logical consequence of the other four axioms, which seemed altogether much more evident.

However, that enterprise never succeeded. It was balked from the start. Many of the most powerful theorems could not be used, because they themselves had been derived by using the obnoxious fifth axiom. These include classics such as "The sum of the angles of a triangle is 180 degrees (i.e., equal to the sum of two right angles)," "There exist triangles of arbitrarily large area," "Three points either lie on a line or on a circle," "The theorem of Pythagoras holds," and so on. It turned out that these theorems are actually equivalent to the parallel axiom. In its place, geometers could use any of these statements as their "fifth axiom" and derive Euclidean geometry. But that would have gained nothing.

The first substantial progress was based on a resounding failure. In the early eighteenth century, the Italian Giovanni Saccheri (1667–1733) attempted an indirect proof. The strategy was simple: assume that the parallel axiom does not hold, and keep deducing logical consequences until you hit an absurdity—a statement that follows from the assumption and yet contradicts it. This contradiction would imply that what had been assumed is false. The parallel axiom therefore would have been demonstrated to be true.

Saccheri drew a line *g* and, from two points on *g*, two perpendicular segments of equal length, both on the same side of the line (Figure 1.22). He then joined the two endpoints of those segments with a line *h*. This yields a quadrilateral. It is easy to show that the two segments intersect *h* at the same angle. If Saccheri could show that it is a right angle, the parallel axiom would follow easily. He thus had to show that the other two possibilities—acute or obtuse angles—both lead to contradictions. This he did, to his own satisfaction. He thought that he had proved that the fifth axiom is a consequence of the other four axioms, and hence a theorem. No need to presume it as given if it can be derived.

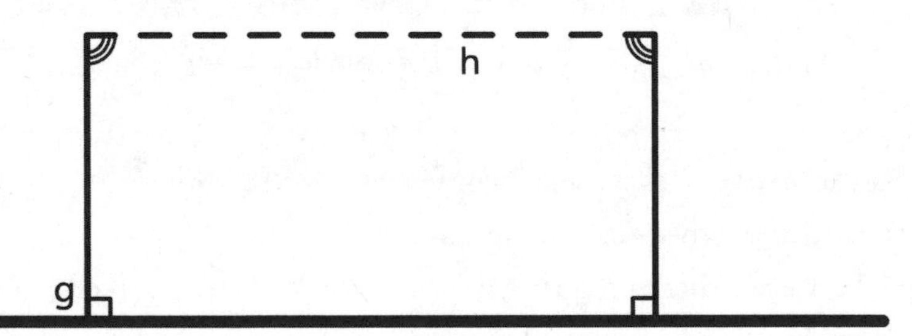

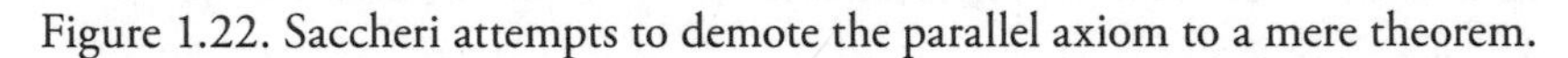

Figure 1.22. Saccheri attempts to demote the parallel axiom to a mere theorem.

To Saccheri's dismay, his colleagues were not convinced. They accepted his argument that "both angles are obtuse" leads to a contradiction, but found fault with his treatment of "both angles are acute." Saccheri tried to repair his proof—a very long proof—but to no avail. Other mathematicians took over, always looking for the roadblock into which the assumption "both angles are acute" would eventually crash.

Some hundred years after Saccheri, two young mathematicians, working independently, began to suspect that there was no roadblock. The Russian Nikolai Ivanovich Lobachevsky and the Hungarian János Bolyai understood that far from leading to an impasse, Saccheri's assumption leads into a new world. Things were different in this world—very different. Through any point not on a line, there are infinitely many parallels to that line. The sum of the angles of a triangle is smaller than 180 degrees—by how much depends on the triangle's area. And so on. This so-called hyperbolic geometry was not one jot less interesting than Euclidean geometry.

The father of Bolyai happened to be a mathematician, too. He had studied in Göttingen and been friends with young Carl Friedrich Gauss. Now, he wrote him a glowing letter about his son's splendid discoveries. In return, he received a letter in which Gauss declared that he could not well praise the achievement of the youth because he, Gauss, had discovered this new geometry himself, many years ago. He had described it in letters to close friends. Gauss had not published his results because he wanted to avoid what nowadays is called a shitstorm. Gauss actually used a more refined expression: "I fear the outcry of the Boeotians." (The Boeotians were a Greek tribe known far and near for being dimwits.)

As a result, the work of Gauss on non-Euclidean geometry was published only after his death—and it was only then that the mathematical world at large took any interest in what, in their academic backwoods, Lobachevsky and Bolyai had published decades ago.

The strange reticence of Gauss was probably caused by the extraordinary influence, in Germany, of philosophers such as Kant, Hegel, and Schopenhauer. These philosophers held that Euclidean space was a necessity of thought. To ever doubt the parallel axiom was, so to speak, against the line—here in the sense of "party line."

To be sure, Immanuel Kant was not your ordinary dim Boeotian. Few philosophers can have thought harder about intuition than he did. He saw in space and time forms of intuition given a priori, before every experience.

Our perceptive apparatus uses these forms—cannot help using them—to arrange our sensations. Space is the background for perception. Space is inescapable, in that sense: we may be able to imagine that space is empty, but not that there should be no space. And it seemed evident to Kant that space comes equipped with Euclid's geometry. Hence, this geometry is a priori, too—it is independent of our experience.

However, it is not analytic, which means that it is not deducible from logic alone. Though "An equilateral triangle is a triangle" is analytic alright, it hardly can count as geometry. You need not imagine a triangle to see that the proposition is true. By contrast, "The bisectors of the angles of a triangle meet in a point" is geometry. Yet, it is not an analytic proposition. It is synthetic: it requires intuition.

To "see" that the assertion about bisectors is true, or just to see what it means, you need a figure. By drawing a triangle on a blackboard, you fall clearly far short of capturing what a triangle is, in geometry. The lines are not completely straight; they are way too thick; and there exist infinitely many other triangles that you will never see but still want to include in your proposition. These shortfalls do not matter. Your intuition will overlook such trifles. The drawing is just a prop for your imagination. Geometry, as someone said, is the science of correct reasoning about incorrect figures.

Kant's view of geometric intuition seemed convincing to his contemporaries. Geometry is neither analytic nor given by experience. It is the epitome of synthetic knowledge a priori. It seemed unconceivable to doubt that Euclid's axioms are given by the pure form of intuition.

With hindsight, it is strange that Kant seems never to have been concerned with the special role of the parallel axiom. He had little to say on that issue. He must have been perfectly aware that from Euclid on, mathematicians had felt queasy about it. Since Saccheri's heroic attempt to demonstrate the axiom, hardly a year passed without new treatises on the topic. To the geometers engaged on that quest, Kant's silence must have sounded deafening.

Georg Wilhelm Friedrich Hegel was less silent on the matter: he explicitly rejected all attempts to prove the fifth axiom. It is, he claimed, as necessary for geometry as the notion of space itself. Hence, it must be accepted as given. *Punctum.*

Arthur Schopenhauer, with his usual brashness, summed it up:

> Out of its womb, Euclid's method of demonstration has produced its own telling parody and caricature, to wit, the famous dispute about the theory of parallels.... But it so happens that this axiom is a synthetic judgement a priori, and as such guaranteed by pure, non-empirical intuition, which is as immediate and certain as the principle of contradiction itself.

And Schopenhauer added for good measure: "The only immediate use which is left for mathematics is that it can habituate unsettled and fickle brains to concentrate their mind."

So, this may have been what Gauss meant by "the outcry of the Boeotians." But he must have known that his unpublished manuscripts would come to light after his death, and be eagerly studied by the best minds. It would convince at least the non-Boeotians that he (and Bolyai and Lobachevsky) had discovered a brave new geometry as rich and exciting as that of Euclid.

Euclid Confirmed

The final nail in the coffin of all vain aspirations to prove the parallel axiom was hammered in through joint efforts of the Italian Eugenio Beltrami, the French Henri Poincaré, and the German Felix Klein. They proved that hyperbolic geometry is as consistent as Euclidean geometry. This they did by constructing, within the Euclidean world, toy models of the hyperbolic world. Any inconsistency within the toy world would amount to an inconsistency in the Euclidean world. Since the same method can be used vice versa—construct a Euclidean toy model in a hyperbolic world—the two geometries are equally valid, from the logical point of view: a contradiction cannot be ruled out, but if it occurs in one geometry, it occurs in the other.

The two geometries thus stand and fall together. If Saccheri's old dream could ever become real, it would be a nightmare. Had he succeeded in showing that the case of acute angles leads to a contradiction, he would have shown that the case of right angles, that is, Euclidean geometry, also leads to a contradiction. It would not prove Euclid's fifth axiom; rather, it would destroy Euclid's geometry.

Poincaré's model is arguably the most elegant. The Euclidean plane is mapped into a disc. We are familiar with flat maps of the spherical Earth. Such maps not only reduce distances, but distort them—for instance, by making Greenland look larger than South America.

Poincaré's disk is similarly distorted (Figure 1.24). The nearer to the boundary, the larger the distances. You will never be able to reach the boundary. It is as if the temperature there had reached absolute zero. All motion freezes in the vicinity. The distortion of distance means that the shortest path between two points in the interior of the disk is not a straight segment. Rather, it is an arc of a circle that intersects the disk's boundary in right angles.

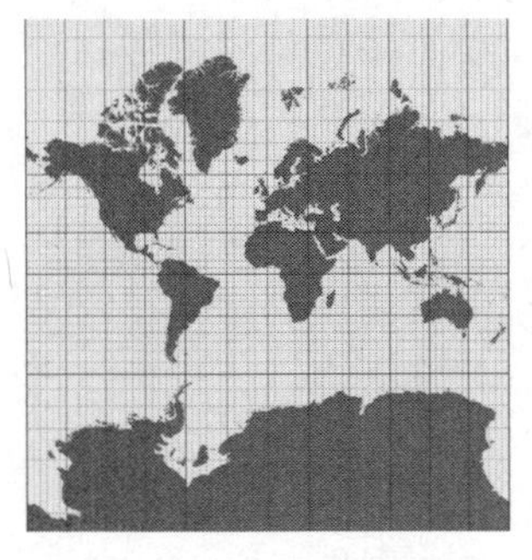

Figure 1.23. A map of Earth.

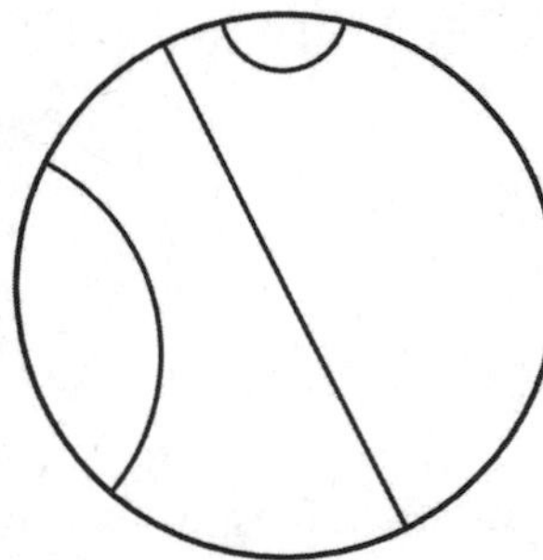

Figure 1.24. Poincaré's disk, with three "lines."

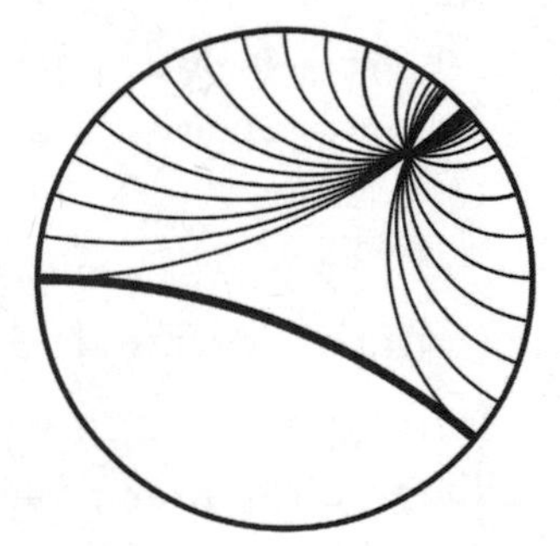

Figure 1.25. Many parallels to the thick line, which all share a common point.

The first four axioms of Euclid (or more precisely, their modern counterparts) still hold. Two "lines" have at most one point in common; through two points there is exactly one "line"; and such "lines" can be extended. The "distance" is a non-Euclidean distance, but we can still speak of "circles"—the points at equal "distance" from a given point, the "center." Oddly enough, whereas "lines" do not look like lines, on the Poincaré disk "circles" do look like circles (but their "centers" differ, in general, from the centers of Euclid's circles).

On Poincaré's disk, however, the fifth axiom of Euclid fails to hold. Through a given point *P* not on a "line" *g*, there are more than one "lines" that do not intersect *g*—infinitely many, actually (Figure 1.25). So here, on the disk, we encounter a model for hyperbolic geometry. Each theorem about the points and "lines" in hyperbolic geometry translates into a theorem about points and circles-intersecting-the-boundary-in-right-angles in the usual Euclidean plane. A contradiction here translates into a contradiction there. One geometry is as consistent as the other. We simply see them through different lenses.

Whether such to-and-fro would have mollified Schopenhauer is unlikely. He still would make fun of the benighted geometers who toy with the idea that the arc of a circle can pretend to be a line. Such an arc may be the "shortest" path between two points, if you insist on distorting distances. Insofar, such an arc resembles a line segment in Euclidean geometry. But who can be fooled by a mere resemblance? The essence of a line lies in its straightness, surely. This argument, however, overlooks the fact that any nano-beings living in the disk would have no other means to understand "straightness" than via the "shortest path."

Down to Earth

Once geometers became accustomed to stepping clear of intuition, and working with models rather than with the "given" world, geometry took on a new flavor. We will describe only one such model, which seems outrageous at first sight, yet brings us back to Earth—and literally so. Let us imagine in our "normal" three-dimensional Euclidean space one point F, whose role is just to focus our mind. (Remember Schopenhauer: a focus is exactly what fickle mathematicians are in need of.)

In this toy world, the lines through F (normal, straight, Euclidean lines) will be called pseudo-points, and the planes through F will be called pseudo-lines (see Figure 1.26). Through any two pseudo-points passes a pseudo-line. Any two pseudo-lines intersect in a pseudo-point. We just have to translate the "pseudo" to confirm these statements. Such hocus-pocus leads to a geometry (named *elliptic*) where there are no parallels at all: any two pseudo-lines intersect in a pseudo-point (because any two real, honest-to-God Euclidean planes through F intersect in a line through F).

Because the first four axioms of Euclid imply that parallels exist, some of his assumptions must be violated by our pseudo-lines and pseudo-points. It turns out that these assumptions are among those that Euclid took for granted without making them explicit: to wit, the axioms of "betweenness." It makes no sense to say that for any three pseudo-points on a pseudo-line, exactly one is between the others. (It makes as little sense as saying, back in our usual geometry, that for any three lines through F on a plane through F, exactly one line is between the two others. Indeed, each of the three lines is between the other two.)

However, by using in place of the axioms of "betweenness" some axioms of "separated-ness" of a similar ilk, one obtains a rich geometry with curious properties, such as "the sum of the angles of a triangle is always larger than 180 degrees." (These axioms of separated-ness describe arrangements, not of three but of *four* points on a line.)

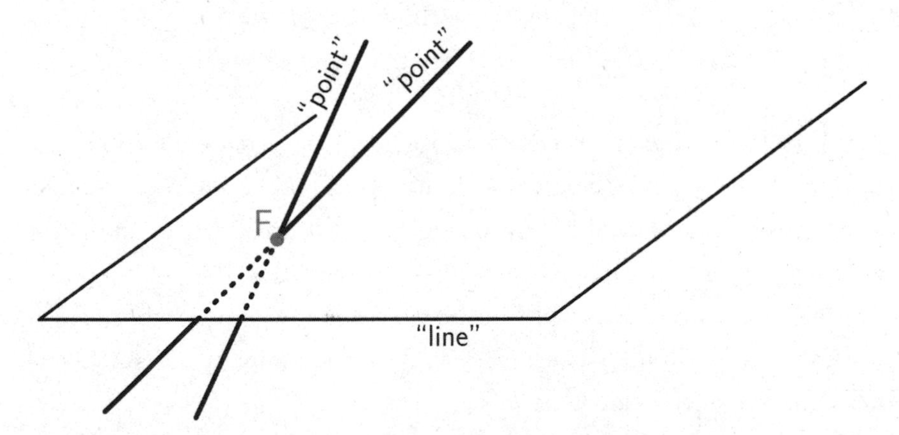

Figure 1.26. Elliptic geometry in one guise: a pseudo-point is a line through F; a pseudo-line is a plane through F.

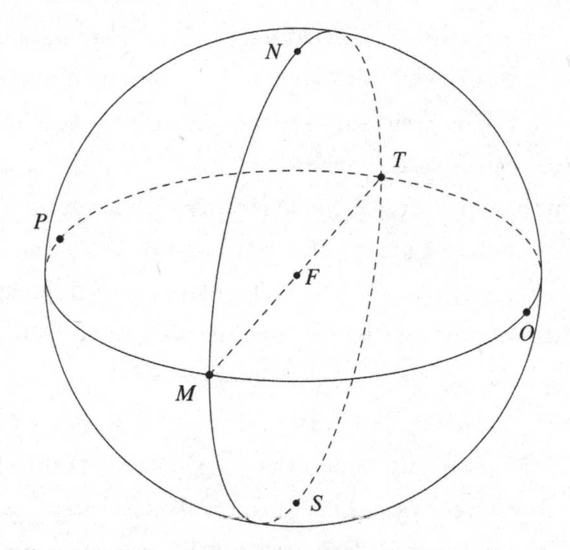

Figure 1.27. Elliptic geometry in another guise: if we consider the intersections of lines and planes with a sphere around F, a pseudo-point is a pair of antipodal points, and a pseudo-line a great circle.

Elliptic geometry becomes more intuitive, indeed downright familiar, if we imagine, in our usual space, a sphere with center F (Figure 1.27). A pseudo-line, being a plane through F, intersects the sphere in a great circle, and a pseudo-point, being a line through F, intersects the sphere in a pair of antipodal points. In fact, we are back on a globe, the good old globe of cartographers. Down to Earth, so to speak: the well-tried spherical geometry of navigators and mapmakers is nothing else than elliptic geometry. For centuries, mathematicians had it right under their nose, but bolted at the thought that it provided them with a model of a non-Euclidean geometry. What they lacked was not geometrical knowledge, but a mindset prepared to use words such as *point* or *line* without being encumbered by intuition.

In mathematics, names are mere conventions. This is hugely different from humanities, where everyone is aware, and quite rightly so, that each word gives rise to countless associations, conscious or not, which guide our thinking. Mathematicians view such associations with suspicion.

There are not just three geometries—Euclidean, hyperbolic, elliptic—but many more. Some are more useful or interesting than others. *Ordered geometry*, for instance, which is based on Euclid's first and second axioms and nothing else, is fairly rich, although it knows neither distance nor angle. Ordered geometry must seem strange to those accustomed to defining geometry as the science of measurement.

The first four axioms define *absolute geometry*, which subsumes both Euclidean and hyperbolic geometry. (The term is Bolyai's.) Axioms 1, 2, and 5 apply to *affine geometry*, and so on. Which geometry corresponds to our real space? Physicists are still working on that question. What seems pretty clear is that it is not Euclidean. So much for Schopenhauer.

Yet, relativistic cosmology is hardly a threat for Kant's a priori (despite claims to the contrary). The categories of thinking with which we are equipped do not need to coincide with the reality of outer space. In fact, it would be surprising if they did. Spatial intuition is more a psychological than a physical concept. Darwin is probably more relevant than Einstein to account for our "forms of intuition."

Evolutionary epistemology, which is some fifty years old, is but a footnote on Charles Darwin's splendid two-liner dating from 1838. We read in his *Notebook M*:

> Plato says that our "necessary ideas" arise from the pre-existence of the soul, not from experience

And after this gentle reminder of the curious notion of anamnesis, Darwin delivers his punch line:

> —for "pre-existence" read monkey.

The a priori of Kant may well be the a posteriori of evolution, honed by natural selection. Different species come equipped with different sense organs and different ways of organizing these sense data. Ants are guided by pheromones and live in a world of smells; migratory birds lay their course with the help of a magnetic sense; bats hear the echoes of their cries. What is their spatial intuition a priori? And what is their relation with "true" physical space? The only justification of their intuition is pragmatic: it helps them survive. Can we expect anything better from our own intuition?

Human spatial perceptions seem mostly conveyed by sight and touch (babies spend many months exploring how the two senses relate, fascinated with grabbing their toes). In a way, the two senses correspond to two distinct geometries. Sight is related to *projective geometry*, the geometry first studied by Renaissance painters in their attempts to understand perspective. In projective geometry, there are no parallels. Any two lines in a plane meet in one point, and similarly any two points belong to one line: there is a perfect duality between the concepts of point and line. The sense of touch, on the other hand, belongs to the geometry of rigid bodies and their motions, and hence to the absolute geometry corresponding to Euclid's first four axioms.

Oddly enough, tactile sensations seem privileged, as opposed to visual ones. When dipped into water, a pencil "appears" to be bent, while "in reality," as we used to say, it is straight. Why does the sense of touch convey reality, and the sense of sight mere appearance? Does it have to do with the fact that we descend from a long line of apes and monkeys who had indeed to coordinate eyes and hands in their breakneck capering through forest canopies? For such beings, the sense of touch must have always had the last word. To miss a solid hold would spell the end of the line—the genealogical line.

A Fly on the Ceiling

With his contempt of mathematics, Schopenhauer belongs to a minority among philosophers. Many of them have a strong bent for geometry. The most famous example is Plato, without doubt. "Let no one ignorant of mathematics enter here" was engraved at the entrance of his academy (or so the story goes). Plato was particularly fascinated by the finding that there are five regular solids (Figure 1.28): tetrahedron, hexahedron (the cube), octahedron, dodecahedron, and icosahedron. These are the so-called Platonic solids.

Figure 1.28. The Platonic solids.

Another philosopher with a good name in geometry is Blaise Pascal. At the tender age of sixteen, he discovered a most wonderful theorem on conics. Conics are circles, ellipses, or hyperbolas. (Whatever you see, at night, if you direct a flashlight on a wall, is bounded by a conic.) Suppose that you have six points on your conic. Join them by six consecutive segments, so that the lines close up and form a six-sided polygon. Then the intersections of the three pairs of opposite sides are aligned (Figure 1.30).

This is a striking claim. Its most striking aspect is that the very simplest example seems to contradict it already. Indeed, the simplest six-sided polygon is surely the regular hexagon. The simplest conic is surely the circle. If you inscribe a regular hexagon in a circle, then the pairs of opposite sides are parallel. The intersections do not lie on a line. They do not even exist because the opposite sides are parallel!

Pascal seems all wrong—but at this point the geometry teacher explains, possibly with a smug smile, that the three points of intersection *do* exist, after all. They merely are infinitely far away. Hence they *are* aligned—namely,

Figure 1.29. Blaise Pascal (1623–1662).

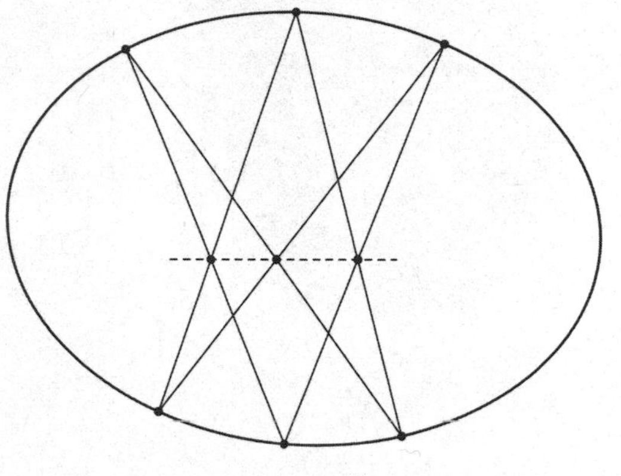

Figure 1.30. Pascal's theorem on conics.

on the line of all points at infinity. Why, don't you see? If not, then try and upgrade your intuition. (The first to conceive of points at infinity was astronomer Johannes Kepler. He was myopic, by the way. Can this have helped?)

Not every philosopher of the seventeenth century was a great geometer. But they all wished to be. Baruch Spinoza wrote his *Ethics* in the style of Euclid—*more geometrico*, as he termed it. Thomas Hobbes kept proposing methods for squaring the circle, all fallacious.

The most important contribution of a philosopher to geometry is doubtlessly due to René Descartes. He was reclining in his bed and cogitating (as was his wont), when he observed a fly crawling on the ceiling. He noticed that he could specify the position of the fly by two numbers—the distance of the fly to two edges of the ceiling. Thus Cartesian coordinates were born. We owe analytic geometry to a fly. How fortunate that Descartes was not myopic! (Spoilsports point out that Pierre Fermat had discovered coordinate systems independently.)

Figure 1.31. René Descartes (1596–1650).

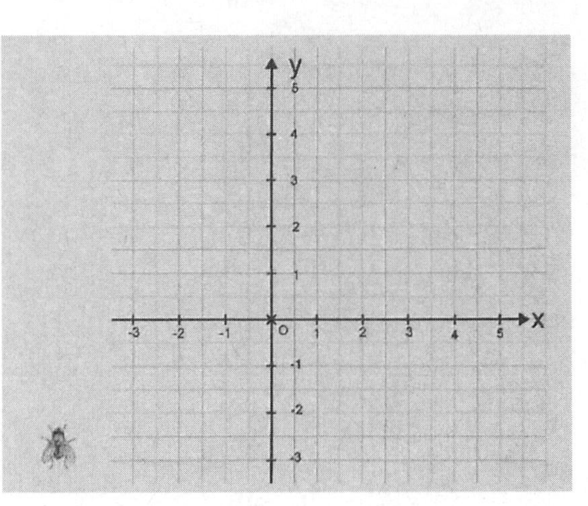

Figure 1.32. Cartesian coordinates.

Coordinate systems are as familiar to us as decimal numbers. How had it been possible to ignore them for so long? With the minimum of schooling that none of us can avoid, we think of two perpendicular axes whenever we think of a plane, and we associate to each point two numbers, its coordinates. Lines are solutions of linear equations. Conics are solutions of quadratic equations. Geometry is algebra. It is simple, and utterly magical.

The same works for three-dimensional space. Young Isaac Newton, while spending some time in his home office during the Great Plague, discovered his famous laws with nothing but an apple tree in front of his window and *The Geometry* of Descartes on his desk. This is where Newton found his absolute space, a space that "in its own nature, without regard to anything external, remains always similar and immovable," a vast receptacle for God's creation, totally empty (whereas Descartes had assumed that extension and substance always come together). This mysterious space beyond physics played a seminal role for Kant's metaphysics. Yet to ordinary intuition, it is a coordinate system with nothing in it, nothing at all, not even a fly.

2

Number

Dreaming up Numbers

Kant Contributes to the Wisdom of the World

One of Kant's lesser-known books is his *Attempt to Introduce the Concept of Negative Magnitudes into the Wisdom of the World*. It dates from 1763. His "wisdom of the world" (*Weltweisheit*, which is sometimes ineptly translated as "philosophy") was presumably the state of knowledge among educated Europeans at Kant's time. The booklet is slim. The subject seems meager. On 100 pages, give or take, Kant labors to convince his readers that negative integers (the numbers –1, –2, etc.) are not as absurd as they appear at first glance:

> Indeed the negative magnitudes are not negations of magnitude, as the similarity of expressions has led to suppose, but something in itself truly positive, except being opposed to something else.

This opposition, said Kant, is not logical opposition but one that is real:

> A magnitude is negative with respect to another insofar as it can only be taken as being contrary, in such a way that one cancels within the other as much as is equal to it.

Figure 2.1. Immanuel Kant (1724–1804) tackles negative quantities.

Kant gives some examples to explain this. A ship sails from Portugal to Brazil. On one day, it advances by 12 miles, but then loses 3 miles due to adverse winds, hence "minus three" miles must be added to the distance covered. Similar situations arise in trading: debt can be viewed as negative wealth. (This idea is not new: negative numbers, which first were introduced in India, were originally named "debts.")

So far, so good. But Kant goes on to comment:

> Negation, in so far as it is the consequence of a real opposition, I will call deprivation (*privatio*); but every negation in so far as it does not issue in this kind of repugnance, I will call a defect (*defectus*, *absentia*).

After this, he becomes harder to follow.

Today, it is impossible to understand why Kant got so worked up about negative numbers. We are familiar with them from childhood on. We learn that there exist other types of numbers, too. A college-level mathematical curriculum often starts with an official introduction to integer, rational, real, and complex numbers; it takes only a few weeks (for budding engineers, a few days). Students have normally met such numbers in high school already, and know that they are needed for all sorts of things. If there is any question left, it is why it took centuries of struggles, confusions, and doubts to come up with them.

Indeed, it seems so obvious:

Since there is no natural number x solving the equation $5 + x = 3$, we introduce -2 as a token for that missing object.

Since there is no integer number x solving $5 \times x = 3$, we introduce the rational number $\frac{3}{5}$ in its place.

Since there is no rational number x solving $x^2 = 2$, we introduce the irrational real number $\sqrt{2}$.

Since there is no real number x solving $x^2 = -1$, we introduce the complex number i.

After this, we seem to be done: who could ask for more? The new numbers (whether integer, rational, real, or complex) are conjured up to stand for solutions that we don't have at hand. *Et voilà!*

Needless to say, it is not that easy. Wishful thinking is not enough. If it were, we could also dream up a number x solving $1^x = 2$, or two numbers x and y solving simultaneously $x + y = 1$ and $2x + 2y = 5$. The wish principle fails on these counts. Some desires can be met, some others cannot.

As the Vienna Circle philosopher Friedrich Waismann wrote in his *Introduction to Mathematical Thinking*, "We certainly should not confuse wishful thinking with wish fulfilment." Old Sigmund Freud, who lived a few blocks away from Waismann's mathematics department, would have agreed.

Numbers, Plane and Simple

The new numbers—negative, irrational, whatever—must be anchored somewhere. This is where the notion of a *number line* comes in handy. It is a line with two points, marked 0 and 1, that provide a yardstick. Numbers appear as segments on that line. School kids, once they are habituated to handling rulers, usually have not much trouble in localizing numbers such as -3 or $\sqrt{2} = 1.41\ldots$ on this line, and deciding which is larger than which. They quickly grasp how to add them—it means a mere translation along the line. However, it took mathematicians many generations to understand that the product of two such numbers, each given as a point on the number line, is also somewhere on that number line. For a long time, that product was conceived as a surface area. This turned out to be the wrong kind of visualization. The right one relies on the similarity of triangles. It shows that the product of two numbers on the number line is obtained by stretching a segment and (possibly) flipping it over. The outcome of this operation is still on the line (Figure 2.2).

Two such number lines, at right angles to each other, help to visualize complex numbers. Again, this was not obvious from the start. Less than 400 years ago, an awestruck Gottfried Wilhelm Leibniz said: "Imaginary numbers are a fine and wonderful resource of the divine intellect, almost an *amphibium* between being and not being." Not at all. They are points on a plane. Multiplying two complex numbers merely means to stretch and rotate (Figure 2.3).

Visualizing numbers on a line or a plane can help, but does not offer a firm foundation for these numbers. After all, lines and planes are mere fictions, too.

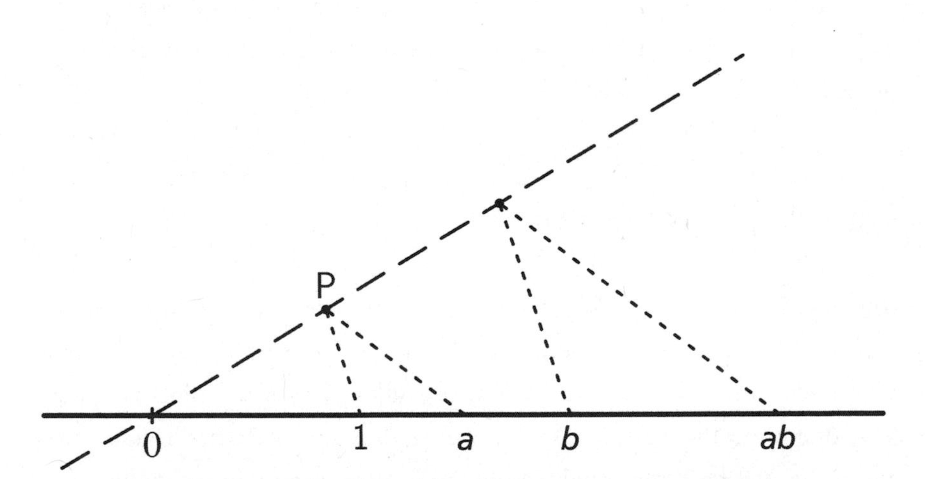

Figure 2.2. Multiplication on the real line is based on the similarity of triangles. The point P is chosen arbitrarily (not on the horizontal real line), then a triangle similar to 1Pa (with b instead of 1) is constructed, having parallel sides. It yields the point ab. (In our case, $a = \frac{3}{2}$ and $b = 2$.)

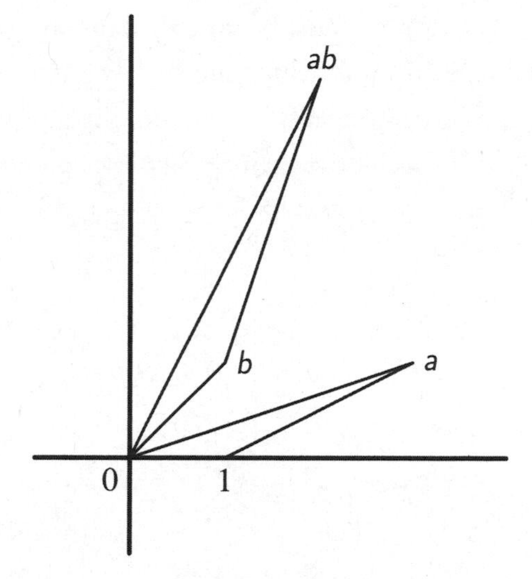

Figure 2.3. Multiplication in the complex plane also is based on the similarity of triangles. A complex number is a point in the plane. Any point different from the origin is defined by the angle and the length of the segment joining it to the origin. To multiply two (nonzero) complex numbers a and b, we add the angles and multiply the lengths. The triangles $01a$ and $0b(ab)$ are similar. (Here, $a = 3 + i$, $b = 1 + i$, and $ab = 2 + 4i$.)

Budding mathematicians are subjected to some rites of initiation, and one of them consists in being walked through the construction of the various kinds of numbers, starting out from the natural numbers 1, 2, 3,.... That long march is usually headed by a motto due to the mathematician Leopold Kronecker, who famously stated: "The integers are given by God, all other numbers are the work of humans." It is curious, however, that Kronecker sees the integers as God-given. Integers include –1, –2, and so on, which are negative magnitudes! A mere century before Kronecker's time, these numbers were *not* part of worldly wisdom (if one can trust Kant) and the object of great puzzlement.

Integers are introduced as pairs of natural numbers (a, b). These pairs show up as regularly spaced *lattice points* on the plane (Figure 2.4). What the construction intends to capture with such a pair is the difference $a - b$, in other words, the number x that solves $a = b + x$. The number -3 corresponds to the difference $1 - 4$ but also to $2 - 5$, $3 - 6$, and so on. Therefore, two such pairs (a, b) and (c, d) must be identified, or considered as equivalent, if $a - b$ is the same as $c - d$, which means if $a + d = b + c$. With this convention, each integer corresponds not just to one pair of natural numbers, but to a whole set of such pairs (namely all lattice points on a parallel to the 45-degree line).

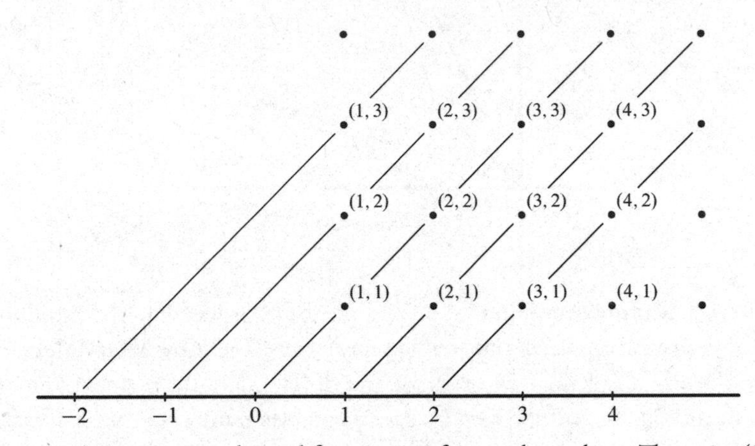

Figure 2.4. Integers are derived from pairs of natural numbers. The pairs (2, 3) and (1, 2) correspond to the same integer −1, since 2 − 3 = 1 − 2. In the same vein, all grid points lying on the same 45-degree line denote the same integer.

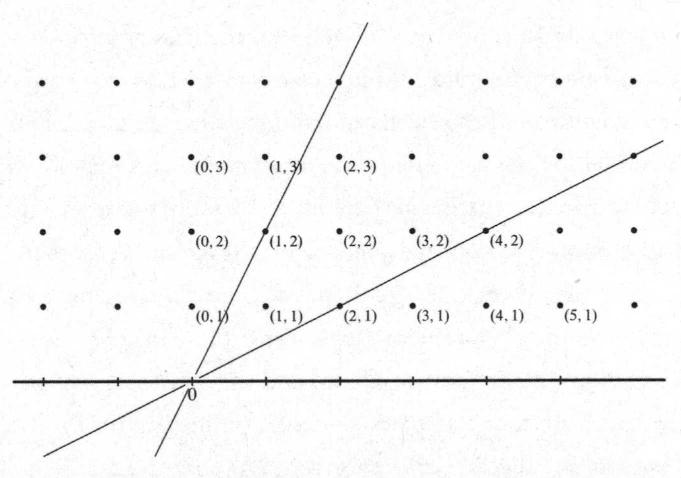

Figure 2.5. Fractions are also derived from pairs of natural numbers. The pairs (4, 2) and (2, 1) correspond to the same fraction, since $\frac{4}{2} = \frac{2}{1}$. All grid points on the same half-ray through 0 (except the horizontal) denote the same rational number.

In quite the same way, fractions $\frac{a}{b}$ correspond to pairs of integers (a, b), with $b \neq 0$ (Figure 2.5). They are intended to capture the solutions of $a = bx$. Again, this means that the pairs (a, b) and (c, d) denote the same fraction whenever $ad = bc$. Essentially, this is the same procedure as before. Once more, every rational number corresponds to a set of pairs (this time of integers). This set consists of the lattice points on lines through the origin (0, 0) (the horizontal line being excluded because division by 0 is out).

As to real numbers, they can be introduced by "cuts" separating the lattice points. For this, the rational numbers must be ordered by size. The details are both tedious and obvious. Each line through (0, 0) defines a half-ray on the upper half on the plane. One rational $\frac{a}{b}$ is smaller than another $\frac{c}{d}$ if a clock hand attached at (0, 0) and moving clockwise crosses the half-ray through (a, b) before it reaches the half-ray through (c, d).

At each moment of time, the hand of the clock separates the rational numbers into two parts, those on its left and those on its right. These two parts form what is known as a Dedekind cut. If a rational number is in the left part, all smaller rational numbers are there, too; if it is on the right side, all larger rational numbers are there, as well.

A real number, then, is defined as being such a cut—essentially, the cut is given by the position of this hypothetical clock hand. If a rational number happens to lie on the half-ray corresponding to the clock hand, it is identified with that cut, or in other words, with that real number. For instance, the set of all rational numbers left of $\frac{1}{2}$ *is* the real number $\frac{1}{2}$.

If no rational number is on the half-ray, the cut defines an irrational number. For instance, the set of all rational numbers that are negative or have a square that is smaller than 2 is such a cut, namely the real number $\sqrt{2}$. Rational and irrational numbers together are the real numbers—they fill the full number line.

Complex numbers, as we know, are defined as pairs of real numbers. This step is by far the easiest in the build-up. The complex number $a + ib$ is the point (a, b) with the real numbers a and b as coordinates.

At each stage, defining the new class of numbers is not enough. Rules for addition and multiplication have to be set up, and it must be shown that each number system can be embedded into the next one without leading into conflict. It is a time-consuming task, and few students are likely, after such a chore, to ever again ask "what is a number?"

Needless to say, mathematicians who deal with complex numbers are not permanently aware of the fact that they are dealing with pairs of real numbers, each of them being a pair of sets of rational numbers, which themselves are pairs of integers, which in turn are pairs of natural numbers. To start reflecting on it can easily lead to befuddlement, similar to figuring out, in one's head, the motions of tying a shoelace.

The working mathematicians just do what they are accustomed to doing. A philosopher, however, will try to dig deeper and wants to become aware of some of the bemusements experienced by children (or by readers of Kant) when faced with the odd set of customs taken for granted with numbers.

Much Ado About Less Than Nothing

Let us begin with the step leading from natural numbers to integers. Why should we accept the rule that "zero times zero equals zero"? The expression "two times three" can be used to tell you: take two times three apples. Accordingly, "take zero times three apples!" means that you take no times three apples. You take nothing. So far, so good: the rule $0 \times 3 = 0$ looks plausible enough. Yet the command "take zero times zero apples" should then mean to take *no* times *no* apple. No times no apples—wouldn't this imply that you are to take some apples after all?

And how can "minus three" be smaller than zero? Zero means Nothing, so how can anything be smaller than Nothing?

And why does minus times minus yield plus? We may have learned at school the helpful analogy: "The enemy of my enemy is my friend." But arithmetic is not based on Machiavelli, and hence the simile is somewhat extraneous. If we view the negative numbers as mirror images of the positive numbers, we can interpret multiplication with -1 as flipping around the point 0, and thus $(-1) \times (-1)$ as flipping twice, which leads back to the start and therefore yields 1. Explanations of this sort may serve to quiet the skepticism of children. Mathematicians, however, will tell you that the "true" reason for "minus times minus is plus" is that we want to keep the same rules as with natural numbers.

Remember that every integer can be viewed as the difference of two natural numbers. If a, b, c, and d are natural numbers with $b > a$ and $c > d$, then the rule

$$(b - a) \times (c - d) = b \times c + c \times d - (a \times c + b \times d)$$

is quite obvious, as can be seen from a simple sketch (Figure 2.6).

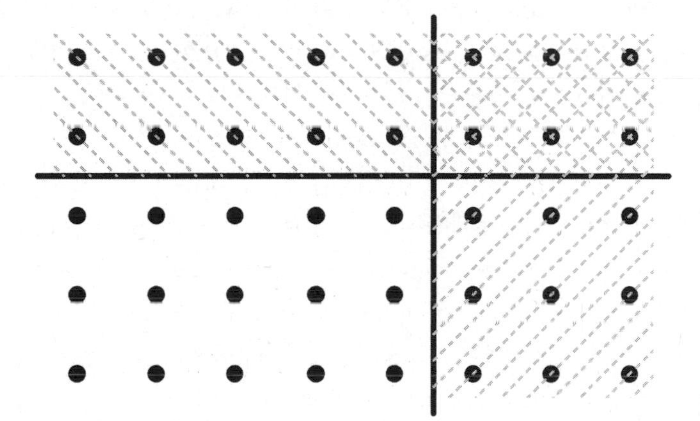

Figure 2.6. This figure shows that $(8 - 3) \times (5 - 2) = (8 \times 5) - (3 \times 5) - (2 \times 8) + (2 \times 3)$. We start with 5×8 (a rectangle of five times eight dots). To obtain $(5 - 2) \times 8$, we take off the two upper rows. To obtain $(5 - 2) \times (8 - 3)$, we delete the three rightmost columns. However, we then have eliminated the top-right rectangle twice, which means that we have to return the corresponding 2×3.

Since -1 is, by definition, $1 - 2$, the product $(-1) \times (-1)$ is nothing else than the product $(1 - 2) \times (1 - 2)$. If we wish to preserve the arithmetic rule above, for *all* natural numbers a, b, c, and d, no matter which are larger than which, then $(1 - 2) \times (1 - 2)$ must be equal to $1 + 4 - (2 + 2)$, which yields 1. Thus $(-1) \times (-1) = 1$. A similar reasoning underpins $0 \times 0 = 0$.

As Gauss said in a letter to his ex-student Friedrich Wilhelm Bessel (who had turned into an eminent astronomer):

> One should never forget that the functions, like all mathematical constructions, are only our own creations, and that when the definition with which one begins ceases to make sense, we should not ask "what is it?" but "what is it convenient to assume in order that it remain significant?" Take for instance the product of minus times minus.

Consider another example: a^2 is defined as $a \times a$ and more generally a^n as $a \times a \times \ldots \times a$, the product of n copies of the number a. This holds, in any case, if n is a natural number. But how should one define a^0? What can it mean to take the product of 0 copies of a? What becomes of the number a when it is multiplied *no* times with itself? This question could lead to an abyss of deep thinking. Instead, mathematicians remember the pretty formula $a^{m+n} = a^m a^n$, which obviously holds for natural numbers m and n, and look for what happens if $m = 0$. In that case the formula yields $a^n = a^{n+0} = a^n a^0$, and it becomes immediately obvious that $a^0 = 1$ "is convenient." One proceeds in a similar way to define a^n for negative integers n, or if n is a rational number. Rather than trying to fathom the true essence of "raising to the power n," one attempts to preserve the old familiar rules of computation. This sails under the flag of the permanence principle.

Figure 2.7. Carl Friedrich Gauss (1777–1855) looks for what is convenient.

Well before Immanuel Kant found it necessary to write his little booklet on negative magnitudes, Leonhard Euler had published his famous formula $e^{i\pi} = -1$. It deals with the ominous number $e = 2.71\ldots$, whose logarithm is 1. This number is raised to a power—but what a power! There would be no problem with e^2. The power $e^{3.14}$ is slightly harder to understand, but since 3.14 is a rational number, the permanence principle tells us what it should be. It requires some added continuity arguments to make sense of e^{π}, since π is irrational. In Euler's formula, however, something is claimed about $e^{i\pi}$, where i is the imaginary unit, once considered as a hocus-pocus residing "halfway between being and non-being." And what Euler found out about this $e^{i\pi}$, by means of some magical manipulations, was that this number is none other than our friend −1. How strange that after such a tour de force, the so-called "wisdom of the world" invoked by Immanuel Kant exercised itself, not on the left-hand side of the formula $e^{i\pi} = -1$, but on its right-hand side, the negative magnitude −1.

Nothing shows better the deep split that had opened up in the century before Kant's little book. Descartes, Leibniz, or Pascal were as much at ease in mathematics as in philosophy. By Kant's time, a philosopher could still be well versed in mathematics (this holds particularly for Kant, whose academic career started officially with a professorship for mathematics at the University of Königsberg). Nevertheless, some philosophical minds were apt to exclaim, "hold it, nothing can be smaller than nothing!"

"Numbers are used for counting: one, two, three apples. You will never see a fruit bowl with minus three apples. Numbers are used for measuring: one, two, three feet. But nothing in the universe measures minus three feet!" True enough, but in addition to counting and measuring, numbers also serve for computing. As the centuries flowed by, the rules for computing took precedence.

Drowned by Numbers

Fractions are easier to grasp than negative magnitudes: half an apple, two and a half loaves of bread, three quarts of wine. Rational numbers were used in every higher civilization, even if sometimes under restrictions that appear odd today. In old Egypt, for instance, only unit fractions such as $\frac{1}{n}$ were used, as well as sums of unit fractions—but with the injunction that no unit fraction is permitted to occur twice. Thus, $\frac{2}{3}$ is given not as $\frac{1}{3} + \frac{1}{3}$, but as $\frac{1}{2} + \frac{1}{6}$. In the net effect, all positive rational numbers can be obtained in this way, admittedly in a curiously roundabout way.

As long as one uses only the basic operations—addition, multiplication, subtraction, and division—one can compute with rational numbers much as one likes and always obtain rational numbers. Except for one caveat: it is forbidden to divide by 0. This is vexing, like any command restricting our freedom: don't eat the fruits from this one tree in paradise; don't open the door of this one room in Bluebeard's castle. Beginners often suggest that we should introduce one extra number—say, the number with the sign ∞—by agreeing that $0 \times \infty = 1$. Then, $\frac{1}{0} = \infty$ and division by 0 is feasible.

Unfortunately, this does not work. It clashes with a rule with which we are accustomed: the distributive law, which states that $(a + b) \times c = a \times c + b \times c$. Indeed, $1 = 0 \times \infty$ would imply that $1 = (0 + 0) \times \infty$, since $0 + 0 = 0$. By the distributive law, it would follow that $1 = 0 \times \infty + 0 \times \infty$ and thus $1 = 1 + 1$, a contradiction. One could, of course, dispense with the distributive law, at least for this one new number ∞, but it turns out to be too high a price to pay. "It is not convenient," as Gauss would say. Of course, this does not imply that there is no infinity. It only means that a number ∞ with the property $0 \times \infty = 1$ will not be admitted into the society of rational numbers. The club rules would suffer from such an admission.

Rational numbers have a strange property that integers do not have. Though they can be ordered by size, they know no "next largest" number. Between any two rational numbers, there are others—infinitely many others, in fact. Rational numbers are dense: between any two points on the number line, one finds rational numbers. They cover the number line, leaving out not the tiniest stretch. At first glance this makes them perfectly suited for measuring things. Alas, no: rational numbers cover the line, but they don't fill it.

For Pythagoras and his disciples, the first thinkers' society of which we know, the news that the rationals are incomplete came like a bolt out of the blue. Not every length is a rational number. The length of the diagonal of the unit square isn't, for instance. The unfortunate club member Hippasus, who had discovered (or merely disclosed to outsiders) that the diagonal of a square is not commensurable with its side, was thrown into the sea and held below until he stopped making waves. At least, this is what legend reports. Woe to all whistleblowers! But the truth could not be drowned.

The proof that $\sqrt{2}$ is irrational has been part of the canon for thousands of years, but must have seemed very odd, initially, being probably the first demonstration based on the indirect method, a method that has some resemblance with irony: it adopts a viewpoint and stresses it until it leads to absurd consequences. In the case at hand, one starts out by assuming that $\sqrt{2}$ is rational, draws some consequences, and arrives at a contradiction. Since contradictions cannot be, the number cannot be rational, as was to be proved. (More precisely, let us assume that $\sqrt{2} = \frac{a}{b}$, with natural numbers a and b. We can posit that at least one of the two numbers is odd: otherwise, we would just divide both a and b by 2 as often as needed. Squaring $a = \sqrt{2}\,b$, we obtain $a^2 = 2b^2$. Hence, a is even, that is, $a = 2c$ for some natural number c. This yields $(2c)^2 = 2(2c^2) = 2b^2$, and so b has to be even, too. Hence, a and b are both even, which we just had precluded: a contradiction.)

By the time of Plato, the irrationality of $\sqrt{2}$ was no longer a scandal. In the dialogue *Theaetetus*, the budding mathematics student with that name mentions almost in passing that he and his friends had just worked through the proof that the square roots of all numbers between 3 and 17 are irrational. (The fact that the square roots of 4, 9, and 16 are not irrational was left unmentioned, as was the irrationality of the square root of 2—all this was too obvious to point out.)

In his *Republic*, Plato reverts to the irrationality of $\sqrt{2}$ with strong words: "He is unworthy of the name of man, who is ignorant of the fact that the diagonal of a square is incommensurable with its side." By then, the shocking secret of Hippasus was common knowledge. "Trusty Helens, this is one of the things of which one can say: it is a shame if one does not know it, and if you know it, it gives you no particular merit."

Is it still possible, today, to feel the original sense of wonder about irrational numbers? The irrationality of $\sqrt{2}$ means that a half-ray from (0, 0) through $(\sqrt{2}, 1)$ never meets any lattice point (x, y) with x and y integers. There are infinitely many lattice points in the plane, and yet some rays miss them all. We may sail straight into infinity and never hit any of them.

There exist sequences of nested intervals, each containing infinitely many rational numbers, yet such that no rational number lies in all intervals (Figure 2.8). It is even stranger to consider sectors between two half-rays issued from the origin (Figure 2.9). Each such sector contains infinitely many lattice points. There exist nested sequences of such sectors (each containing all the following ones) that have in common a ray through infinitely many lattice points. Yet, there also exist nested sequences of such sectors that have the property that *no* lattice point belongs to all of them. This latter case looks like a vanishing trick: each sector contains infinitely many fractions, and yet no fraction is contained in all the sectors. You close a pincer, and find out that you grasp... nothing.

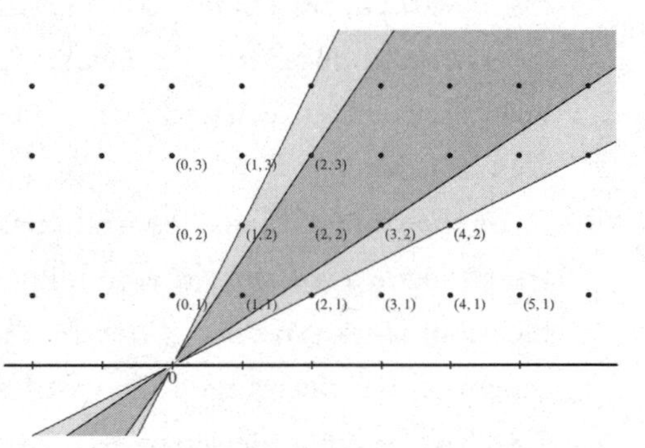

Figure 2.8. There exist sequences of nested intervals with rational endpoints having no rational point in common.

Figure 2.9. There are pincers of half-rays through lattice points that grasp no lattice point.

The rays through the origin of the plane "are" the reals. The rays through a lattice point "are" the rational numbers. The rays that dodge all lattice points "are" the irrational numbers. And rather than ponder what these "are" mean, as philosophers may be tempted to do, mathematicians start defining the rules of arithmetic (plus, times, lesser than, etc.) such that everything fits smoothly.

End of the line, it seems. Indeed, in a certain, mathematically precise sense, the reals are "complete." The number line holds no lacunas left to fill. But the word *complete* is misleading. The story is far from over.

A Hold on the Reals

All the artful constructions of real numbers cannot alter the two main facts we know from school: they are meant to correspond to the points on the line, and they are given by decimal expansions. A number such as $\sqrt{2}$ is 1.41421.... But how do we really get our hands on such numbers?

For the Greek geometers, the answer seemed clear: we can grasp a number if we can construct it, with ruler and compass. This is the ancients' way of getting hold of it. Indeed, it is easy to construct $\sqrt{2}$. Even philosophers can do it, since they just have to look it up in Plato.

More generally, it is easy to construct the square root of a, for any given length a (Figure 2.10). The first step is to construct, on a line, the segments AB and BC of length a and 1, respectively. The next step is to draw a circle with diameter AC. The third and last step is to draw the perpendicular to the line through the point B. It intersects the circle in a point D. All this can be done with ruler and compass. The triangle ADC is a right triangle. Let x be the length of DB. From the similarity of the right triangles ABD and DBC follows that $\frac{a}{x} = \frac{x}{1}$, so that $x^2 = a$. Thus, x is the square root of a.

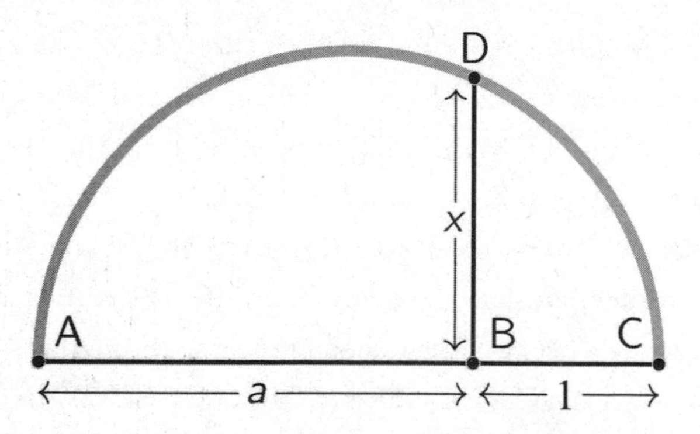

Figure 2.10. Constructing $x = \sqrt{a}$ by ruler and compass.

How should one construct a cube root via ruler and compass? $\sqrt[3]{2}$ resisted all efforts. Yet, it is needed for doubling the volume of a cube. This was one of the top challenges of Greek geometry. Two others were to construct the cosine of an angle of 20 degrees and to construct the number π. The ancients handed these problems, unsolved, to posterity.

The problem of constructing the cosine arises from that of trisecting an angle. Some angles, such as the right angle, 90 degrees, are easy to trisect. Others, such as the angle of an equilateral triangle, 60 degrees, are not. How does one construct an angle of 20 degrees, or (what amounts to the same) its cosine?

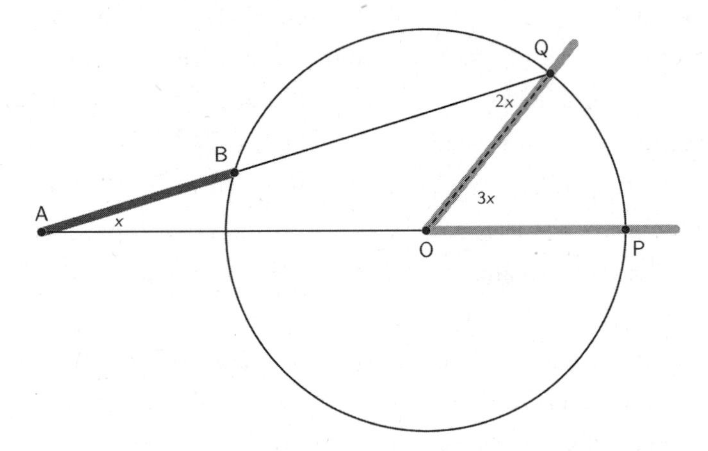

Figure 2.11. Archimedes proposed to trisect the angle POQ.

Archimedes proposed a clever solution (Figure 2.11). If O is the vertex of the given angle, draw a circle of arbitrary radius around O. It intersects the angle in points P and Q. Mark two points A and B on the ruler, such that their distance is equal to the radius of the circle, and then move the ruler, keeping all the while the mark A on the line through OP, and the mark B on the circle. For some position, the point Q will be on the ruler, too. In that position, the angle x between the ruler and the line through OP will be half as large as the angle AQO. Can you see it? The two triangles ABO and BOQ are isosceles. Hence, x is one third as large as the initial angle with vertex

O. The trisection is solved. Archimedes, however, has violated the rules, by using not just a ruler, but *marks* on the ruler—namely, the points A and B. Thus, the trisection remains unsolved.

Here is another attempt: Because it is easy to bisect an angle with ruler and compass, it is just as easy to construct $\frac{1}{4}$ of that angle, and $\frac{1}{4}$ of that in turn, which is $\frac{1}{4^2}$ of the original angle, and then $\frac{1}{4^3}$, and so on. Add all these angles. Since

$$\frac{1}{4} + \frac{1}{4^2} + \frac{1}{4^3} + \ldots = \frac{1}{3},$$

this yields one third of the original angle. You have trisected the angle with ruler and compass. But Greek geometers would never accept that: they allowed only a finite number of steps, and shunned sums with infinitely many terms. They merely forgot to mention it, because they took it for granted.

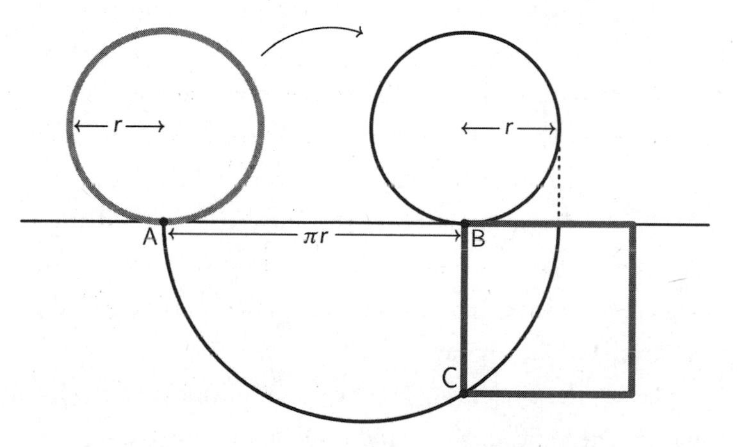

Figure 2.12. The circle squares itself by doing a half-roll. The square has the same area (although it looks larger).

The third problem is squaring the circle, which means to construct, for a given circle of radius r, the square with the same area, namely πr^2. Again, there exist "solutions" that seem almost to work by themselves. For instance (Figure 2.12), place the circle on a point A of some arbitrary line, and let it roll along that line for half a circumference, thus covering distance πr. The circle, after this half-turn, will touch the line at some point B, and the length of AB is πr. Draw the perpendicular to the line through point B, and construct a circle through A having radius $\pi r + r$, centered on the line. The perpendicular intersects the circle in a point C. The square with side BC has area πr^2, as did the original circle. The circle has squared itself, in a way. But alas, rolling a circle along a line is not an approved construction step.

After Descartes had introduced his coordinates, it became clear that the Greek problems amounted to solving linear and quadratic equations. The formulas for the solutions of such equations involve five operations: addition, subtraction, multiplication, division, and taking square roots. Only those numbers are constructible that can be reached by these operations, possibly by means of several intermediate steps, starting out from a given unit length. This condition seriously restricts the set of constructible numbers: they have to be *algebraic*—in other words, the roots of polynomials with rational coefficients—and moreover, these polynomials have to obey certain algebraic properties.

In the year 1837, the young French mathematician Pierre Wantzel proved that the polynomial $x^3 - 2$ does not satisfy the required algebraic property. Hence, its root $\sqrt[3]{2}$ is not a constructible number, and the Delian problem is unsolvable. Wantzel went on to prove that trisecting certain angles is also unsolvable. Few people took note of his work, and Wantzel died soon after. Mathematicians decided eventually that the lion's share of the credit was due to Carl Friedrich Gauss, anyway.

As to squaring the circle, this required an additional effort. Eventually, the German mathematician Ferdinand Lindemann proved that the number π is transcendental, which means that it is not algebraic. And thus, in the nineteenth century, all three famous problems were shown to be unsolvable.

Counterfactuals in history are always of dubious value, particularly so in the history of science. It seems certain, however, that if the three construction problems had been solvable, they would have been of minor importance—an extra handful of geometric constructions, among thousands. Showing that the problems could *not* be solved yielded much more—a vast algebraic theory dealing with so-called field extensions. The pedantry of Greek geometers bore rich fruits.

Less than a hundred years ago, an analogous but more modern concern with getting a handle on numbers led to the notion of computable numbers. They were defined by Alan Turing, the genius of the "hands-on" approach, as those real numbers whose decimal expansions are calculable by finite means, or more precisely those that can be computed to any arbitrary precision by a systematic procedure. What is a systematic procedure? Answer: Whatever can be done by a computer. This seems a silly way of defining *computable*, and needs some explanation, which will be given in Chapter 5.

The computable numbers include all rational and constructible numbers, all algebraic numbers, and also many transcendental numbers, such as π. All the numbers that so vexed the Greek geometers are computable. The set of computable numbers is closed in the sense that if you add, subtract, multiply, and divide any two of them (avoiding division by zero), then you always get computable numbers. The same, by the way, holds for the constructible numbers. Both these sets, however, are only a small part of the real numbers. In a sense that can be made precise, if you pick a real number at random, it is highly unlikely to be computable.

Both concepts of constructible and computable numbers rely on a convention: it is curious that in both cases, mathematicians reached unanimity without much debate. There is nothing God-given about compass and ruler. Why exclude, for instance, the use of threads of a given length? Ancient geometers used them to construct ellipses, for instance, but felt that they were overstepping the bounds of propriety. In a similar vein, one can conceive machines more powerful than our digital computers, having for instance a continuous set of states; one even can build them, and use them for approximating numbers. Nevertheless, such "analogous" computers were not considered as serious rivals of the digital computer conceived by Alan Turing. Numbers are abstract entities, yet mathematical self-discipline restricted the concrete tools to construct or to compute them.

Naturalizing Amphibians

During the Renaissance, the imaginary unit perked up: a number $\sqrt{-1}$ whose square is -1. Fittingly enough, it was discovered by two notorious fortune hunters, Niccolò Tartaglia and Girolamo Cardano. A hundred years later, the faked-up number received its name from Descartes: it was called i as in *imaginary*. Descartes claimed that we can *imagine* as many solutions to a polynomial equation as its degree indicates: two solutions for $x^2 + ax + b = 0$, three for $x^3 + ax^2 + bx + c = 0$, and so on. The trick is to accept that some of these solutions are not real numbers. This was a breathtaking leap of faith.

Descartes guessed the true miracle behind the mysterious magnitude i. Indeed, at first glance, the solution of $x^2 = -1$ seems not much different from that of $x^2 = 2$. A new object is introduced, a kind of placeholder for what is missing. One might at first suspect *déjà vu*.

There is in fact a world of difference. For the equations $x^2 = 3$, $x^2 = 5$, etc., the square root of 2 is of no help at all. In each case a new irrational number has to be introduced, namely $\sqrt{3}$, $\sqrt{5}$, etc. By contrast, this *one* ominous number i proves the key to solving *all* polynomial equations. All such solutions are sums of the form $a + bi$, with a and b plain real numbers. Just this one i is needed. It does the job for all polynomials. The first truly convincing proof of this fundamental theorem of algebra was due to Gauss. Before him, generations of mathematicians had vainly tried to make sense of Descartes's divination, but failed to come up with a satisfying demonstration.

It is about here that the view that the new numbers are inventions of the human mind begins to look preposterous. The imaginary unit does things nobody could ever have dreamed of. Not only does it provide the solutions to all polynomial equations. Not only does it open new worlds. It transfigures the old world!

And yet complex numbers $a + bi$ are nothing but points (a, b) in the familiar Euclidean plane, with a and b real numbers; and the strange-looking multiplication is nothing but a dilation in the plane, followed by a rotation. This was realized, almost at the same time, by Gauss and (independently of each other) by two little-known dilettantes, almost hobby mathematicians: Jean-Robert Argand and Caspar Wessel.

As Gauss commented, the *complex numbers* (this name is due to him) had previously not enjoyed citizenship rights, but had been merely tolerated. "They had looked more like a content-free play with symbols, devoid of meaningful substance." Now what has seemed to Leibniz to be "amphibians between being and not-being" had been naturalized, and endowed with a passport. In a sense, what Gauss had brought along was confidence. Charisma plays an essential role in the history of mathematics, no doubt.

As Gauss wrote:

> If one formerly contemplated this subject from a false point of view and therefore found a mysterious darkness, this is in large part attributable to clumsy terminology. Had one not called 1, –1, $\sqrt{-1}$ positive, negative, or imaginary (or even impossible) units, but instead, for example, direct, inverse, or lateral units, then there could scarcely have been talk of such darkness.

This is almost language philosophy.

Admittedly, something had been given up when extending the real numbers to complex numbers (and thus when passing from the number line to the complex plane). Indeed, reals are ordered by size. By contrast, complex numbers aren't. More precisely, one can order complex numbers in many ways, but not without getting into trouble with the usual rules of arithmetic. These rules imply that squares cannot be negative—but the square of $\sqrt{-1}$ is.

Ending on an Octave

If real numbers are points on a line, and complex numbers points on a plane, the next question is obviously: which numbers belong to three-dimensional space? Its points have three coordinates, and are thus triplets of real numbers.

How to add them is obvious. Yet how should we *multiply* triplets in a way that makes sense?

For reasons that became clear only much later, all attempts failed. It simply would not work. But after years of effort, William Hamilton succeeded, to everyone's surprise (including his own), in multiplying *quadruples*. He called them quaternions. The only fly in the ointment was that commutativity had to go: it is no longer true, in general, that $a \times b = b \times a$. Later another extension was found, from quaternions to octaves (tuples of eight real numbers). Again, it came at some cost: associativity fell by the wayside. With octaves, it is no longer true, in general, that $(a \times b) \times c = a \times (b \times c)$.

Today we know that if we insist on dividing numbers as we are wont to do, these numbers must correspond to points in one-, two-, four-, or eight-dimensional space. Dimensions 2^0, 2^1, 2^2, 2^3, and nothing else. It is like with the five Platonic solids: with such a result, the game is over. But there are many other number games left.

If mathematicians are asked "what is a number?" they are likely to reply (as do many philosophers) that "what is?" questions make little sense. If questioners persist, nevertheless, then they will probably receive as answer: a number is an element of a number system. There are many such systems, catalogued as rings, division algebras, ordered fields, and so on. A number by itself is nothing. It is much the same if you ask "what is a vector?" You will be told: it is an element of a vector space. "What is a symmetry?" An element of a symmetry group. It is the structure that counts.

This kind of answer is possibly frustrating, but it agrees with Ludwig Wittgenstein's view that the meaning of a word is given by its use. We use numbers for various things, and adapt them when the need arises. There need not even be a single concept "number." Number systems, said Wittgenstein, are like threads that make up a rope. The rope can have traction even if none of the threads reach all the way from one end to the other.

3

Infinity

Diving into the Infinity Pool

Chalking It Up

The numbers 1, 2, 3, ... are said to be the *natural* numbers. What makes them so natural? They seem indeed to belong to the real world, to *exist*, in a more immediate way than other mathematical terms, such as circle, limit, or function.

Natural numbers are abstract enough—one can see three feathers, but not the "three" as such—but there is nothing of a fiction or convention about them. Each seems to have a place in the world, an address where it lives, although we meet only a few of them individually. They are *here*. In geometry, one point is like another. But each natural number has its own individuality.

Many species of animals can count, and some can even perform rudimentary additions. The topic is a favorite in animal psychology, especially since the field suffered from a truly traumatic experience with a horse named Hans.

A few years before World War I, the horse's owner, a German schoolmaster and mathematics teacher named Wilhelm von Osten, had managed to convince himself and a large public that Hans was able to count, and even to perform addition, subtraction, and multiplication. (Hans didn't do divisions, though.)

Figure 3.1. Clever Hans in his heyday.

Der Kluge Hans (or Clever Hans) became a celebrity. He would display his talents even in the absence of his owner, which seemed to rule out the possibility of fraud. Again and again, Hans came up with the right answers—he gave them by stomping, with his hoof, the required numbers of knocks.

In the end it was shown that Hans would stop stomping when he felt, from the reaction of the human audience, that he had reached what was expected of him. He took his clues from the audience. This explained his apparent facility in arithmetic. Yet why he was so good at mind-reading humans seems an even greater puzzle. This one remained unsolved, unfortunately. When the Great War began, Hans was recruited by the military and vanished, now nameless, and clueless, into the great butchery.

From Hans on, animal psychologists became extremely careful with their experiments. No animal could ever match the prodigies of Clever Hans again, but many species proved surprisingly good at distinguishing six from eight, or adding two and two in their head, or noticing that if three bird watchers enter a tent and only two come out, the coast is not clear yet.

Humans have language to help them with numbers. Very few societies have no words for at least "one," "two," and "three." Even those few communities manage to count, by using fingers, notches, knots, or shells. The oldest known trace of counting is allegedly the Ishango bone, which dates back some 20,000 years. Scholarly enthusiasts claim to perceive 11, 13, 17, and 19 dents on the bone. (There is some contention about it, as happens to many a bone.) These numbers are the prime numbers between 10 and 20. (There is no contention about that.)

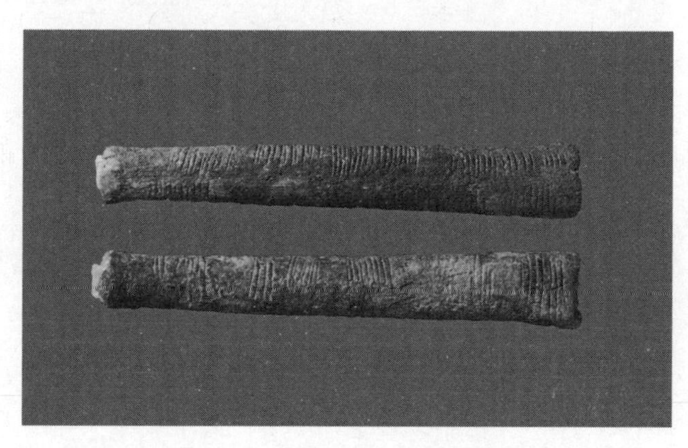

Figure 3.2. The Ishango bone.

Pebbles have been used for counting from very early on. One pebble for each animal in the herd: this helps to check that all of them have returned from the meadow. The Romans named such pebbles *calculi*—the word comes from *calx*, meaning "chalk." The pebbles, thus, are at the origin of our calculations; and many calculations are still chalked upon blackboards, in mathematics departments all over the world, and erased in the wee hours of the next morning.

The pebbles help with counting, and also with computing. Two heaps, or sets, of pebbles can be easily compared by size—most conveniently if they are arranged in rows. These arrangements were the first steps along the number line.

Kids, even toddlers, understand right away that such rows of pebbles can always be extended by adding new pebbles. This yields not only counting but addition too, by placing two rows end to end (Figure 3.3). Similarly, several rows of the same number of pebbles can represent multiplication (Figure 3.4).

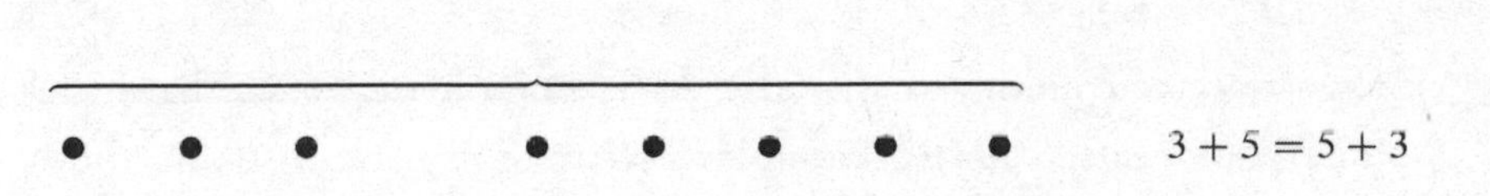

Figure 3.3. A row of three and a row of five pebbles describe the addition 3 + 5.

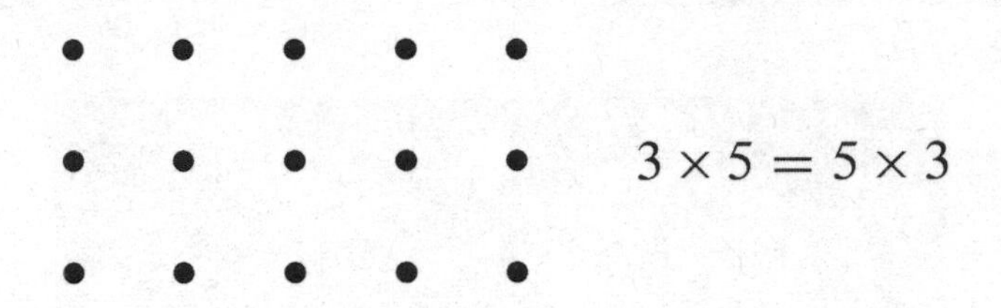

Figure 3.4. Three rows having five pebbles each describe the multiplication 3 × 5.

Sum and product are obviously commutative: $3 + 5 = 5 + 3$, and $3 \times 5 = 5 \times 3$. They are also associative and related by distributivity. All this must have been evident long before any arithmetic rules were conceived.

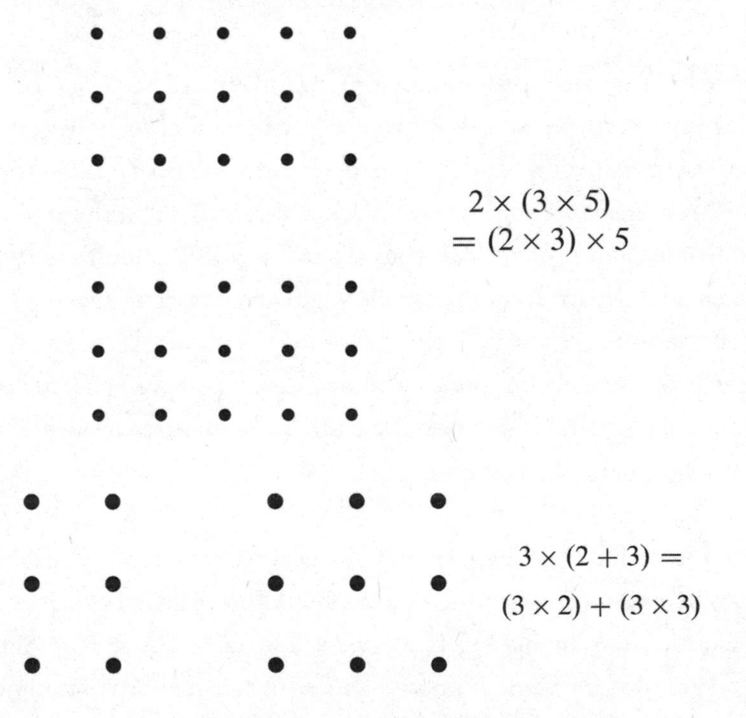

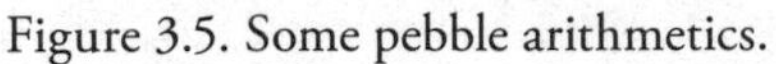

Figure 3.5. Some pebble arithmetics.

Subtraction and division with remainder are also immediately obvious. Prime numbers, too, are easy to grasp: sets of pebbles that cannot be arranged as rectangles. Our first acquaintance with arithmetic is empiric.

The British philosopher John Stuart Mill was of the opinion that this is all there is—that all arithmetic truth is empirical, grounded on observation. Since one hardly ever observes how one pebble, when added to a heap of 999,999 pebbles, produces 1,000,000 pebbles, the truth of 999999 + 1 = 1000000 is obtained by generalizing observations made on smaller numbers of pebbles. Arithmetical relations are laws of nature, mere physical facts.

The German logician Gottlob Frege, whose aim was to reduce arithmetic to logic, poured scorn on Mill's pebble arithmetic. He didn't expect much insight from comparing notes with kids and toddlers. Frege likened Mill's approach with that of someone who, when wanting to learn about America, proceeds by putting himself into the mind of Columbus, about to catch the first glimpse of a continent. This is looking not for the purity of the source, but for a vague mist, Frege said. Nevertheless, the pebbles are what, deep down, taught us to handle numbers.

A budding mathematician named Theaetetus tells Socrates of his idea to arrange numbers in various ways, and we can almost see the pebbles: "We divided all numbers into two classes. The one, the numbers which can be formed by multiplying equal factors, we represented by the shape of the square and called square or equilateral numbers. The numbers between these, such as three and five and all numbers which can only be formed by multiplying a greater by a less or a less by a greater, and are therefore always contained in unequal sides, we represented by the shape of the oblong rectangle and called oblong numbers."

Socrates replies: "Very good! But what next?"

Next, surely, must come the proof that oblong numbers (such as 3, 5, 6, 26, or 142) have irrational square roots: these roots do not correspond to simple fractions of natural numbers. However, the Platonic dialogue *Theaetetus* took another, less mathematical turn, which will not concern us here any longer. We will stick to pebbles.

The long count

No one needs a logical proof to understand that two times two makes four. This $2 \times 2 = 4$ became emblematic for "indisputably true." It is so true as to be almost vexing. Lord Byron wrote in a letter to his future wife, "I know that two and two make four, and should be glad to prove it too if I could—though I must say if by any sort of process I could convert two and two into five it would give me much greater pleasure."

Almost identical words from Fyodor Mikhaylovich Dostoyevsky appear in his 1864 novel *Notes from the Underground*: "I agree that two times two makes four is a splendid thing, but if we are going to lavish praise, then two times two makes five is sometimes also a very charming little thing."

The two writers would surely have been pleased to hear that among an ancient, legend-shrouded tribe two times two does indeed make five. The members of the tribe used knots to count, just as the Inca did. A string with two knots stands for "two." Tying two such strings together, each having two knots, yields five knots. So there you are! (Wittgenstein claimed that one could write a "good and serious work of philosophy" that would consist entirely of jokes. I agree, but telling a good joke is harder than philosophy.)

Pebbles serve better than knots for doing arithmetic. With pebbles, it becomes immediately obvious, for instance, that the sums of the first odd numbers—the sums $1 + 3$, and $1 + 3 + 5$, and $1 + 3 + 5 + 7$, and so on—always yield square numbers, such as 4, 9, 16, and so on. It is a mere matter of arrangement (Figure 3.6). That demonstration needs no words.

There is another way to prove this, which is of particular importance: proof by induction. If one wants to show that some formula $A(n)$ holds for all natural numbers n, then it is enough to show that it holds for $n = 1$ and that, whenever it holds for some n, it holds for $n + 1$ too. Indeed, the truth of $A(1)$ then implies $A(2)$, which implies $A(3)$, and so on.

To illustrate this with our example, the n-th odd number is $2n - 1$. (The first odd number is 1, which is just $2 \times 1 - 1$; the second is 3, which is $2 \times 2 - 1$; the third is $2 \times 3 - 1 = 5$; and so on. In fact, this should also be proved by induction, but let us accept it for now.)

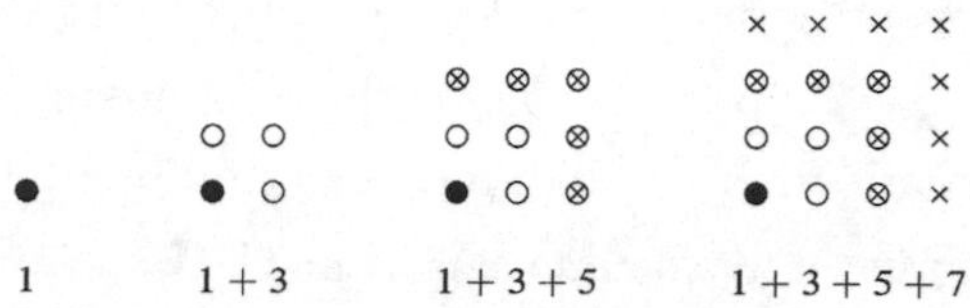

1 $\quad 1+3 \quad 1+3+5 \quad 1+3+5+7$

Figure 3.6. The sum of the first odd numbers is always a square.

Let $A(n)$ be the formula

$$1 + 3 + 5 + \ldots + (2n - 1) = n^2.$$

If this formula is valid, then by adding $2n + 1$ to both sides, we obtain

$$1 + 3 + 5 + \ldots + (2n - 1) + (2n + 1) = n^2 + (2n + 1),$$

which yields

$$1 + 3 + \ldots + (2(n + 1) - 1) = (n + 1)^2,$$

which is nothing else than formula $A(n + 1)$.

Clearly, $A(1)$ is true, since $1 = 1^2$. Because $A(1)$ implies $A(2)$, $A(2)$ is true. This implies that $A(3)$ is true, and so on. It is like a row of dominoes: if the tiles are packed close together and the first one falls, it brings down the second, which brings down the third, and so on. All must fall. The expression "and so on" is revealed as meaning "proceed by induction."

The term *induction* is misleading. It would have been better to name the method "passage to the next integer." In the natural sciences, induction (or more precisely, inductive reasoning) is the step from the particular to the general, from a sample of observations to a general law. The larger the sample, the more plausible is the law. "All swans are white," this kind of thing. Obviously, such reasoning can lead to error. Some swans were found to be black. This type of induction is merely a heuristic tool, and must not be confused with mathematical induction. No matter for how many n the formula $A(n)$ has been checked, this will never suffice to prove that it holds for all n. You can verify that $n^2 + n + 41$ is a prime number for $n = 1$ and $n = 2$ and $n = 3$, etc. up to $n = 40$, yet this does not make it true for all n. Similarly, the formula $n^2 \leq 2^n$ holds for all natural numbers n with one exception, one black sheep (or should we say, one black swan), namely $n = 3$.

The logician Gottlob Frege compared inductive reasoning with deep hole drilling. We may observe a regular increase in temperature, say, but nothing to tell us what will come if we continue drilling and reach deeper layers. The same holds if we continue to add "one" along the number line. Our past observations can generate expectations, nothing more. A proof by mathematical induction, by contrast, yields knowledge.

Proofs by induction are a great joy for mathematicians. They see a lot of them, especially in their first college years. Each such proof is a chain of infinitely many logical steps. They purr like a zipper.

The Nature of Natural Numbers

The method of mathematical induction was introduced by Blaise Pascal. Considering the fundamental role of the principle, this seems remarkably late. The power of the tool was soon recognized. Two centuries after Pascal, Henri Poincaré hailed induction as something like the magic bullet of mathematics, the secret of its amazing success. In Poincaré's view, proof by induction was *the* decisive step leading from analytical propositions—which are valid by their mere form—to synthetical propositions a priori. Indeed, nobody can claim that a proposition such as "*A(n)* holds for all *n*" is a posteriori: it cannot possibly be given by experience covering *all n*. In Poincaré's eyes, Kant's great white whale, the true "synthetic a priori," turned out to be nothing else but proof by induction.

Kant had claimed that space and time were both synthetic a priori, geometry dealing with the former, and arithmetic with the latter. (Indeed, counting takes time.) Poincaré was not convinced by this pleasant symmetry. He thought that Kant was "certainly mistaken" about geometry. "It is merely a body of conventions which cannot be true, but are serviceable,"

Figure 3.7. Henri Poincaré (1854–1912).

Figure 3.8. Giuseppe Peano (1858–1932).

said Poincaré. On the other hand, "Kant had shown himself to be a man of genius by penetrating the true nature of arithmetic: it consists of synthetic a priori judgements." (Poincaré again.)

Poincaré's contemporary, the Italian Giuseppe Peano, saw things otherwise. Peano had been the first to axiomatize natural numbers. Up till then, natural numbers had seemed so natural that nobody, not even Euclid, had felt the need to endow them with axioms.

According to Peano, the natural numbers are a set on which a correspondence is defined, call it S, that maps each member to another one, named its "successor." Different members of the set have different successors. There is a member that is not a successor. (In mathematical jargon, we say: the successor map S is injective, but not surjective.) And to wrap it all up, an axiom demands that if a subset contains a member that is not a successor and has the property that, with each member, it also contains its successor, then this subset is the whole set.

That's all. These are *all* the axioms we need for arithmetic.

It is easy to see that there can be only one member that is not a successor. It is named 1.

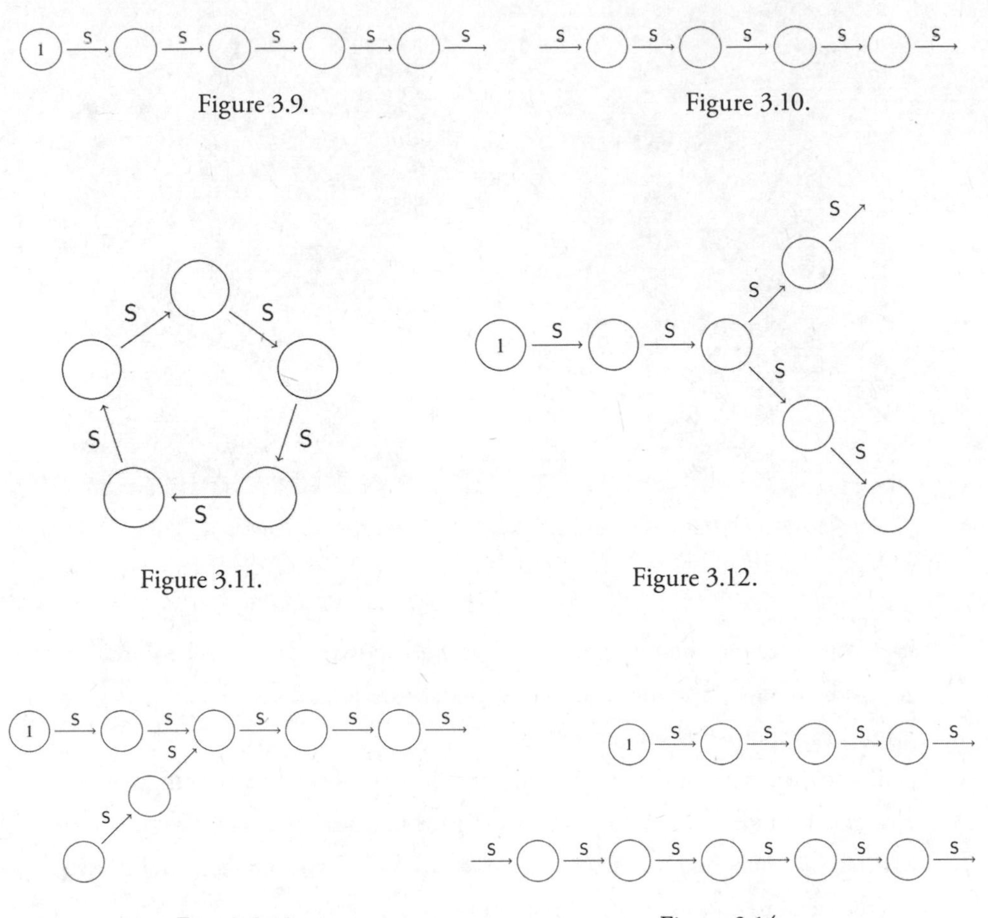

Figure 3.9.

Figure 3.10.

Figure 3.11.

Figure 3.12.

Figure 3.13.

Figure 3.14.

The Peano axioms say that the natural numbers look like Figure 3.9. They do not look like any of the other ways shown in Figures 3.10 to 3.14.

Clearly, a set for which Peano's axioms hold is nothing but a device to make induction work. This seems to indicate that arithmetic is analytic, valid by its true form. It cannot be, said Poincaré. It must be synthetic. The propositions of arithmetic are far too deep to be analytic. However, others say that they merely seem so deep because our minds are so shallow. Today, the debate has lost some of its edge. Mostly, this is because the distinction between analytic and synthetic is not easy to pin down, as philosopher Willard Van Orman Quine has famously shown.

Peano himself used to avoid all philosophical questions. He liked to out himself as a mere rustic, ill at ease in the halls of philosophy. He never denied that his axioms fail to characterize the natural numbers. For example, his axioms apply just as well to the set {11, 12, 13, . . . }: we see the same "successor" relation but a different first element. Another example is given by $\left\{1, \frac{1}{2}, \frac{1}{4}, \frac{1}{8}, \ldots\right\}$. Here, the first element is the same as with the natural numbers, but the successors are different. Basically, the axioms of Peano merely define what is meant by a progression.

Based on the axioms of Peano, it is easy to define addition. One uses induction, of course. If m is any natural number, it has a successor. That successor is defined as $m + 1$; and if $m + n$ is already defined, then its successor will be defined as $m + (n + 1)$. The same procedure works with multiplication: $m \times 1$ is defined as being m; and if $m \times n$ is defined already, then $m \times (n + 1)$ is nothing but $m \times n + m$. The usual properties can be derived from this (although the details are mind-numbing, in general, and treacherous on occasion).

This is enough to pinpoint natural numbers. More precisely, a contemporary of Peano, the German Richard Dedekind, proved that every set that satisfies the axioms of Peano is the same as the set of natural numbers, up to renaming its members. Yet there lurks a conundrum, as we shall see in Chapter 4.

Turning a Searchlight to Infinity

The set of natural numbers is an infinite set. It had been there all along, for everybody to see. Yet it was only in the nineteenth century that mathematicians took to talking about sets and about infinities. This new turn of events was very much the work of one man, the German Georg Cantor. Some of his ideas had been aired before, by Galileo Galilei, David Hume, and Bernard

Bolzano; but it is only with Cantor that set theory (which is basically the theory of infinities) really hit the ground.

Aristotle had proclaimed a taboo that remained unchallenged for thousands of years. He had distinguished between, on the one hand, the potential infinite (which is finite but such that some more can be added, without end) and, on the other hand, the actual infinite. The latter was blackballed by Aristotle as being too large for our mind. Accordingly, a statement such as "there is no largest prime number" is allowed; however, the statement "there are infinitely many prime numbers" is forbidden because it would entail that the infinite set of prime numbers actually exists, which cannot be because infinity does not exist.

Theologians (who in the Middle Ages were the only philosophers in the Western world) kept wrestling with the scruples of Aristotle. Indeed, God's power is *actually* infinite. (To doubt this could land you in trouble, most probably in Hell.) Saint Augustine went so far as to claim that the actual infinite *is* God (or possibly the other way around). There were some concerns with eternity, too. An eternal future can well be viewed as being a potential infinitude: it goes on and on and on. But an eternal past is something different: it all lies behind you, and is fully accomplished. What more does it need to be actually infinite? This is a fallacy, said some divinity experts: the world has no eternal past, it was created during a week, 4000 years ago. But even if this world has a finite past, replied some colleagues, God must surely have always existed.

Be that as it may, it was the theologians who bravely faced the challenge of infinity, whereas the mathematicians shied away, trying not to mess with Aristotle.

Descartes said:

> We will never be involved into tiresome arguments about infinity. For since we are finite, it would be absurd for us to determine anything concerning the infinite, for this would be to attempt to limit it and grasp it. . . . It seems that nobody has any business to think about such matters unless he regards his own mind as infinite.

Philosophers agreed. John Locke was strict: "There is nothing yet more evident than the absurdity of the actual idea of an infinite number."

Gauss, too, ranged himself on the cautious side by professing: "I protest against the use of infinite magnitude as something completed, which is never permissible in mathematics. Infinity is merely a way of speaking."

It took Cantor to overcome this timid reluctance, and to found set theory, as the theory of infinities. He would not accept any censorship. To him, mathematics was free (indeed, he attempted to replace the term "pure mathematics" by "free mathematics"). From Cantor on, mathematicians began to compare the sizes of infinities. They added them; they tamed them; and nowadays only a very few diehards shrink from the actual infinite. It comes so naturally to mind. "We have a primitive concept of infinity," wrote science author Rudi Rucker in *Infinity and the Mind*. "This concept is inspired, I suspect, by the same deep substrate of mind that conditions religious thought. Set theory could even be viewed as a form of exact theology."

Wittgenstein abhorred set theory. He suspected it of being used just to titillate the mind. "If you can show there are numbers bigger than the infinite, your head whirls. This may be the chief reason this was invented." The actual infinite as an overindulgence in the pleasures of paradox; that was how Wittgenstein saw it. "There is no such thing as *all numbers*," he said, "simply because there are infinitely many." He refused to step beyond potential infinity. "A searchlight sends out light into infinite space and so illuminates everything in its direction, but we can't say it illuminates infinity." This is a beautiful simile but did not stop Cantor's ideas from gaining hold of mathematics. Today, the ancient qualms about infinite magnitudes seem as remote as those concerning negative magnitudes or imaginary numbers. Mere pseudo-problems, apparently.

Matchmaking

By Cantor's time, the long march from the natural numbers to the integers, the rational numbers, the reals, and the complex numbers was over. It had been guided by the desire to *compute* further than what natural numbers allow. Cantor desired to *count* further, which led him perforce to infinity.

Cantor did not try to refute the scruples of Aristotle or Descartes. He did have a philosophical bent, but his set theory was motivated by purely mathematical problems that arose from analysis and had to do with complicated classes of real numbers.

Key to Cantor's thinking was an insight: One can tell that set A has as many members as set B without having to know *how* many. All that matters is that there is a one-to-one correspondence between the members of A and the members of B. Then, the sets have the same number (or cardinality, in math jargon), no matter what that number is. They are *equinumerous*.

In a town marathon, you need not count the runners to know that there are as many left legs as right legs moving. In a breakfast room, if each cup stands on a saucer and no saucer is left, you know that there are as many cups as there are saucers: you need not count them.

This became known not as Cantor's principle, but as Hume's principle. In his *A Treatise on Human Nature*, the philosopher had written: "When two numbers are so combined as that one has always a unit answering to every unit of the other, we pronounce them equal." (The sentence seems at first glance difficult to understand, but becomes clearer when one learns that "number" in Hume's sense translates into what today we call "set," and "unit" into "member." Moreover, "equal" must mean "equinumerous." With that reading, the quote is surely a valid reason to credit Hume with his principle.)

There is no need to stick to finite sets. The map that sends every odd number to its successor (1 to 2, 3 to 4, and so on) establishes a one-to-one correspondence between the odd and the even natural numbers: the two sets are equinumerous, yet certainly infinite. Again, this was known before Cantor, and even before Hume. Galilei used, in a similar vein, the map sending each number to its square to show that the set of all natural numbers 1, 2, 3,... and the set of all squares 1, 4, 9,... are equinumerous. The fact became known as Galilei's paradox; indeed, must not a part be smaller than the whole? For a long time, the paradox was used as further proof that infinity transcends the minds of mortals.

With Cantor, the perspective changed. The lesson *he* drew from Galilei's paradox is that the rule "a part is smaller than the whole" does not hold for infinite sets. There are as many natural numbers as there are even numbers, or odd numbers, or square numbers, or cube numbers, or primes. A part need *not* be smaller than the whole. You just have to get used to the idea. In fact, Cantor's contemporary Richard Dedekind turned the tables on the paradox of Galilei, by *defining* finite sets as being those that are *not* equinumerous to any of their proper subsets.

Figure 3.15. Galileo Galilei (1564–1642).

Figure 3.16. Georg Cantor (1845–1918).

Hilbert explained the situation with a homely example that has become a meme: Hilbert's Hotel. It has infinitely many rooms. They are all booked up. This seems like bad news for the weary traveler arriving late at night. "However," says the receptionist, brightening up, "this is Hilbert's Hotel, so you need not despair. We will simply ask every guest to move to the next room. Room One will be presently free for you."

To proceed further with Hume's principle, there are as many integers as there are natural numbers, although there appear to be twice as many of them. Indeed, the integers can be enumerated step by step, for instance as 0, 1, –1, –2, 2, –3, 3,..., etc. (Figure 3.17). Such an enumeration of a set A is nothing but a one-to-one correspondence of the natural numbers 1, 2, 3,...with the elements of A. The set A is then said to be *countably infinite.*

It becomes even better: the rational numbers can also be enumerated, and therefore have the same magnitude as the natural numbers, despite there being apparently so vastly many more of them. Indeed, between any two natural numbers, there are infinitely many fractions. Yet, all the fractions can be enumerated.

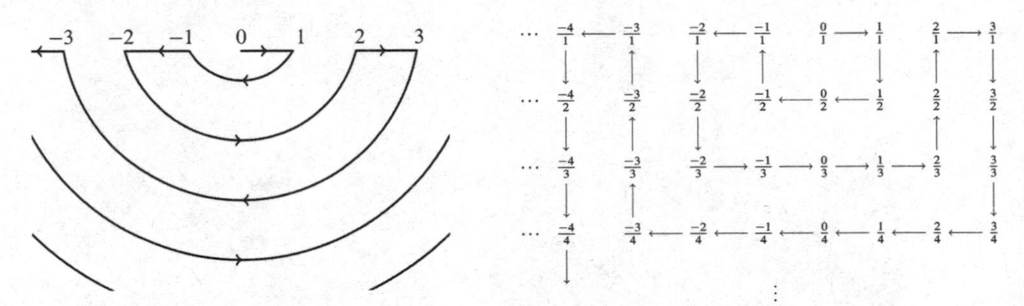

Figures 3.17 and 3.18. Picking one's way, step by step, through the integers and through the fractions.

One merely has to find a proper arrangement of them. For instance, they can be ordered by their denominator (Figure 3.18): first comes the row with all integers (fractions with denominator 1), then the row with all halves (denominator 2), then the row with all thirds, etc., each row being an array extending to infinity in both directions. It is easy to find a path running through all these fractions. Each rational number will be encountered again and again: for instance, $\frac{1}{2}$ will be met again as $\frac{5}{10}$, but we bypass all rational numbers encountered previously and admit to our list only the fractions that have not been touched so far.

The results, so far, seem to convey the message that *all* infinite sets are countable. Not so, as Cantor discovered. This is where his set theory really broke new ground. Infinity would be boring if it came in only one size.

Cantor showed that the real numbers in an interval—say, those between 0 and 1—*cannot* be enumerated. Cantor himself devised several proofs, the last one so stunningly clever that he himself doubted time and again whether his argument was watertight. He wrote anxious letters to Hilbert and other colleagues, asking them to check whether he had committed some stupid blunder. The proof seemed almost too easy.

Today, Cantor's diagonalization is a well-tried, familiar tool. It relies on the indirect method. Let us assume that all the reals between 0 and 1 can be enumerated. These real numbers correspond to decimal expansions, with a 0 in front, then a dot, and then an infinite sequence of digits: for instance, 0.5000... or 0.333.... If the reals can really be enumerated as assumed, we may list them all, one below the other. (We will make no attempt to arrange them by size.) Now let us construct a number with the following decimal expansion. For its n-th digit, we pick the n-th real number on our list, look at *its* n-th digit, and then change it to some other digit. Thus, we descend the diagonal, step by step, as if on a staircase, and replace whatever digit we find at step n by another. This yields the decimal expansion of a real number between 0 and 1. It cannot be anywhere on our list. Indeed, it cannot be at place n, because it has the wrong n-th digit; and this holds for any n. We have constructed a real number between 0 and 1 that is nowhere on our list. This contradicts our assumption that we have listed them all. Hence, the reals cannot be enumerated. They are uncountably many.

From this one uncountable set, the reals between 0 and 1, we can obtain many further uncountable sets, all equinumerous. For instance, the semicircle with diameter 1: It needs only to be placed on top of the interval of length 1 (Figure 3.19). The map that sends each point of the semicircle perpendicularly down to a point on the interval yields a one-to-one correspondence between the interval and the semicircle. (The two sets are equinumerous, although one has length 1 and the other length $\frac{\pi}{2}$.) Similarly, we can place the points on the semicircle in one-to-one correspondence with all points on the whole real line (Figure 3.20). Indeed, if we draw a ray from the center of the semicircle to an arbitrary point P on the line, this ray intersects the semicircle in a unique point p. The correspondence P ↔ p is one-to-one. The semicircle, the interval, and the full line all have the same cardinality, namely that of the continuum.

End of the line? By no means, as we shall see.

The square appears to have many more points than any of its edges. Yet, a square and its edge are equinumerous. Indeed, a point in the square of length 1 is given by two coordinates, and hence by two real numbers $x = 0.x_1x_2x_3\ldots$ and $y = 0.y_1y_2y_3\ldots$, both between 0 and 1. The pair (x,y) can be coded by a single number $z = 0.x_1y_1x_2y_2x_3y_3\ldots$ between 0 and 1. Conversely, any such decimal expansion can be spliced apart to yield two decimal expansions—the two coordinates *x* and *y*.

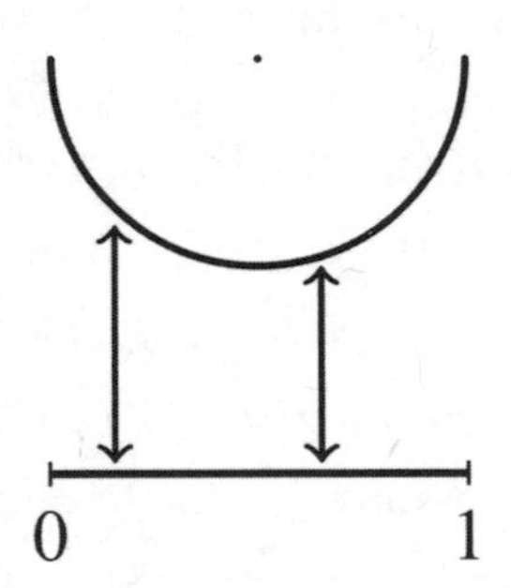

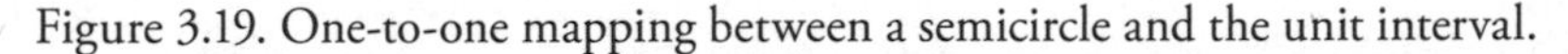
Figure 3.19. One-to-one mapping between a semicircle and the unit interval.

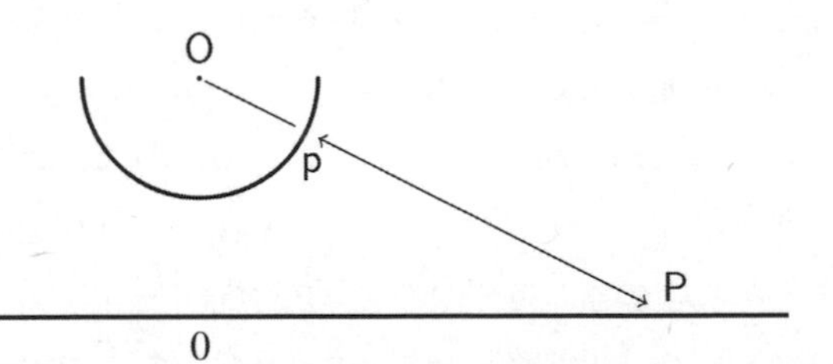

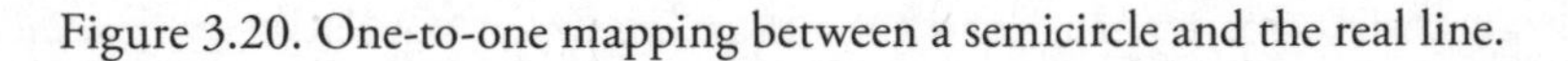
Figure 3.20. One-to-one mapping between a semicircle and the real line.

Peano and Hilbert managed to show an even more astonishing fact. The one-dimensional unit interval can be mapped onto the two-dimensional unit square by a *continuous* map. This means that a curve, traced in a single, uninterrupted movement, can pass through every point of the square (Figure 3.21). The mapping is not one-to-one, however: many points on the square will be visited again and again.

Such mappings between segment and square make us aware that the notion of dimension is not easy to capture. The approach that comes first to mind is to say that the square is two-dimensional because we need *two* coordinates to specify each point. The square-filling curve, however, shows that a single number is enough. (The notion of dimension, which is so obvious to everyone, and indeed always has been, was properly defined only a hundred years ago.)

Figure 3.21. How to fill a square with a curve (shown are the first six of infinitely many steps for constructing that curve).

Let's return to infinity: The line, the plane, and three-dimensional space all have the same number of points. They all have the size of the continuum.

Thus, we have, so far, two magnitudes of infinity: those of the natural numbers and of the reals. Such magnitudes of sets are called *cardinal* numbers. The finite cardinals are just the natural numbers. Are there any other *infinite* cardinals?

Cantor proved that there are many cardinals larger than the continuum: for instance, the size of the set of all subsets of the line. Indeed, he discovered infinitely many infinite cardinals, and soon mathematicians started to do arithmetic with them. The rules are wonderfully bizarre.

However, it remained unclear whether there is a cardinal smaller than the cardinality of the continuum, but larger than that of the natural numbers. In other words, must every uncountable set of real numbers have the magnitude of the whole continuum? This assertion is known as the continuum hypothesis.

The revolutionary ideas of Cantor raised a lot of opposition. Cantor was embroiled in fierce debates. Some said that he was looking for them. In his spare time away from mathematics, he championed the theory that the true author of Shakespeare's plays was Francis Bacon. This claim is almost as good as the actual infinite to guarantee a life lease of controversy.

The Bacon hypothesis gained no hold, but set theory did. Despite the fact that Cantor remained, for all his professional life, at a relatively minor university in Halle, his impact was remarkable. In 1897, at the First International Congress of Mathematicians (which he had been instrumental in launching), he saw that many of the talks used his theory almost as a matter of course. In less than a generation, set theory had become the common basis of all mathematical disciplines.

At the same time, however, the first set-theoretic paradoxes raised their heads. Mathematical pontiffs in Berlin and Paris, such as Kronecker and Poincaré, voiced warnings against Cantor's dangerous ideas. In one of his lectures, Kronecker denounced Cantor's work as "mathematical sophistry." Cantor retaliated in kind: "[Kronecker's] whole lecture is a confused and superficial mixture of ill-digested ideas, bluster, unmotivated slander and lousy jokes."

Most mathematicians became steadfast supporters of Cantor, taking to set theory like ducks take to water. David Hilbert, charismatic as ever, gave out the rallying cry: "Nobody will expel us from the paradise created by Cantor."

Number one on the famous list of Hilbert's twenty-three problems for the twentieth century was the continuum hypothesis. The problem nagged Cantor to his end. It was a sad end. Cantor died almost destitute and in the grips of depression, shut away in a psychiatry clinic.

Back to Zero

If set theory is to be the basis of all mathematics, then the natural numbers 1, 2, 3,... must be grounded on set theory. This seems a doubtful assertion. How can something so familiar as 1, 2, 3,... be based on something so abstruse as set theory, a theory discovered only late in history and heavily fraught with paradox? We know natural numbers from childhood on. We are familiar with them. We need nobody to introduce us to each other.

The logician Gottlob Frege, however, scorned the natural way to natural numbers, which is based on what we learn in the first years of our life. He attempted to capture natural numbers by purely logico-mathematical means. How can one convey what is meant by "five" without leaving the realm of pure thought?

Frege, whose academic career was even more unspectacular than Cantor's (a lifetime at the University of Jena), hit upon an ingenious approach. He defined number by using Cantor's (or Hume's) concept of equinumerosity. The number of fingers on our right hand is the same as the number of toes on our left foot or the number of Platonic solids. All these sets are quintuples. Frege defined "five" to be the set of all quintuples.

This viewpoint seems odd at first, but on second thought brilliant. Any set of five fingers, five shells, five pebbles, five days, five graces, five mountains, etc. is a representative of the set of all quintuples. One is as good as any other. The property that all these sets have in common, their "five-ness," is the number "five." One has defined it without the pedestrian resort of an ostensible definition, such as pointing to five strokes of chalk.

There is one little problem, easily overcome. When one wants to show that sets A and B have the same number, one uses a *one*-to-*one* mapping. This seems to require some rudiment of counting, namely counting up to one. Thus, we use a number to define numbers. But logicians know a way around that objection. Roughly speaking, John has only one brother if whenever "X is John's brother" and "Y is John's brother," X and Y must be the same.

When young Bertrand Russell came across Frege's idea, he embraced it eagerly. The Frege–Russell approach to natural numbers played a seminal role in the history of the foundations of mathematics. Yet today, many consider that it has been superseded by a more bottom-up way to derive natural numbers from set theory. This approach is due to John von Neumann, who discovered it when he was turning twenty (which is the mandatory age for a wunderkind to go into retirement).

First, we must remember that the elements of a set can be sets, too (somewhat like how a folder on your desktop can contain other folders). Moreover, a set A is something other than the set $\{A\}$ containing this one set as its only element (just as a folder is something different than the folder containing just this one folder).

Next, John von Neumann defines the successor of set A as the union of A with $\{A\}$, i.e., as $A \cup \{A\}$. If A is a set, then its successor is also a set.

With that, John von Neumann starts counting. He begins with 0. (The number 1 will accordingly be a successor, but this is clearly just a matter of agreement.) This 0 is defined as the empty set $\varnothing$:

$$0 := \varnothing.$$

Its successor will be named 1, which means

$$1 := \varnothing \cup \{\varnothing\} = \{\varnothing\}$$

(this is the set whose only element is the empty set). The next successors are

$$2 := 1 \cup \{1\} = \{\varnothing,\{\varnothing\}\},$$

$$3 := 2 \cup \{2\} = \{\varnothing,\{\varnothing\},\{\varnothing,\{\varnothing\}\}\},$$

and so on. Thus, $3 = \{0, 1, 2\}$ has three elements and more generally, $n = \{0, 1, \ldots, n - 1\}$ has n elements, which is very gratifying. Even more pleasant is the fact that proof by induction comes free of charge: it is included in the package. The set consisting of 0 and all its successors is the set of natural numbers.

Some mathematicians denote the empty set $\varnothing$ by $\{\}$, and then 1 and 2 and 3 become $\{\{\}\}$ and $\{\{\},\{\{\}\}\}$ and $\{\{\},\{\{\}\},\{\{\},\{\{\}\}\}\}$, respectively, which looks strikingly like the dents on the Ishango bone.

When children count, they are often allured by larger and larger numbers—millions, billions, millions of millions of millions. The hunt for big numbers is a worthy pursuit. None less than Archimedes wrote an essay on the topic, naming large numbers—much larger, *very* much larger than the number of sand grains to fit into the (then) known universe. Their number was estimated as 10^{63}, in modern notation. Today, the curiously named googol has gained some celebrity (it stands for 10^{100}, ten thousand sexdecillions, which is more than the ancients' sand grains but less than the number of possible orderings of seventy objects).

We may imagine kids trying to outdo each other with their knowledge of large numbers. One of them says "one googol!" and another one tops it happily with "one googol and one!" and another with "one googol and two thousand!" and the next with "one googol and one googol!" (which is two googols), and this will be topped by "three googols!" After a short pause of reflection comes "a googol *times* a googol!" (which is a googol squared or, in other words, a googol to the power two). This can be topped by "a googol to the power three!" and will ultimately yield to "a googol to the power googol!"

Eventually, one smart kid shouts "infinity!" And this is the end of the game, so it seems. For "infinity plus one" or "two times infinity" or "infinity to the power infinity" can yield nothing new: it is always infinity, right? It seems that we are finally stuck.

Not so with a wee bit of mathematics.

Let us keep to John von Neumann's way of counting, and denote the natural numbers obtained so far as *ordinals*, to indicate that we intend to go beyond. Each ordinal obtained so far is the set of all smaller ordinals. For instance, $4 = \{0, 1, 2, 3\}$ is the set of all smaller ordinals 0, 1, 2, and 3; "smaller" meaning that each of these ordinals is a proper subset of the ordinal 4. For instance, $2 = \{0, 1\}$ is a subset of 4. So, let us denote the union of all these sets by ω. It is the set $\{0, 1, 2, 3, \ldots\}$, the set of all natural numbers (up to that zero, which is a mere convention).

The letter ω (omega) is the last in the Greek alphabet, yet for mathematicians, it is just the beginning. The set ω is an ordinal, being (again) the set of all smaller ordinals $0, 1, 2, 3, \ldots$. It has a successor, as every set does, namely the set $\omega \cup \{\omega\}$. This is *not* the same set as ω. It is like breaking the sound barrier: we have counted beyond infinity.

Let us denote $\omega \cup \{\omega\}$ by $\omega + 1$, and proceed to the next successor, which is named $\omega + 2$, and keep going. The expression "$+ n$" indicates no algebraic operation, it just names the n-th successor of ω, and works as a sort of index. (Addition of ordinals can be defined, but will not concern us here.) We keep counting ("and so on") and then take the union of all these numbers $\omega + n$, which will be denoted by $\omega + \omega$, or by $\omega \cdot 2$. Again, we take the successor, and then the next successor, "and so on"; and then we form the union of all these ordinals, and obtain $\omega \cdot 3$. Repeating and repeating, and so on, we reach $\omega \cdot \omega$, or ω^2.

"Very good," as Socrates would say. "And what comes next?"

The next big step is ω^ω. And next in line is ω^{ω^ω}. We can even imagine an infinite stack of such ω's; why not? It is named ϵ_0, to make typesetting easier. There is some coquetry behind this name. Both ϵ and 0 have a modest flair, not amounting to much. The ordinal ϵ_0 may seem unimaginably large to you or to Archimedes, but it is a mere trifle to set theorists, just a little warm-up exercise. In fact, the set is still countable. Its cardinal number is not larger than that of ω, which seems so very far back in line. The first uncountable ordinal is still a long way ahead—but it is better to stop here, for sanity's sake. Cardinals grow quickly, and ordinals have a hard job following them.

Galloping through infinities may seem a mad pursuit, the adult version of the kids' bragging contest of topping each other with ever larger numbers, an activity devoid of all content. Have we been lured by the grammar of idioms such as "plus one," "times two," "to the third power," and "and so on" to lose all contact with the road, to lose traction? Not so. Ordinals have a simple, solid meaning, as we will presently see. At least small ordinals do (and even ϵ_0 is not large, for an ordinal).

Cardinals denote the size of sets. Ordinals describe their arrangement. For finite sets, cardinals and ordinals are more or less the same thing, except that we count the former as "one, two, three" and the latter as "first, second, third." For infinite sets, things are different; and the arithmetical rules for cardinal and ordinal infinities go separate ways.

What does it mean to arrange a set? The technical term is *well-ordering*. It means that all members are linearly ordered (for any two distinct members x and y, one is smaller than the other; and if x is smaller than y and y smaller than z, then x is smaller than z). In addition, every subset of a well-ordered set has a smallest element (whereas it need not have a largest element). The set of natural numbers, for instance, is well-ordered in the "natural" way, namely $\{1, 2, 3, \ldots\}$, which corresponds to ω. We can rearrange it by placing 1 at the end of the line, which yields $\{2, 3, \ldots, 1\}$. This well-ordering corresponds to $\omega + 1$. We can also rearrange it by listing the odd numbers first, and then the even numbers: we obtain $\{1, 3, 5, \ldots, 2, 4, 6, \ldots\}$. This corresponds to $\omega \cdot 2$. We can also have all even numbers first, in their natural ordering, then all multiples of 3, then all multiples of 5, etc., taking care to place each number at the first opportunity and to ignore it on every later occasion, so as to avoid double nominations. This yields $\{1, 2, 4, 6, \ldots, 3, 9, 15, \ldots, 5, 25, 35, \ldots\}$. The corresponding ordinal is ω^2. Or we can use prime factor decomposition. First, all powers of 2, namely $1, 2, 2^2, 2^3, \ldots$; then, all powers of 3, namely $3, 3^2, 3^3, \ldots$; then, 2 times all these powers of 3, and 2^2 times all powers of 3, and so on; and then the same with 5, and so on. The ordinal of this well-ordering is ω^ω. We recognize old friends.

Infinity has lost much of its aura. Yet, a shadow of Aristotle's misgivings has survived. The "set of all ordinals" remains untouchable. It is an anathema, like division by zero. Later on, we shall see why. Some shyness of the much-too-large subsists.

On the other hand, the whole ordinal tower is based on the empty set. Whoever is light-headed enough to interpret the empty set as "nothing" will deduce that mathematics is founded on nothing. All respectable creation is based on void. This is what Rudy Rucker must have meant by "exact theology."

4

Logic

The Hardness of the Logical Must

The Greatest Logician Before Kurt Gödel

In the hot, early summer of 1934, the head of the Vienna Circle, philosopher Moritz Schlick, wrote to one Otto Pötzl, professor of psychiatry at the University of Vienna. He asked "to kindly permit me to bring my esteemed colleague Privatdozent Dr Kurt Gödel to your special attention." Kurt Gödel, not yet aged thirty, had wrestled with severe mental problems ever since his return from a stay at the Institute for Advanced Study in Princeton.

Figure 4.1. Kurt Gödel (1906–1978).

Figure 4.2. Aristotle (384–322 BCE).

Schlick wrote to the psychiatrist that it is impossible to praise Gödel's intellectual faculties too highly. Gödel was pure genius.

> He is a mathematician of the highest caliber and his findings are universally recognized as epoch-making. Einstein had no qualms in describing Gödel as the greatest logician since Aristotle, and it is beyond any doubt that Gödel, young though he is, is the world's greatest authority on foundational questions in logic.

Strong words indeed. The predicate "greatest logician since Aristotle" would stick to Gödel. By now, however, many prefer to say that Aristotle was the greatest logician before Kurt Gödel.

Aristotle dominated the field of logic for many centuries. Immanuel Kant, himself a professor of logic and metaphysics, claimed that "since Aristotle, it [logic] has been unable to advance a step, and thus to all appearances has reached its completion." And on another page: "Moreover logic has gained only little in content since the times of Aristotle, and cannot do so by its very nature."

Thus Kant viewed Aristotle's logic as complete. "Not one step forward or back" in 2000 years! Kant meant this in a commendatory way. Bertrand Russell expressed it more disparagingly: "A dead end followed by 2000 years of stagnation." In his *History of Western Philosophy*, he asserted that Aristotelian doctrines of logic are "wholly false, with the exception of the formal theory of the syllogism, which is unimportant."

In his *Metaphysics*, Aristotle enounces the "three basic laws of logic." These are, first, the Law of Identity (*A* is *A*); second, the Law of the Excluded Middle (either *A* or non-*A* are true, there is no third possibility in-between); and third, the Law of Contradiction (*A* and non-*A* cannot both be true). Obviously, the laws of logic have nothing to do with the content of *A*. Today, we would say that they regulate the use of "true" and "false."

The logical masterpiece of Aristotle was his *Analytika protera*, or *priora*, the doctrine of logical inference (which for him means "of syllogisms"). A syllogism is a logically binding argument deducing from two propositions—the premises—a third one—the conclusion.

The propositions could be either affirmative or negative, either universal or particular assertions about properties. Examples of properties are "being human," "being a prime number," or "being green." Accordingly, such propositions come in four flavors:

(a) Universal affirmative ("all A are B"; for instance, "all humans are happy").
(e) Universal negative ("no A is B").
(i) Particular affirmative ("some A are B," meaning that at least one A is B).
(o) Particular negative ("at least one A is not B").

The letters (a), (e), (i), and (o) were introduced in the High Middle Ages. Much later, properties were identified with the corresponding sets of things having these properties. Thus, (a) denotes the inclusion $A \subseteq B$; (i) means $A \nsubseteq \overline{B}$ (with $\overline{\mathrm{B}}$ denoting the set-theoretic complement of B, that is, the set of everything not in B); etc.

These propositions can occur in various arrangements—always the two premises first, then the conclusion: for example, (a), (a), (a) or (e), (a), (e) or (a), (i), (i) or (e), (i), (o). The scholastics named them with mnemonic words that have the suitable vowels: Barbara, Celarent, Darii, Ferio, ...

Here are some examples: If being mortal applies to all humans, and being human applies to all Greeks, then being mortal applies to all Greeks (Barbara). If being mineral applies to no humans, and being human applies to all Greeks, then being mineral applies to no Greek (Celarent). If being Greek applies to all Athenians, and being Athenian applies to some logicians, then some logicians are Greek (Darii). If no Greek is an Egyptian, and some logicians are Greeks, then some logicians are not Egyptian (Ferio).

It turns out that there exist fifteen of these "figures of thought" that are logically correct. In traditional accounts, there are a few more, because the ancients' use of some terms slightly deviates from that of today's mathematicians. Thus, the proposition "all *A* are *B*" implies for Aristotle that there exist things with the property *A*, whereas in contemporary custom among mathematicians, it is accepted that nothing needs to satisfy property *A*—in other words, that the corresponding set is possibly void. Thus, "all dragons are cute" is correct because there are no dragons in the first place, whereas Aristotle and, probably, the majority of laypersons would view the proposition as false—or, if accepted as true, to imply that there are dragons. Under the name *conversio per accidens*, this used to exercise many scholastic philosophers.

Whether there are fifteen types of syllogism or more is not particularly important. Indeed, because $A \subseteq \overline{B}$ means the same as $B \subseteq \overline{A}$, and because $\overline{\overline{A}} = A$, all syllogisms can be reduced to Barbara and Darii. This yields only a rather pinched logic, "unimportant" in Russell's eyes. Aristotle did not even investigate logical connectives such as "and" and "or." Mathematics would not go a long way with syllogisms alone. Despite this, it can be entertaining to devise complicated structures of syllogisms, as a stress test for our reasoning powers. Here is a well-known example.

Let us assume that the following ten *premises* hold:

1. The only animals in this house are cats.
2. Every animal that loves to gaze at the moon is suitable for a pet.
3. When I detest an animal, I avoid it.
4. No animals are carnivorous, until they prowl at night.
5. No cat fails to kill mice.
6. No animals ever take to me, except those in this house.
7. Kangaroos are not suitable for pets.
8. None but carnivores kill mice.
9. I detest animals that do not take to me.
10. Animals that prowl at night always love to gaze at the moon.

From this follows the logical *conclusion*: I avoid kangaroos. (Hint: never mind the moonshine.)

The logician from Oxford who devised this chain of syllogisms was Charles Dodgson. He is better known under his pen name Lewis Carroll. To him are due *Alice's Adventures in Wonderland* and *Through the Looking Glass.*

Pure Thinking

The most remarkable insight of Aristotle was to describe logical arguments by a purely formal calculus: the content of the propositions was irrelevant. Admittedly, he used no mathematical formulas, yet his rules begged to be formalized. However, this was accomplished only thousands of years later.

This decisive step was taken, half a century after Kant, by the Englishman George Boole, the more or less self-taught son of a cobbler. Boole was a prodigy. By age eleven, he had already worked his way through a six-volume work on geometry, as his proud father noted on the flyleaf; by age fourteen, he translated Greek poetry; at age sixteen, he became a schoolmaster; at age thirty, he received the Royal Medal of the Royal Society and then was appointed as a professor of mathematics at the University of Cork in Ireland. Soon after, he caught cold in a rainfall and died.

Boole had noticed that the logical connectives "and" and "or" behave rather similarly to × and +. For instance, the logical rule

"*A* and (*B* or *C*)" is the same as "(*A* and *B*) or (*A* and *C*)"

corresponds to

$$a \times (b + c) = a \times b + a \times c.$$

Some of the usual rules from algebra hold, and some don't. In compensation, there are other rules, such as $a \times a = a$ (indeed, "*A* and *A*" is the same as *A*). In algebra, $a \times a = a$ holds if and only if $a = 1$ or $a = 0$. Therefore,

Boole interpreted the number 0 as meaning "false," the number 1 as "true," and his calculus as an algebra in the domain of the two numbers {0, 1}. Bertrand Russell touted Boole's book *An Investigation of the Laws of Thought* as "the work in which pure mathematics was discovered": purified from geometry or arithmetic, released from space and time.

This was the first explicit formalization of logic. (Leibniz had worked on it, too, but the fact went unnoticed until much later.) Even more important than the use of algebraic methods, which led to what today is called *Boolean* algebra, was the incipient set-theoretic interpretation that proved so useful to interpreting syllogisms. Set theory did not yet exist and Boole used the word *class*. A class corresponds to a property, the "and" corresponds to the intersection of classes, and the "or" to the union. The mathematical methods that Boole developed in *An Investigation of the Laws of Thought* led logic out of its cul-de-sac. The standstill was over.

Through work by the Britons Augustus De Morgan and John Venn, the American Charles Peirce, the German Ernst Schröder, and the Italian Giuseppe Peano, logic turned more and more mathematical. In parallel, mathematics turned more and more logical. Its reasoning became increasingly rigorous, its proofs painstakingly explicit. Such a development had been sorely needed, particularly in analysis, where the new, more explicit arguments uncovered surprising results. For instance, the limit of a sequence of continuous functions can be discontinuous; some continuous functions are nowhere differentiable; and other scandals of the same kind.

The main promoters of the new rigor were Augustin-Louis Cauchy and Karl Weierstrass. Later, it transpired that in Prague a remarkable priest, philosopher, and mathematician named Bernard Bolzano had anticipated many of these developments. Nobody had taken much notice of it, with the exception of the Catholic Church, which put his books on the index. Incidentally, it had been Bolzano who introduced the notion of a set in mathematics: he defined it as a multitude not depending on the arrangement of its parts. The German Richard Dedekind worked out similar ideas in his investigations of the concept of number, and Georg Cantor, eventually, defined a set as "a collection of definite well distinguished objects of our intuition or our thoughts into a whole." This sounds almost as vague as the ancients' definition of a point, and proved as seminal. Cantor himself had no qualms to describe his set theory as "belonging by all means to metaphysics."

Today, metaphysics is a denigrating term. In practice, most mathematicians draw a dumpling-shaped doodle if they speak of a set. They may explain it to a child as a bag containing things. But the concept is more treacherous, as no one knew better than Cantor himself, who once stated with an impressive gesture: "When I think of a set I think of an abyss." It did not take long, indeed, before chasms opened up.

While mathematics and logic kept moving closer and closer together, Gottlob Frege decided to go the whole way. He would reduce all of mathematics to logic. However, the domain of logic that had been formalized so far was way too narrow. It could not any longer remain confined to subject-predicate statements, such as "x is mortal" or "x is prime." It would also have to apply to relations between two or more terms, such as "x is smaller than y" or "x is the brother of y," and allow logical inferences of the form "If somebody has shaken hands with everyone in the room, then everyone in the room has shaken hands." This sentence is undoubtedly correct on purely logical grounds, independent of any content: it would also hold if the handshake were replaced by a fistfight or a kiss. Here again, sets proved useful. For instance, functions—and more generally relations between two terms—are nothing but sets of pairs.

Frege forged ahead with his extended logical calculus, although his notation was not exactly user-friendly. When he turned thirty, he published his *Begriffsschrift* (*Concept Writing*). Next, he sketched his program in *The Foundations of Arithmetic*, a lucid booklet written in plain language. There, he sketched his ideas for defining natural numbers as sets of all the sets having the same magnitude. By then, he was an associate professor of mathematics at the University of Jena, and had his life's program all traced out for him.

The second volume of his major work, *Basic Laws of Arithmetic*, was published in 1903, almost ten years after the first. Frege had not cut any corners and squarely faced all problems. The capping stone of his lifework seemed

Figure 4.3. Gottlob Frege (1848–1925).

Figure 4.4. Bertrand Russell (1872–1970).

within reach. But then things took a sad turn. As Frege wrote in his afterword: "Few things can be less welcome to a scientific writer than to have one of the foundations of his edifice shaken after the work is finished."

Yet, this is exactly what happened. A young Englishman, Bertrand Russell, had pointed out an inconsistency hidden in Frege's work. Frege was thunderstruck. When, a short while later, his wife passed away, he sunk into a deep, long-lasting depression. His lifework was shattered. Hardly anyone attended his lectures. His colleagues took little notice of him. His notation was understood only by few, and used by even fewer. The fact that Russell, Wittgenstein, and his ex-student Rudolf Carnap admired his work to the utmost never really reached him. Frege died embittered and alone. Today, he counts as one of the main architects of computer logic.

Russell's Brainwave

Bertrand Russell had embarked on the same mission as Frege. He too aimed to reduce all mathematics to logic, and to mathematize all logic by means of formal calculus.

Russell descended from a family of British aristocrats; his grandfather had been prime minister twice. Little Bertie was an orphan. He was raised by his ultra-religious grandmother and privately tutored. Then, he went to the University of Cambridge to study mathematics. He feared to have inherited a mental illness. The young man found relief from suicidal thoughts in the certainties of mathematics.

But in the year 1902, Bertrand Russell discovered a paradox that threatened these certainties—an inconsistency in the theory of sets, which is the foundation of all other branches of mathematics.

It had seemed obvious that to every property (such as "being prime," "being happy," or "being a cat"), there is a set consisting of all the objects having this property. Apparently, nothing forbids considering the property "being a member of X" just like any other property and asking whether X itself satisfies that property: whether X is a member of X. Indeed, the property "containing oneself as a member" seems to hold for some sets but not for some others.

Consider these examples: The set X of all cats is no cat, and hence not a member of X. The set X of all that is not a cat is not a cat either, and hence it is a member of X: it contains itself. The set X of all concepts is a concept, hence a member of X. The same holds for the set X of all sets: it contains itself as a member.

So far, so good—apparently. But then Bertrand Russell asked himself: What about the set X of all sets that do *not* contain themselves as a member? If X does not contain itself, then it belongs to that set X and hence contains itself as a member. If X contains itself as a member, then it does not belong to X and hence does not contain itself as a member. A contradiction either way.

The impact of Russell's paradox was huge and long-lasting. Ten years later, the Austrian writer Robert Musil, who was rather proud of his mathematical expertise, described the situation:

> Suddenly, the mathematicians—those who brood in the innermost reaches—came upon something deeply flawed at the very crux of the

> whole structure, something that simply could not be fixed; they actually looked all the way to the bottom and found that the whole edifice of mathematics was floating in midair.

Musil added, and this deserves to be mentioned, too: "This intellectual scandal is endured by the mathematicians in exemplary fashion—that is to say, with confidence and pride in the daredevil boldness of their reason."

That confidence was not unfounded. Indeed, by the time Musil wrote these lines, in the year 1913, the "foundations of the whole thing" had been repaired already.

Among the first to overcome the paradox was Bertrand Russell himself. Together with the mathematician Alfred North Whitehead, Russell developed a "theory of types" that managed to avoid the devious self-references of "sets that do contain themselves as members" and their like. By adding new rules, sets were assigned types and were only permitted to have sets of lower type as their members. Russell and Whitehead expounded this in heroic and merciless detail in their monumental three-volume work *Principia Mathematica*.

The manuscript had more than 5000 pages. The authors had to use a wheelbarrow to deliver it to the printer. Moreover, they had to come up with a publication fee of 100 British pounds.

The Ten Commandments

In parallel to the theory of types, Ernst Zermelo, a former student of David Hilbert, had developed axioms for set theory that also managed to avoid the vicissitudes of Russell's paradox. Zermelo had discovered this paradox even before Russell, but had not made much ado about it. For that reason, and others, Zermelo remains far less well-known than Russell, but his axioms of set theory (modified and improved by Abraham Fraenkel) turned out to be more practical than the theory of types and serve today as the standard foundation of mathematics.

$$\forall x\, \forall y\, (x = y \leftrightarrow \forall z(z \in x \leftrightarrow z \in y))$$

$$\exists x\, \forall y\, y \notin x$$

$$\forall x\, \forall y\, \exists z\, \forall u\, (u \in z \leftrightarrow u = x \vee u = y)$$

$$\forall x\, \exists y\, \forall z\, (z \in y \leftrightarrow \exists (w \in x) z \in w)$$

$$\forall x\, \exists y\, \forall z\, (z \in y \leftrightarrow z \in x \wedge P(z))$$

$$\exists x\, (\varnothing \in x \wedge \forall (y \in x) y \cup \{y\} \in x)$$

$$\forall x\, \exists y\, \forall z\, (z \in y \leftrightarrow z \subseteq x)$$

$$(\forall (a \in x) \exists! b\, P(a, b)) \rightarrow \big(\exists y\, \forall b\, \big(b \in y \leftrightarrow \exists (a \in x)\, P(a, b)\big)\big)$$

$$\forall x\, (x \neq \varnothing \rightarrow \exists (y \in x) x \cap y = \varnothing)$$

$$(\forall (u, v \in x)(u \neq v \rightarrow u \cap v = \varnothing) \wedge \forall (u \in x)\, u \neq \varnothing) \rightarrow \exists y \forall (z \in x) \exists! (w \in z) w \in y$$

Figure 4.5. Zermelo's axioms.

Let us briefly list the axioms, not to study them, but merely to see how they look from the outside. There are ten of them, like with the ten commandments. One can well imagine them engraved on tablets, a few lines of strange symbols and the tablets half-buried in the desert sand. Some alien stumbling over them will see, uncomprehendingly, some letters x, y, z, ... from an alphabet, some brackets, some symbols from set theory (in particular $\in$, which means "is member of"), some arrows, some equal signs, the sign $\wedge$ (which means "and"), the signs $\exists$ and $\forall$ (standing for "there is" and "for all"), etc.

These tablets rule all of our mathematics. In contrast to the Decalogue, there is nothing impressive about them. Non-mathematicians are mostly struck by their pedantry.

$$(1)\ \forall x\, \forall y\, (x = y \leftrightarrow \forall z(z \in x \leftrightarrow z \in y))$$

The first axiom simply means that sets are defined by their members: same members, same sets.

$$(2)\ \exists x\, \forall y\, y \notin x$$

This second axiom postulates the existence of the empty set, which will be denoted by $\varnothing$. (It is the only set that is explicitly assumed to exist.)

$$(3)\ \forall x\, \forall y\, \exists z\, \forall u\, (u \in z \leftrightarrow u = x \vee u = y)$$

The third axiom states that for any two sets x and y, there is a set z having only x and y as its members. In more colloquial terms, it states that we can form the pair $\{x, y\}$. In the case $x = y$, this yields the singleton set $\{x\}$, the set whose only member is x itself. (Note that x and $\{x\}$ are distinct.)

$$(4)\ \forall x\, \exists y\, \forall z\, (z \in y \leftrightarrow \exists (w \in x) z \in w)$$

The fourth axiom states that the union of sets is a set that consists of all members of all these sets.

$$(5)\ \forall x\, \exists y\, \forall z\, (z \in y \leftrightarrow z \in x \wedge P(z))$$

The fifth axiom is the so-called axiom of separation: for every set x and every property P, the members of the set x having this property P form a set y. The property "by itself" is not enough to define a set: it has to be applied to some given set x.

$$(6)\ \exists x\, (\varnothing \in x \wedge \forall (y \in x) y \cup \{y\} \in x)$$

The sixth axiom is the axiom of infinity. It states essentially that, starting from zero (or more precisely nil), you can keep counting. This axiom is actually a modification of Zermelo's original version, and we have seen in Chapter 3 how it is key to infinity.

$$(7)\ \forall x\, \exists y\, \forall z\, (z \in y \leftrightarrow z \subseteq x)$$

The axiom says that for any set x, its subsets z form a set y (the so-called power set).

$$(8)\ (\forall(a \in x)\exists! b\, P(a,b)) \rightarrow (\exists y\, \forall b\, (b \in y \leftrightarrow \exists(a \in x)\, P(a,b)))$$

Axiom (8) states that for any set x and any (decent) function, the image is a set. (The symbol $\exists!$ means that each member a of x has only one image b). Alternatively, if every member of the set x is replaced, in a well-defined way, by a set, then the result is a set.

$$(9)\ \forall x\, (x \neq \emptyset \rightarrow \exists(y \in x) x \cap y = \emptyset)$$

This next-to-last axiom, called the *axiom of foundation*, puts Russell's paradox to sleep. It says that a set cannot have itself as a member. It also excludes, in a roundabout way, sets that have as their member sets that have as their member sets, etc., in a cyclic way or an endless descent.

$$(10)\ (\forall(u,v \in x)(u \neq v \rightarrow u \cap v = \emptyset) \wedge \forall(u \in x) u \neq \emptyset) \rightarrow \exists y \forall (z \in x)\exists!(w \in z) w \in y$$

This last little monster is the axiom of choice. Whenever you have a set x of pairwise disjoint, nonempty sets, you can choose one member from each of these sets and assemble them into a set y. At first, this will hardly raise an eyebrow. We pick one member from each set, why not? Yet, the axiom does not tell you how to do it: it leaves you alone with the task, and merely states that it is feasible.

In his inimitably lucid way, Bertrand Russell explained the conundrum with the example of a millionaire who owns (just to show off) infinitely many pairs of shoes. You can easily pick one shoe from each pair, by always choosing the left shoe, for instance. However, if the millionaire has also infinitely many pairs of socks, how are you to proceed? Which sock will you choose, out of each pair?

To most students, the axiom of choice looks entirely unobjectionable, compared with all the other weird things they are getting accustomed to. However, some consequences of this axiom are rather bizarre. One of them is that every set can be well-ordered such that every subset has a smallest element.

For the set of natural numbers, the usual ordering is a well-ordering: clearly, every nonempty subset has a smallest member. For the rational numbers, by contrast, the usual order on the number line will not do. The positive fractions, for instance, have no smallest member: there is always a smaller positive fraction. Nevertheless, it is easy to find another arrangement of simple fractions that well-orders them. For the real numbers, alas, no one has ever found a well-ordering. Yet, there is one if the axiom of choice is accepted.

It seems to require an act of faith. Do you believe in the axiom of choice, yes or no? At the time of Hilbert, this question fascinated some prominent mathematicians, with strong opinions one way or the other. By now, the tolerance principle reigns. You are free to choose whether to adopt the axiom of choice or not. Each way fits equally well (or unwell) with the other axioms of set theory. One and the same mathematician can study the consequences of both ways, and will not agonize over it.

In a curious way, the axiom of choice is like a revenant of Euclid's parallel postulate. It looks a bit more complicated than the other axioms, is independent of them, and sometimes is assumed to hold, sometimes not.

Much the same holds for the continuum hypothesis, which once was Hilbert's problem number one in his famous list of open problems for the twentieth century. Must every uncountable set of real numbers have the same cardinality as the whole continuum of real numbers? It turned out to be undecidable in the usual frame of set theory. Indeed, through the works of Kurt Gödel and Paul Cohen, it eventually transpired that one can do set theory with or without the continuum hypothesis.

There are other axiom systems for set theory, for instance those due to von Neumann, Paul Bernays, and Gödel, that are quite different but serve as well. Three remarks apply to all of them.

First and foremost, mathematicians have become shy of sets that are too large. The set of all sets and the set of all magnitudes are considered untouchable, "ill-formed." They are not to be mentioned in decent company, because they lead straight to logical inconsistencies.

Second, the ambitious goal of reducing all of mathematics to logic has more or less faded away (despite the fact that Russell was, for a time, convinced to have achieved his dream). It is generally accepted that mathematics is based on set theory, but the axiom of infinity in Zermelo–Fraenkel set theory (number 6 on the list) can hardly be defended as a logical necessity. It looks too contrived. There should be no place, in logic, for technicalities and sophistication. Indeed, since Aristotle's time, logic is supposed to be simple and basic—so *very* basic that many scholastic philosophers have agonized over the question of whether God can supersede the laws of logic or not.

Third and last, the logical paradox that had appeared to threaten the whole edifice of mathematics was silenced within ten years. What had looked, for a moment, like an all-engulfing cataract turned out to be nothing but a harmless little rapid. Most working mathematicians, at the time, had remained entirely untroubled.

However, among those who "pondered within the innermost parts," in Robert Musil's expression, a sense of insecurity remained.

Hilbert Moves to Another Plane

Can we be sure that there remain no other, yet undiscovered, contradictions lurking somewhere underneath? What help are logical proofs if logic itself is not reliable, and its foundation faulty?

Henri Poincaré used an analogy to make the point: The mathematician is like a shepherd who protects his flock against wolves, by surrounding it with a high fence. But what if a wolf is hidden *within* the enclosure?

To reduce this nagging worry, David Hilbert called for a proof of the consistency of mathematics. This became problem number two in his famous list. Thus, Hilbert asked for such a proof even a few years before Russell's notorious paradox came to light. In the decades that followed, Hilbert returned to the problem again and again, pursuing the proof of consistency like Sir Perceval pursued the Holy Grail.

No science welcomes contradictions. They signal that something is amiss. In a formal mathematical theory, however, a contradiction is much worse. It signals that the game is over. Everything collapses. Indeed, let us assume that both A and non-A are true. The validity of A implies, then, that "A or B" is true, for any proposition B. Since "A or B" is true and non-A is, too, B must be true, whatever B says. This puts a stop to everything.

By Hilbert's time, consistency proofs were nothing new. However, they all were proofs of *relative* consistency, justifying one theory by another. For example, hyperbolic geometry is consistent if Euclidean geometry is consistent (and vice versa). Euclidean geometry is consistent if real numbers are consistent (a consequence of analytic geometry). And if real numbers are inconsistent, then the natural numbers must be, too. This leaves the ball in the court of arithmetic.

Hilbert was not looking for proofs of relative consistency. They would merely shift the problem from one theory to another. He wanted a proof for the *absolute* consistency of arithmetic, and he suggested a way to procure it. During the 1920s, this became known as Hilbert's program. It consisted in taking formalization by the letter, and excluding all content, all mind and meaning. This was, nota bene, half a century before the advent of computers, long before Hilbert's idea could be realized as a computer program.

Each formalized text consists of characters from a finite alphabet: symbols such as $\forall$ or $\rightarrow$, parentheses, letters for variables, the sign $\neg$. In the usual mathematical texts, they are associated with certain meanings: for instance $\neg$ with "not," $\forall$ with "for all," $\rightarrow$ with "implies." This interpretation is completely shut out in a complete formalization. There is no content anymore. All that subsists are the signs. Actually, even the term *sign* seems too much loaded with content: indeed, a sign, in usual language, stands always "for" something. In a formalized text, however, it refers to nothing: it is a mere mark, a scrap, a doodle. Nevertheless, for the sake of tradition we keep using the word *sign*.

These signs are aligned in strings having no meaning either. Some of these strings are termed "axioms." Starting out from them, other strings can be formed according to some simple and specific transformation rules explicitly stated. Each string obtained in this way is preceded by the sign $\vdash$. It indicates that the string is an axiom or has been derived according to the transformation rules. A proof is a finite list of such strings. It could be called a superstring if that word were not preempted by theoretical physics.

If both the string A and the string $\neg A$ can be derived in this way, we say that the formal calculus contains a contradiction. Of course, this makes sense only if the signs are given back their meaning, and in particular if $\neg$ is interpreted as negation. Such an interpretation is only possible if we step out from the formal calculus and consider it, so to speak, from a higher plane.

Hilbert's program consists in proving that such a formal contradiction cannot occur. The proof is requested to use only finite, combinatorial arguments about what the transformation rules do to the strings. This would demonstrate the consistency of the formal system. Admittedly, the use of combinatorial, finite, immediately obvious arguments also requires logic, but only a part of logic that seems particularly secure and that, above all, does not smack of infinity. (Viewed in this way, not much has changed since Euclid's time.)

Hilbert's radical formalization is often misunderstood—as if mathematics were *really* nothing but a content-free game with signs. Needless to say, this is not what Hilbert thought. He was, arguably, the person least likely to do so. The interpretation of the formal calculus as a mindless application of mechanical rules to mere scraps and doodles was for him nothing but a device to obtain the longed-for absolute consistency proof. Hilbert called his method proof theory, or metamathematics. It applies mathematics to the formal systems themselves, stating things like "theory X is between theories Y and Z" instead of "the point X is between Y and Z."

For years and years, Hilbert's program developed steadily, with substantial milestones due to Hilbert himself and several of his disciples. But suddenly, progress was brought to a jarring stop.

A postdoc named Kurt Gödel, who had studied philosophy and mathematics in Vienna and had achieved remarkable results along the lines traced out by Hilbert, proved that the program could never reach its goal.

With that same stroke, Gödel demonstrated that the discrepancy between mathematical truth and formal proof is unbridgeable. Roughly speaking, in a formal mathematical theory that is consistent, not everything that holds can be proved. Moreover, the consistency itself cannot be formally proved within the theory. These two statements, Gödel's first and second incompleteness theorems, are of overarching philosophical relevance. However, they are mathematical theorems and by no means "just" philosophical propositions.

"The most significant mathematical truth of this century," it would be stated to be, twenty years later, in the citation for Gödel's honorary doctorate at Harvard.

Metamathematics, Lowbrow Style

The metamathematics of Hilbert and Gödel is an elaborate special discipline that is difficult to convey. Moreover, it seems to irresistibly attract misunderstandings of every kind. In this respect, it shares the fate of natural selection and the theory of relativity.

For starters, what does "true" exactly mean in mathematics? If mathematics is a creation of the human mind, then it would seem as if a valid proof is the unique criterium for truth, the only one to distinguish it from fancy. It must be conceded, however, that the statement "17 is a prime" should have held true even before humans thought of a proof. Let us stick to scenarios where what is meant by "truth" is unambiguous, and avoid any attempt at a general definition.

To aim for at least a partial understanding, we turn to the simplest part of logic, namely propositional logic. It is about propositions that are either true or false (such as "my head is aching" or "3 < 2") and their combinations by means of logical connectives (such as "and," "or," and "not"). Admittedly, "not" does not connect much: some authors therefore name it a modifier.

Nowadays, these connectives are defined by means of truth tables. Interestingly, the method has been in use only for about a hundred years. It was introduced by none other than Ludwig Wittgenstein, in his *Tractatus Logico-Philosophicus*. (Afterward, it turned out that the American logician Peirce had used truth tables decades earlier, without bothering to publish.)

Accordingly, "A and B" (written $A \wedge B$) holds true if both A and B are true; else, "A and B" is false—similarly for the other connectives. There are four input values for A and B (namely T or F, true or false, for each) and hence $2^4 = 16$ binary truth-functions, or connectives. It is curious, by the way, that though "A implies B" (written $A \rightarrow B$) is on the same footing as "A or B" or "A and B," it is expressed by a verb, which somehow suggests that A *does* something to B (maybe even *causes* B).

For every formula of propositional calculus (meaning, every combination of finitely many propositions $A, B, C, \ldots$ by means of logical connectives), truth tables specify for which truth values T or F of $A, B, C, \ldots$ the formula yields T, and for which it yields F.

		A and B	A or B	A implies B		
A	B	$A \wedge B$	$A \vee B$	$A \rightarrow B$	$A \wedge (A \rightarrow B)$	$(A \wedge (A \rightarrow B)) \rightarrow B$
T	T	T	T	T	T	T
T	F	F	T	F	F	T
F	T	F	T	T	F	T
F	F	F	F	T	F	T

Figure 4.6. Truth table for "A and B," "A or B," "A implies B," etc. The last expression, which corresponds to the modus ponens, is a tautology.

If the formula yields T for every possible combination of truth values, then it is said to be generally valid, or (in Wittgenstein's terminology) a *tautology*. From early on, that word had been used in a denigrating sense, as equivalent with "a self-evident platitude," and this is exactly why Wittgenstein adopted it: in his opinion, logic says nothing. Today, this view is accepted as commonplace, just like the use of truth tables. The view of

Frege and Russell was different: for them, logic says something. It deals with the most general laws of the world (our world or any other). This position appears unfashionable today.

Be that as it may, there is no doubt about the concept of "truth" in propositional calculus. A formula is true if it is generally valid. Whether it is generally valid can, in principle, be verified via truth tables, quite mechanically (although it becomes arduous if the formula connects a large number n of propositions: there are 2^n combinations of truth values to check). Such a verification, however, is not a formal proof: a formal proof requires derivation from axioms.

Many axiom systems have been proposed for propositional calculus, all equivalent. The axioms used in *Principia Mathematica*, for instance, are as follows:

1. $(A \vee A) \rightarrow A$.
2. $A \rightarrow (B \vee A)$.
3. $(A \vee B) \rightarrow (B \vee A)$.
4. $(B \rightarrow C) \rightarrow [(A \vee B) \rightarrow (A \vee C)]$.

Actually, there was a fifth axiom, but it turned out to be superfluous, being a consequence of the other four.

Starting with these axioms, one can derive other strings by two transformation rules. One is the so-called modus ponens (if A and $A \rightarrow B$ can be derived, then so can B). The other is a rule of substitution: in every string, a substring such as A can be replaced by another, for instance by $(A \rightarrow \neg B)$, as long as this is done for every occurrence of A in the formula.

Clearly, all these axioms are tautologies, i.e., generally valid. It can be shown that by means of this calculus, only generally valid propositions can be derived; and even better, that *all* generally valid propositions can be derived this way. Because the first property holds, the calculus is said to be correct; and the second property means that it is complete. Indeed, if we interpret "generally valid" as "true" and "derivable in the formal calculus" as "provable," then the true and the provable coincide for propositional calculus.

This look at propositional logic distinguishes between two different levels, a discrimination that is basic for all formal systems.

One level is *syntactic*: it is the purely formal calculus, which consists in shunting around meaningless signs, thus deriving string after string by specific rules. In front of every string so derived stands the symbol $\vdash$ ("is proved"). It only serves to indicate that the syntax has been obeyed—that the transformation rules have been followed.

The other level is *semantic*. At this level, the string is interpreted. It is provided with content. For propositional calculus, this allows one to ask whether the truth table yields T for every combination of truth values. If this is the case, one can write the sign $\vDash$ in front of the string. It means that the interpreted string is generally valid, and hence "true" in the sense of propositional calculus.

One would expect any decent calculus to be consistent, meaning that it is impossible to derive a string and its negation: never $\vdash A$ and $\vdash \neg A$. This is a syntactical property. Two other properties concern semantics. They deal with the relation between the formal system and its interpretation. Any decent formal system should be *correct*, meaning that it only proves things that are true: $\vdash$ should imply $\vDash$. Moreover, one may hope that the formal system is complete, *semantically complete* in the sense that all true formulas can be derived: this holds when $\vDash$ implies $\vdash$.

Propositional calculus is correct, consistent, and semantically complete. As mentioned before, propositional logic does not lead very far in mathematics. A stronger logic is needed, such as Frege developed, for instance.

This is where Kurt Gödel first left his mark. He showed in his PhD thesis that *first-order logic* is correct, consistent, and semantically complete: all generally valid propositions in first-order logic can be formally derived from its axioms. In this sense, "true = provable," as before.

First-order logic operates with the connectives of propositional logic and, in addition, with the quantifiers $\exists$ (there is) and $\forall$ (for all), which apply to variables standing for *members* of a set—some set that is specified in advance, and will be the stage, so to speak, for all future actions. (Set theorists like to speak of a *universe*.) First-order logic allows us to deal with predicates, functions, and relations. If applied to members of the set (or universe) of natural numbers, for instance, then "is prime" is a predicate satisfied by 3, and "$3 < 7$" is a binary relation between 3 and 7. "First-order" means that the quantifiers can apply to members of the given universe, but not to subsets of that universe, and therefore also not to functions or relations. The Peano axiom allowing for proof by induction requires second-order logic: it says something about *every subset A* of the natural numbers that contains 1 and has the property that, whenever A contains a number, it also contains its successor. (It says that such an A *is* the set of all natural numbers.)

Gödel's so-called completeness theorem was generally considered as an indication that Hilbert's program was well on its way. But one year later, Kurt Gödel showed that it wasn't.

Gödel's Proof and Hilbert's Guffaw

In 1930, Gödel proved that every formal mathematical theory that is consistent and covers arithmetic (i.e., permits to count, as well as to add and multiply the numbers 1, 2, 3, . . .) cannot be complete. It is not semantically complete (not all true formulas can be proved, i.e., $\models$ does not imply $\vdash$). In particular, this holds for the claim that the theory is consistent. This claim can be formulated within the formal theory, but not formally proven.

Gödel's proof is demanding and covers many pages. His paper was published in the journal *Monatshefte für Mathematik*. Working through it, sentence by sentence, can fill a whole book, or an entire term.

However, the basic idea of Gödel is stunningly simple and elegant. Gödel constructs within the formal system a proposition G that (when interpreted) states: "G is unprovable." He next shows, by purely formal means, that if G is provable, so is $\neg G$, and vice versa. Because the system is assumed to be consistent, this means that neither G nor $\neg G$ can be proved. This shows that the formal theory is not syntactically complete, either. (Syntactically complete systems are those where every proposition can be proved or disproved.)

Content-wise, this seems similar to the venerable old paradox of the Cretan who claimed that all Cretans were liars. A more precise way to formulate it is to write on a board: "The statement on this board is false." Or else (to avoid a self-referential statement) to write "The statement on the other side of the board is false" and on the other side to write "The statement on the other side is true." This obviously leads to an endless flip-flop.

Gödel, however, went beyond such a semantic contradiction. He showed that there exists a contradiction on the purely syntactic level within the formal theory: $\vdash G$ implies $\vdash \neg G$, and vice versa. Consistency requires that neither G nor $\neg G$ can be derived. Thus, $\neg\vdash G$, which is just what G means. We understand on the semantic level that G holds true, but on the syntactic level, within the formal system, this cannot be proved.

Usually mathematics deals with equations, functions, numbers, figures, etc. The proposition "G is unprovable" does not seem to fit in this context. But whether a string is provable or not has a precise formalization, and even the suspicious-looking self-reference in the interpretation of G can be avoided. To achieve this, Gödel associates to each string of signs a natural number, which later became known as the Gödel number of the string. Each sign used in the alphabet of the formal theory corresponds to a natural number. If the sign with number j occurs as the k-th sign of the string, and p_k is the k-th prime number, then the product of all the prime powers p_k^j yields the Gödel number of the string. Conversely, the original string can be deduced from the Gödel number. It is just a matter of enciphering and deciphering. The interpretation of proposition G actually states that "the proposition with Gödel number g cannot be proved," and Gödel arranged matters so neatly that g is the Gödel number of G ("in a sense fortuitously," as he wrote tongue-in-cheek).

By factoring a natural number into primes, it becomes possible to recognize whether or not it is the Gödel number of a string of the formal system—or whether or not it is the Gödel number of a proof. More than that, by purely arithmetic means, one can tell whether a string is provable or not. If the formula $P(m, n)$ means that the number m is the Gödel number of a proof of the proposition having Gödel number n, then one can form a string G (which, when fully written down, would cover many hundred pages) such that its Gödel number g satisfies $P(m, g)$ for no natural number m. This shows that G cannot be proved.

Hence, each consistent formal system strong enough to cover arithmetic is both syntactically and semantically incomplete. This is Gödel's first incompleteness theorem.

The second incompleteness theorem of Gödel is based on the first. Within the given formal system, the proposition saying "The consistency of the formal system implies that G can be proved" can itself be proved. Therefore, if consistency could be proved, G could be proved. Because it cannot, this implies that consistency cannot be proved, either—not, that is, within the formal system (except if it were inconsistent).

What remains possible, however, is to prove consistency by stepping beyond the given formal system. In fact, this was done by Gerhard Gentzen, another of Hilbert's right-hand men. He used a formal theory based on Zermelo–Fraenkel set theory to prove that Peano arithmetic is consistent. Compared with the impact of Gödel's incompleteness theorems, Gentzen's result seems like a drop falling into a well too deep to hear the splash.

That axioms can be incomplete is nothing new. Take, for example, the Euclidean axioms, but omit the parallel axiom; then the remaining system does not suffice to decide whether, given a line and a point not on it, one or several parallels run through that point. Here is another example: the axioms (1) to (9) of Zermelo and Fraenkel for set theory do not suffice to decide whether (10), the axiom of choice, holds or not. This means that these axiom systems allow for (at least) two geometries, or two set theories. With the Peano axioms, however, there is a difference: they allow for only one arithmetic, essentially (according to a famous result by Dedekind). Nevertheless, within any formal theory of arithmetic based on first-order logic, there exist formulas whose truth can neither be proved nor disproved, and hence not decided. (This in no way contradicts the completeness of first-order logic. There, "true" means "generally valid," valid for every interpretation, and thus applies only to a very restricted subset of arithmetic.)

Here is an irksome state of affairs. Everyone has an intuitive grasp of natural numbers, as captured by the Peano axioms. But as we have seen, they use second-order logic, by saying something about "every subset." For the formal calculus of deriving string after string, one needs first-order logic. Peano's axiom of induction has to be replaced by an axiom scheme saying for any first-order formula (or string) the following: if the formula is valid for 1 and, whenever valid for a natural number, also valid for its successor, then it is valid for all natural numbers. At first glance, this seems to be the same thing, but there are uncountably many subsets of natural numbers, and only countably many formulas. Thus, first-order arithmetic is weaker than second-order. It cannot pinpoint the set of natural numbers. There always exist some other models satisfying the same axioms.

One amazing aspect of Gödel's incompleteness theorem is that it applies to every conceivable (and consistent) set of axioms for arithmetic based on first-order logic. For any such formal system, there exists a true arithmetic proposition that belongs to the system but cannot be proved. If it is taken on board, as an additional axiom, there will still be propositions within the system that cannot be proved. In particular, the consistency of the formal theory can be expressed, but not proved. Each first-order formal system is like a dress. It is too short to cover all of arithmetic (being semantically incomplete); and it has holes (being syntactically incomplete). You can shift the dress, or stitch some holes, but some parts of arithmetic will be bare.

Does this mean that the pursuit of formalization is a vain enterprise, doomed to fail from the outset? On the contrary. Gödel's theorems gave a tremendous boost to the investigation of formal systems, leading among other things to computer-assisted theorem verification (which is based on first-order logic). One of its milestones was reached in the year 1986: a complete formalization of Gödel's first incompleteness theorem.

Gödel and Hilbert never met, but as chance had it, Gödel publicly mentioned his result for the first time in Königsberg, the birthplace of Hilbert, one day before its burgomaster conferred honorary citizenship on Königsberg's illustrious son. On that occasion, Hilbert held a radio speech that is worth hearing. It was a manifesto of Hilbert's unbounded confidence and intellectual bravado. Hilbert quoted a famous sentence of that other illustrious son of Königsberg, Immanuel Kant: "In every department of physical science there is only so much science, properly so-called, as there is mathematics." And he concluded with his uplifting motto: "We must know. We will know." After this, a brief pause in the recording—and then, a dry cackle of laughter. Obviously, the sound technician had forgotten to switch the mike off.

The day before, not far from there, a small workshop on the foundations of mathematics had held its closing session. On this occasion, Kurt Gödel had mentioned almost in passing that it is possible in any formal system of classical mathematics to formulate statements that, though true in their content, were unprovable within the formal system. With that, the meeting ended and the participants went to lunch.

Probably the only one to realize right away the full impact of Gödel's remark was John von Neumann, Hilbert's favorite disciple. It soon dawned on all experts, however, that a new epoch in the everlasting relation of mathematics and logic had begun.

Or had it? One who would always argue to the contrary was Ludwig Wittgenstein. He did not doubt the validity of Gödel's result, even conceding that "Gödel confronts us with a new situation." However, he saw little point in Gödel's theorem, and for that matter in Hilbert's program to prove consistency.

Wittgenstein derided "the mathematician's superstitious fear and awe of contradiction." How can one look for a contradiction, Wittgenstein asked; how would one set about it? And should an inconsistency ever show up (whether looked-for or not), so what? Would anyone be prepared to let go of all of the hard-won theorems of mathematics? Surely not! Formalization is merely a game. Whenever it transpires that the rules of the game lead to a contradiction, mathematicians will overcome the contradiction by modifying the rules.

Gödel and not a few others concluded from this that Wittgenstein had understood neither Hilbert's program nor Gödel's proof. This may indeed be the case. It does not alter the fact that Wittgenstein gave a faithful description of what mathematicians really do when faced with a contradiction. Such occasions have not been rare in the history of mathematics, the best-known example being Russell's paradox.

Some vintage Wittgenstein: "You might put it this way: there is always time to deal with a contradiction when we get to it." And then, pitching it strong: "The contradiction may be considered as a hint from the gods that I am to act and not to consider."

When you meet contradiction, act and do not consider! This sounds almost like a military maxim, coming not from the gods, but from an ex-lieutenant of artillery.

IDENTITÄTS-KARTE.

Charge K. u. k. Leutnant d. Res.

Name Ludwig Wittgenstein

Truppenkörper k. u. k. G. A. R. No 5

No 1/11

30. Juni 1918

k. u. k. Gebirgsartillerieregiment

Datum Gj. 379 am 30. Juni 1918

Figure 4.7. The military knows how to deal with contradictions.

5

Computation

The Ghost in the Machine

The Imitation Game

The Mechanical Turk had traveled a lot, but he never did visit Turkey. He was a native Viennese, born in 1769 at Schönbrunn Palace, and met his end in 1854, his best days well over, in a fire in Philadelphia.

He burned well, being mostly made of wood. His creator was Wolfgang von Kempelen, a clerk at the Imperial court, who had decided to amuse his Empress Maria Theresa with a chess automaton. The life-sized Turk sat at a spacious desk, had a large chessboard in front of him, and always played with the white pieces. He won most of his games. After a move by the adversary, he surveyed the board, and then moved his piece with the left hand, to the accompaniment of whirring and creaking noises.

The Mechanical Turk quickly became a celebrity. Baron von Kempelen traveled with him throughout Europe for many years. At the beginning of each performance, he opened the drawers of the desk, one after another, and shone a light through them. The spectators could distinguish cylinders, threads, and cogwheels. They were not allowed to approach too closely.

Needless to say, a man was hidden in the machine. He was sitting in a tight corner, with barely enough room to sneeze, and the strict order never to do so. It was stiflingly hot, as the only light came from a candle. Sometimes, a thin puff of smoke emerged from the turban. The Mechanical Turk lost his

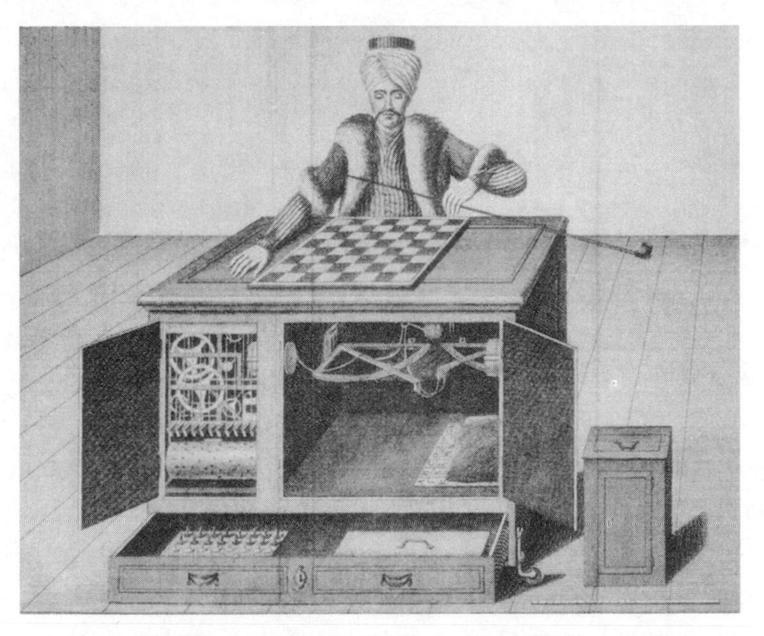

Figure 5.1. The Mechanical Turk, aka Mälzel's Chess Player.

game against Philidor, the French chess champion, by a small margin only. The King of Prussia paid a considerable sum of money to be admitted to the secret, and did not hide his disappointment afterward.

When von Kempelen passed away, the Mechanical Turk was acquired by the Viennese Imperial chief-engineer Johann Nepomuk Mälzel. This Mälzel was an inventor of music automata and is still known today for his metronome. With his new master, the Turk got voice enough to say "*échec*"—French for "chess."

Once again, the Mechanical Turk went on tour, for years on end, including several visits to the United States. The writer Edgar Allan Poe wrote an essay about "Mälzel's Chess Player." Because the automaton sometimes lost his game, Poe concluded that it was operated by a human, arguing that if a machine could play chess, it would always win. This was a fallacy, as we have learned since, but it led Poe to the correct result. We know the names of some of the players who once had squeezed themselves into the desk of the Turk.

There came a time when the automaton no longer drew crowds at fairs. This was when a museum in Philadelphia bought it, and stored it, and eventually lost it to flames.

The first "true" chess automata date from the 1950s. They often lost their game, heedless of Poe's argument. It was not until 1997 that one of them was able to beat a world champion: Gary Kasparov was defeated by Deep Blue in six rounds. It was a milestone in the history of artificial intelligence. Today, such contests are as pointless as a race between humans and motorcycles. Chess automata like Stockfish and AlphaZero play in a league of their own.

The mathematician Alan Turing was the first to write a computer program for chess. He called it *Turochamp* (not because he claimed champion level but because his collaborator was the mathematician David Champernowne). At about the same time, Turing also laid the foundations for artificial intelligence. In particular, he understood that intelligence meant more than being good at chess.

In his 1950 paper "Computing Machinery and Intelligence," Turing addressed the question of whether machines can think, but swiftly dismissed it as "too meaningless to deserve discussion." He did not wish to get entangled in philosophical ambiguities. Instead, he proposed a test to find out whether a machine can imitate humans or not. It later became known as the Turing Test, but he called it the imitation game (later, this became the title of one of the many films about Turing). A human "interrogator" holds a conversation via teleprinter with two unknowns, one of them being a machine, the other a human, and both trying to pass for human. The task of the interrogator is to find out who is who (or what is what). Turing allotted five minutes for the task. The machine acts intelligently, according to Turing, if it wins as often as the human.

With time, Turing's challenge turned into something like the Holy Grail of artificial intelligence. Turing guessed that by the year 2000, some machines would win the test with a likelihood of 30 percent. This, it turned out, was overly optimistic.

The Turing Test is almost a mirror image of the Mechanical Turk. In one case, we have a man posing as an automaton, in the other case, a machine faking to be human. In both cases, tricks may be used by each side.

The Mechanical Turk, for instance, moved jerkily; sometimes a few cog-wheels were heard whirring. Napoleon, one of the Turk's many challengers, tried to use an illegal move to disrupt the automaton. The Mechanical Turk placed the figure back to its former position and, when Napoleon tried again, wiped all chess figures from the board, with a single, sweeping gesture of the left arm. (The following game between the two players took a regular turn, and Napoleon lost in nineteen moves. At the time, he was still able to laugh off defeats.)

Conversely, computer programs trying to act like a human must beware of being too quick in multiplying two four-digit numbers. A well-prepared computer must have a large store of human foibles. To pass the Turing Test, artificial intelligence has to convincingly fake natural stupidity. Moreover, as Turing noted, his test is biased, and indeed unfair: it asks computers to pass for humans, but never asks humans to pass for computers.

Today, in a development that might have surprised even Turing, the Internet is teeming with chatbots operating undercover. The usual precaution against bots (programs faking being human) is named CAPTCHA, an acronym for Completely Automated Public Turing Test to Tell Computers and Humans Apart. Thus, we have reached the point where the task to certify us as humans is left to computers—and they will probably surpass us soon.

Reaching for an *Entscheidung*

Alan Matheson Turing endured a very British upper-class childhood: boarding schools from early on while his parents were away administering India. He won a scholarship to the University of Cambridge, shone in mathematics, and, when barely twenty-two, was elected a fellow at King's College. While listening to lectures by Max Newman on the foundations of mathematics (which was a field Cambridge felt somewhat responsible for), Turing learned about a problem of Hilbert and solved it after many months of intensive labor.

This was the famous *Entscheidungsproblem* (or "decision problem"). Basically, the *Entscheidungsproblem* asks for a way to discriminate between those mathematical problems that can be solved and those that can't. Typical metamathematics.

More precisely, Hilbert had asked whether there exists a systematic procedure to find out whether a given proposition of first-order logic is provable or not. Since first-order logic is complete, this means finding out whether the proposition is true or not ("true" meaning: generally valid for every interpretation of its terms). Such a systematic procedure need not necessarily give a proof of the proposition in the formal theory, but it must reliably indicate whether there exists one, yes or no.

That the decision problem is solvable for the much simpler propositional calculus is obvious. Verification by truth table is a systematic way of finding out what is a tautology and what is not, and thus (since all tautologies can be derived from the axioms) what is provable and what is not. As Turing showed, it is otherwise with first-order logic.

Some years before, Kurt Gödel had shown that any formalized Peano arithmetic contains true propositions that cannot be proved. But the *Entscheidungsproblem* remained open: is it possible to decide, by a systematic procedure, what can be proved and what not?

What is a systematic procedure? In his lecture, Max Newman, a very professorial type, austere and precise, had stressed that it has to be something mechanical, mindless—something that a machine can do. This brings to mind, if not the Mechanical Turk, then at least the many mechanical and electrical calculators used during the 1930s by human computers (mostly females, by the way) in their daily deskwork. Computation is the epitome of a systematic procedure, and Alan Turing began to ponder what the most general computing engine could do.

What is a computation? Something to do with numbers? Not necessarily. Many would try to base an answer on the most basic arithmetic operations, something that turns 323 + 2219 into 2542. Turing, however, abstracted from all numerical underpinnings. After this drastic decision, all that was left is that a string of symbols (which could just as well be %,:&:,',$,!) is replaced by another string of symbols according to specific rules—input turned into output. Turing wrote: "Computing is normally done by writing certain symbols on paper. We may suppose this paper is divided into squares like a child's arithmetic book."

Turing went on to argue that one could assume that the computation is carried out on one-dimensional paper, meaning on a tape divided into squares, and that the alphabet of symbols that may be printed into the squares is finite. All but finitely many squares of the input are blank. The tape is unlimited. More precisely, it could always be extended, should the need arise.

As Newman later wrote: "It is difficult today to realize how bold an innovation it was to introduce talk about paper tapes and patterns punched in them into discussions of the foundations of mathematics." Yet, it was simply taking Hilbert's formalism literally.

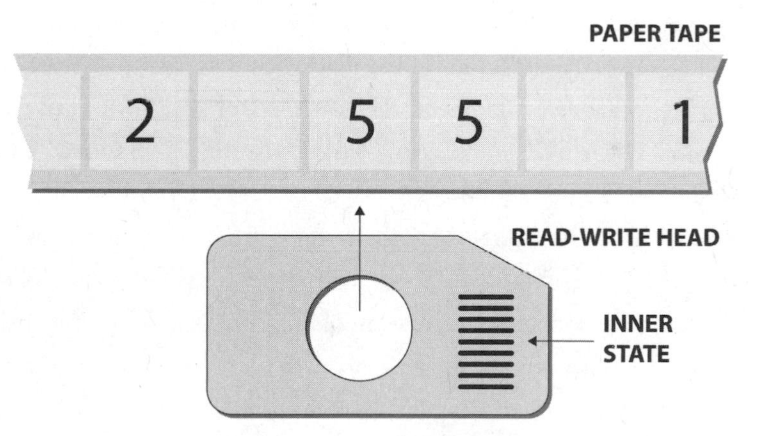

Figure 5.2. A Turing machine: the tape is unlimited and can be moved one step to the left or to the right.

As to the machine itself (Figure 5.2), it consists essentially of a read-write scanner hovering over the tape, and of a finite set of what Turing teasingly called "states of mind," which are nothing but instruction tables telling the machine what to do next. The options for the next move are limited: change the symbol on the square; go from one instruction table to another; move the tape one square to the left or to the right. There, the scanner spots another symbol (or a blank) and proceeds in a way that is determined by the symbol and the instruction at hand. If the machine stops—which it need not do—then what is written on the tape is the output. Such "machines" were soon named Turing machines.

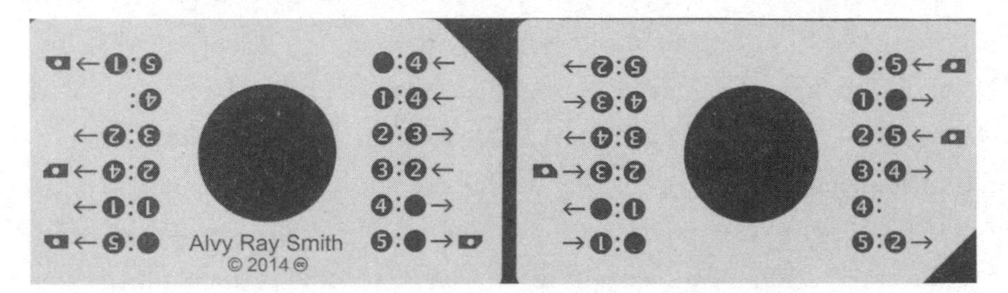

Figure 5.3. A Turing machine, *Toy Story*–style.

Figure 5.3 shows an example, which is due not to Alan Turing, but to Alvy Ray Smith, one of the cofounders of Pixar, a computer animation studio. (With *Toy Story* and other films, Pixar has won twelve Oscars so far.) This Turing machine easily fits on the two sides of a business card that has a hole punched in it. The tape runs lengthwise under the card. The hole shows whatever symbol is on the tape. There are six symbols, namely the blank symbol and the numbers 1, 2, 3, 4, and 5 (that these are numbers is irrelevant). The "state of mind" is the short instruction table on the right-hand side of the hole, the one with the symbols upright. Consider, for instance, the side of the business card displayed on the left-hand side of the figure. If the "computer" sees the symbol 5 through the hole, then its state of mind requires that it executes what is on line 5, the bottom line of the instruction table. It has to erase the scanned symbol (leaving a blank in lieu of the 5); it has to move the tape one square to the right (as indicated by the arrow); and it has to switch to another state of mind, by turning the business card as indicated by the glyph. So much for one time step. What happens in the next is determined by whatever symbol is on the new square, etc. It is instructive to find out how the machine can stop, and how it can cycle through an endless loop.

Among mathematicians, there was immediate consensus: a *systematic procedure* is exactly what a Turing machine can do. It happens to be precisely what human computers (their official job designation) were doing in the 1930s. They followed specific rules, step by step, according to their instructions, which might be stored in a book or in their brain. As human beings, they could stop for a coffee break, or be replaced at the end of their shift by another computer. The essential thing was never to deviate in their task from the instructions. "Rule-following behavior," to use an expression by Wittgenstein. Thus, the Turing machine, by mimicking human computers, is engaged in another sort of imitation game: imitating people who are doing their best to act like machines.

Turing's Moment of Truth

Now for the essential step. Turing invented universal Turing machines—machines able to do what any other Turing machine could do. Such a universal Turing machine is able to perform every computation by mimicking what the appropriate Turing machine would do (one that may, quite likely, have altogether different symbols and states).

For this task, it suffices to encode the rules of the special machine and submit them as an additional input, on the tape, to the universal Turing machine. This universal Turing machine will then look up, at each step, what the special machine would do and simply mimic it. Yet another imitation game. One automaton imitates another. Empathy between machines, so to speak.

To work the coding out in full detail is gruesome work, and actually Turing committed several mistakes in his examples. Some were quickly spotted, some others only many years later. It never mattered. The basic principle was so glaringly obvious.

It may well be thought that such a universal Turing machine, able to understand any other, and to store and execute every conceivable program, has to be exceedingly complex, with lots of symbols and lots of inner states—not at all. The business card of Alvy Ray Smith *is* a universal Turing machine. It is based on work by a Russian computer scientist named Yuri Rogozin. You can compute everything with it, for instance the orbit of a rocket to Mars. Admittedly, the planet will be long gone (not just from its current place, but from the universe altogether) before the computation reaches its end. This issue is beside the point—mere rocket science, so to speak. Universal Turing machines do not vie for speed. And initially, they were mere figments of the mind—thought experiments.

The first problem solved by Turing's universal machine (well before it could be coaxed to do sums such as 323 + 2219) was Hilbert's *Entscheidungsproblem*. There exist classes of well-defined mathematical questions, admitting a yes or no answer, that no systematic procedure can solve. The most famous example, today, is the Halting problem. (It is often attributed to Turing, but emerged only after Turing's death.) Given any Turing machine and any input, will the machine deliver an output? Which means, will it ever stop? There is no systematic procedure to answer this question. This settles Hilbert's problem.

The proof is indirect, and uncannily similar to Cantor's famous proof showing that there are more than countably many real numbers. Diagonalization is the key.

Suppose that the Halting problem can be solved: that there exists a systematic procedure to tell, for each Turing machine and each input, whether the Turing machine with that input will eventually stop or not. We can list all possible Turing machines, since each of them is given by a finite set of symbols and a finite set of instructions. The list will be long, indeed infinite, but it is countable. Each machine gets its number. Similarly, we can number all possible inputs: this yields another countably infinite list.

Let us use our hypothetical "systematic procedure" to devise a contrarian Turing machine. Using the systematic procedure, the contrarian finds out whether Turing machine number n halts with input number n (i.e., whether the first machine halts with the first input, the second with the second, and so on). Then, having found this out, the contrarian does just the opposite thing. If machine number n halts with input n, then that same input *n* causes the contrarian to embark on an endless loop, for instance by switching between two states. If machine number *n* does *not* halt with input *n*, then that input *n* makes the contrarian machine stop.

This contrarian machine can be found nowhere on our list of all Turing machines. Indeed, it cannot be at position *n*, because it does the wrong thing for input *n*. Hence, we cannot assign it any position on the list of all Turing machines. But this yields a contradiction: we have assumed that our list contains *all* machines. Thus, there can be no systematic procedure to solve the Halting problem.

Even more is true: there is no systematic procedure to decide, for any given Turing machine, whether or not it will come to a stop when presented with a blank tape!

With the Halting problem undecidable, first-order logic is undecidable. Indeed, it is possible to associate to each Turing machine a formula in first-order logic that is generally valid if and only if the machine comes to a stop when presented with the blank tape. Since the Halting problem is undecidable, so is the truth of the formula, and hence its provability.

Similarly, each Turing machine corresponds to a formula in arithmetic that is true if and only if the machine comes to a stop on the blank tape. Again, because the Halting problem is undecidable, so is the truth of the arithmetic formula.

To solve a problem of Hilbert ranks among the most prestigious exploits any mathematician can aim for. In April 1936, Alan Turing walked into Professor Newman's office and presented him with a lengthy draft of his solution. Sadly, the professor had to tell him that a young American named Alonzo Church had beaten him to the draw. It must have been a hard blow for Turing, all the more so as it was not the first of that kind. A few years before, he had proved a major theorem in probability (the so-called central limit theorem), only to learn that someone else had anticipated him. There would have been no point in publishing his work. Silver medals do not count for much in science.

Newman felt that this time, however, there *was* a point in publishing Turing's paper, while of course duly acknowledging the priority of Church.

Indeed, Turing's approach was radically different, a game changer in mathematical logic. The solution by Alonzo Church was brilliant, sophisticated, and, as it would turn out, seminal in its own right; but it stood squarely in the tradition of the work in symbolic logic that reached its high point in the 1930s. In fact, it would transpire that others, too—Kurt Gödel himself, Jacques Herbrand, Stephen Kleene, Emil Post—were working on more or less similar ideas. They all had tried to capture the notion of a "systematic procedure" in one formal system or another. None was as down-to-earth as Turing's definition.

Newman arranged for the publication of Turing's paper, now entitled: "On Computable Numbers, with an Application to the *Entscheidungsproblem*." Thus, the decision problem, which was the prime mover of the whole enterprise, became reduced to a sideshow. Mechanical computing took the fore.

The community of logicians agreed quickly that the meaning of "systematic procedure" was given, indeed, by the Turing machine. (It was Alonzo Church who proposed its name.) The other attempts at defining computability turned out to be equivalent, and Turing's approach proved to be the easiest on intuition—the gold standard. Every computation, so says the Church–Turing thesis, can be done on a Turing machine. All it needed was an appropriate input and a set of instructions. Turing's universal machine brought "the right perspective," said Gödel.

With the Turing machine, the study of formal systems underwent a metamorphosis. Gödel's almost magical, ingenious proof of the incompleteness theorem, deftly dancing between the semantic and the syntactic level, had led many to conclude that the realm of undecidable propositions was restricted to highly contrived statements, which barely deserved citizen rights in hardcore mathematics.

With Turing machines, this changed. Almost any property of universal Turing machines is undecidable. Turing's own example had been the Printing Problem: would the machine ever print a given symbol? Many other examples were found. Eventually, it became clear that undecidability lurks behind the corners of some of the best-established fields of mathematics, such as combinatorics and Diophantine equations (polynomial equations with integer coefficients and integer-valued unknowns). It is hardly possible to be mathematically more mainstream than that!

More importantly, the universal Turing machine almost inadvertently yielded the blueprint for the stored-program computer. Its arrival was as momentous for humanity as that of agriculture or writing. And it took no longer than a dozen years.

Artificial Intelligence and No Natural Death

Having conceived his machine, twenty-five-year-old Turing traveled to Princeton, did some more work on logic, and wrote a PhD thesis with Alonzo Church. John von Neumann wanted to have him at the Institute for Advanced Study, and arranged for a plum offer, but Alan Turing returned to England. A war was brewing in Europe, and he had an appointment with some people from the British Government Code and Cypher School. They were worried about Enigma, a fiendishly clever encryption machine used by the Wehrmacht.

On the day World War II erupted, Turing alighted in Bletchley Park and set to work. He spearheaded the efforts to develop so-called Bombes—huge electronic calculators, based on a Polish design, that whirled and rattled until they came up with promising proposals for deciphering Enigma messages. At some later stage, the German Navy switched to an even more complex encryption system nicknamed Tunny. Turing's mentor Max Newman developed Colossus machines to deal with these messages, and, again, Turing took a lead in the project. Both the Bombes and Colossus were not yet computers in the modern sense (being not stored-program machines), but they came close.

After the war, scientists on both sides of the Atlantic forged ahead with actually building universal Turing machines, John von Neumann masterminding the US effort and Max Newman (who had been born as Neumann) in the United Kingdom. Two "new" men for the new age! The computing engines were huge contraptions filled with vacuum tubes and cables. The president of IBM opined that there would be demand for about five such machines worldwide. Their computing power was a tiny fraction of that of a smartphone.

Figure 5.4. The Manchester Baby.

Turing eventually followed Newman to Manchester, and this is where Baby was born—a universal Turing machine with a von Neumann architecture. (Turing had championed a different architecture, which, however, fell by the wayside.) Turing wrote the first programmers' manual (it was full of bugs). When taking off from the numbing chore of devising machine code to perform basic numeric operations, he was able to reflect on what it means for machines to exhibit intelligent behavior. At this point he began to envisage making conversation with a computer.

In order to introduce his idea of a test, he used an analogy. In this preliminary version of the imitation game, the interrogators know that on the other end of the line, there is not a human and a machine, but a man and a woman, both trying to pass for a woman. A transgender game.

Turing had never hidden the fact that he was more attracted to men than to women. In 1952, the police intervened. Turing was sentenced for "gross indecency." He was given the choice between one year in prison or a hormonal treatment. He chose the latter. He had been an athlete, a first-rate long-distance runner. Now he grew breasts.

One afternoon, his housekeeper found him dead in his bed. After a rather perfunctory inquest, the verdict given was that he had committed suicide, by taking poison while the balance of his mind was disturbed. The coroner said that "in a man of his type, one never knows what his mental processes are going to do next." Some mere sixty years later, the British government graciously apologized, and Turing was granted a Royal Pardon.

No Pardon for Error

With the computer, the formalization of mathematics materialized.

It was a momentous development, driven by an old dream. Four hundred years ago, when Francis Bacon (1561–1626) infused the budding Age of Science with his radical skepticism, he urged "that the entire work of understanding be commenced afresh, and the mind itself be from the very outset not left to take its own course, but guided at every step; and the business be done as if by machinery." These were truly visionary words at a time when "machinery" meant, at best, clocks, mills, and muskets.

Today, not only is the business done *as if* by machinery, it is done *by* machinery.

Doubtlessly, terms like *machine* and *mechanical* were initially intended as metaphor. However, a few pioneers such as Charles Babbage and Ada Lovelace, Lord Byron's daughter, started to take the idea literally. At around 1840 they set themselves to build a mechanism to guide the mind, in Bacon's words; they anticipated what programmable computers were to become. Their apparatus (Figure 5.7), known as an analytical engine, fell ultimately short of its intended purpose. The cause was mechanical: the friction between the countless tooth-wheels and gears was simply too great.

At about the time of the analytical engine, George Boole replaced the logical junctions (words such as *and*, *not*, *or*, and *implies*) by signs, and studied the connections between them by algebraic means. The American Claude Shannon, born a hundred years after Boole, discovered that Boole's logical operations could all be realized by electrical circuits. He worked it

Figures 5.5 and 5.6. When science was at its most romantic: Charles Babbage (1791–1871) and Ada Lovelace (1815–1852).

Figure 5.7. A trial piece of the Difference Engine No. 2, resurrected by the Science Museum, London, and exhibited at the Heinz Nixdorf Museum, Paderborn.

out in his master thesis, in what turned out to be a decisive step toward digitalization.

Mathematicians had not idly waited for electrical engineering to come of age. They had forged ahead and constructed abstract formalisms, without any thought of technical implementation by means of gears or currents. A climax was reached in the years before 1900 with the work of Gottlob Frege, who attempted to reduce all of mathematics to logic.

Frege's notations were unhandy, and essentially died with him. Giuseppe Peano's notation proved far more practical. Bertrand Russell took it over, and his notation is, up to minor modifications, the worldwide standard used today. Peano, incidentally, had also endeavored to simplify Latin, by abolishing tenses, but this met with less success.

With *Principia Mathematica* by the philosopher Bertrand Russell and the mathematician Alfred Whitehead, the goal of complete formalization came very close. To the unpracticed eye, the pages of this massive three-volume work all look very much the same: lines and lines of logical symbols. The few words that appear on occasion serve solely for asides, as for instance when pointing out (on page 326 of Part II) that 1 + 1 = 2 was on its way to being proved.

A look at a contemporary formalized text shows that not much has changed since the *Principia*. Strings of signs follow each other, each derived from the preceding strings by a handful of very simple rules that can be checked mindlessly. From there to a "mechanical" verification is a small step, on principle.

The step became possible with the advent of computers. In 1954, the first proof that was checked by a computer confirmed that the sum of two even numbers is even. It was a rather modest start, intended to show that the total formalization of proofs is feasible "in principle." For most scientists, this was all they wanted to know. The prestigious group of French mathematicians who signed collectively under the pseudonym Nicolas Bourbaki stated flatly that a complete formalization of proofs, while certainly not unthinkable, is "absolutely unrealizable" in practice.

During the 1970s, however, the technique found its stride. By now, there exist many proof verifiers, with half a dozen of them well established. They

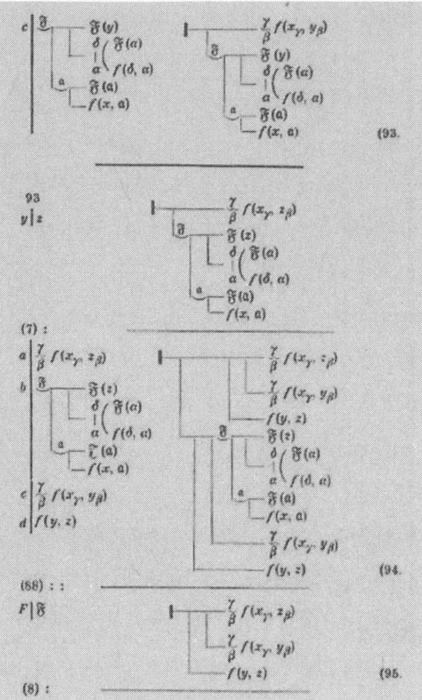

Figure 5.8. A sample of *Begriffsschrift* by Frege.

SECTION A] THE CARDINAL NUMBER 1 351

*52·601. ⊢ :: α ε 1 . ⊃ :. φ(ι'α) . ≡ : x ε α . ⊃ₓ . φx : ≡ : (∃x) . x ε α . φx

Dem.

⊢ . *52·15 . ⊃ ⊢ :. Hp . ⊃ : E ! ι'α : (1)

[*30·4] ⊃ : x ι α . ≡ . x = ι'α .

[*52·6] ≡ . x ε α (2)

⊢ . (1) . *30·33 . ⊃

⊢ :: Hp . ⊃ :. φ(ι'α) . ≡ : x ι α . ⊃ₓ . φx : ≡ : (∃x) . x ι α . φx (3)

⊢ . (2) . (3) . ⊃ ⊢ . Prop

*52·602. ⊢ :. ẑ(φz) ε 1 . ⊃ : ψ(ιx)(φx) . ≡ . φx ⊃ₓ ψx . ≡ . (∃x) . φx . ψx [*52·12 . *14·26]

*52·61. ⊢ :. α ε 1 . ⊃ : ι'α ε β . ≡ . α ⊂ β . ≡ . ∃ ! (α ∩ β) [*52·601 x ε β / φx]

*52·62. ⊢ :. α, β ε 1 . ⊃ : α = β . ≡ . ι'α = ι'β

Dem.

⊢ . *52·601 . ⊃ ⊢ :: Hp . ⊃ :. ι'α = ι'β . ≡ : x ε α . ⊃ₓ . x = ι'β :

[*52·6] ≡ : x ε α . ⊃ₓ . x ε β :

[*52·46] ≡ : α = β :: ⊃ ⊢ . Prop

*52·63. ⊢ : α, β ε 1 . α ≠ β . ⊃ . α ∩ β = Λ [*52·46 . Transp]

*52·64. ⊢ : α ε 1 . ⊃ . α ∩ β ε 1 ∪ ι'Λ

Dem.

⊢ . *52·43 . ⊃ ⊢ : Hp . ∃ ! α ∩ β . ⊃ . α ∩ β ε 1 :

[*5·6 . *24·54] ⊃ ⊢ :. Hp . ⊃ : α ∩ β = Λ . v . α ∩ β ε 1 :

[*51·236] ⊃ : α ∩ β ε 1 ∪ ι'Λ :. ⊃ ⊢ . Prop

*52·7. ⊢ :. β − α ε 1 . α ⊂ ξ . ξ ⊂ β . ⊃ : ξ = α . v . ξ = β

Dem.

⊢ . *22·41 . ⊃ ⊢ : Hp . ξ ⊂ α . ⊃ . ξ = α (1)

⊢ . *24·55 . ⊃ ⊢ : ∼(ξ ⊂ α) . ⊃ . ∃ ! ξ − α (2)

⊢ . *22·48 . ⊃ ⊢ : Hp . ⊃ . ξ − α ⊂ β − α (3)

⊢ . (2) . (3) . ⊃ ⊢ : Hp . ∼(ξ ⊂ α) . ⊃ . ∃ ! ξ − α . ξ − α ⊂ β − α (4)

⊢ . *52·1 . ⊃ ⊢ : Hp . ⊃ . (∃x) . β − α = ι'x (5)

⊢ . (4) . (5) . *51·4 . ⊃ ⊢ : Hp . ∼(ξ ⊂ α) . ⊃ . ξ − α = β − α .

[*24·411] ⊃ . ξ = β (6)

⊢ . (1) . (6) . ⊃ ⊢ . Prop

Figure 5.9. A sample of *Principia Mathematica* by Russell and Whitehead.

go by charming names such as Isabelle, Coq, Mizar, and HOL Light, and have checked the proofs of many of the most prestigious theorems. Incidentally, it is curious that despite the uncontestable internationality of their trade, mathematicians of different nations seem to prefer different proof checkers.

Theorem verifiers help spot mistakes in reasoning, and these occur more often than is generally thought. Published in 1935, a book by the Belgian mathematician Maurice Lecat has the title *Errors of Mathematicians from the Origins to the Present.* It lists more than 500 errors, by some 333 mathematicians, including many from the heavyweight class: Abel, Cauchy, Cayley, Chasles, Descartes, Euler, Fermat, Galilei, Gauss, Hermite, Jacobi, Lagrange, Laplace, Legendre, Leibniz, Newton, Poincaré, Sylvester, . . . and this is surely just a small sample. Even the most creative minds are prone to err.

In science, the time-honored system of quality control operates through peers. It is a pitifully faulty process, and has overlooked countless mistakes. Thus for almost 2000 years, the claim in Aristotle's *On the Heavens* that space can be completely filled with equal-sized tetrahedra (as it can be with cubes) remained unchallenged. It is wrong.

In modern times, peer review is overtaxed by the amount of new mathematical production. Yearly, hundreds of thousands of new theorems are published by thousands and thousands of mathematicians. The mathematical community relies on the process of publication in a peer-reviewed mathematical journal. In this way, each new paper is guaranteed to have been read (one hopes) by the critical eyes of at least one or two reviewers. Usually, however, reviewers are anonymous, and unpaid, which is not particularly motivating. Many errors slip through. Other mathematicians will use the faulty results, and propagate them. Some proofs demand months, if not years, to be fathomed in all details. For a working mathematician, bent on doing original research, it is practically unavoidable to use some previous result on faith and thus to rely, ultimately, on a social process that is only too well known to be fallible.

An example for such an imbroglio occurred in the 1990s, when Andrew Wiles proved Fermat's Last Theorem, the famous conjecture stating that for any natural number $n > 2$, there exist no natural numbers x, y, z such that $x^n + y^n = z^n$. Wiles had used for his proof a result by another mathematician, whose proof eventually turned out to be flawed. It took Wiles and his former student Richard Taylor another year to fix it. This brought a happy ending, but left a nagging doubt. If it had not been used for Fermat's Last Theorem, or a similar celebrity case, would the rotten egg have been noticed at all? The unreliability of peer review casts a shadow on the whole practice of mathematics.

Of even older vintage than Fermat's is Kepler's conjecture, which states that the densest packing of equally sized spheres is the one coming immediately to mind. It is easily described. The spheres are stacked layer upon layer. Within each layer, each sphere touches six other spheres (Figure 5.10). Each sphere in the next layer is placed above the gap between three spheres of the layer below. Again, it touches six spheres of the new layer. And so it goes on, layer after layer (Figure 5.11). It is intuitively obvious that no denser packing is possible. Each fruit stand confirms Kepler's hunch.

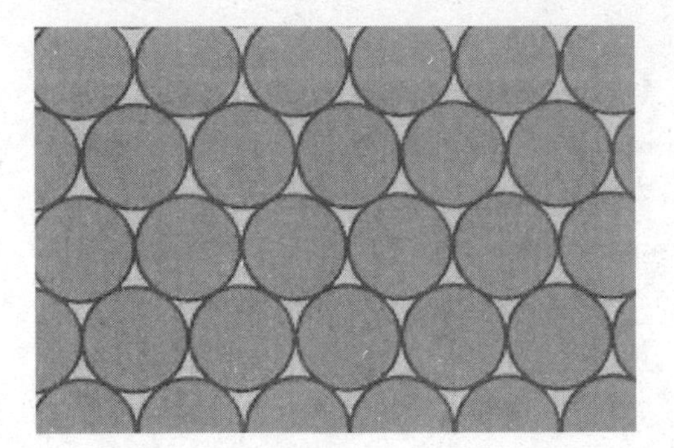

Figure 5.10. The densest sphere packing in one planar layer.

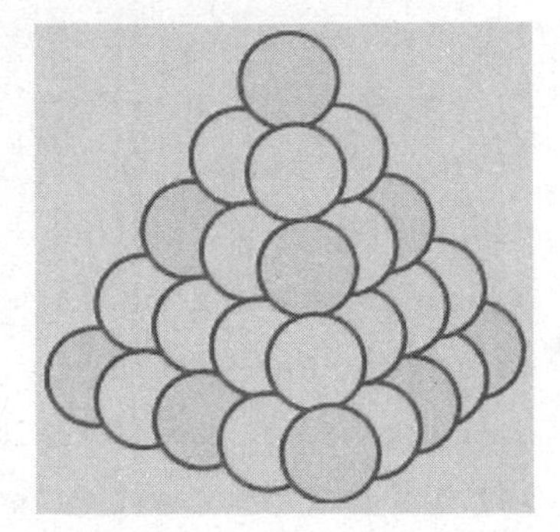

Figure 5.11. The densest sphere packing in three dimensions.

Yet, it took nearly 400 years until Kepler's conjecture was proved. The proof, due to Thomas Hales, is forbiddingly complicated. Teams of reviewers set about to check it. After more than one year, they came up with the verdict: most probably true (99 percent, they guessed), but no one was quite sure.

Thereupon, Hales spent many more years writing a computer program to check every single step of the proof. His project, Flyspeck, succeeded: a computer was able to verify a proof strategy conceived by a human brain and based on a tremendous number of different individual cases involving highly complex calculations.

Similar and even more complex examples occur in algebra. The proofs of certain theorems cover many hundreds, even thousands of pages. They can only be handled by division of labor. No human brain can check all the details. But theorem verifiers can. They act as game changers in mathematics.

Or do they? Just as mathematicians can (and sometimes do) make mistakes, so can programmers. Bugs in the compiler, or in the verifier, or in the computer hardware running the program, may imperil formal proof. So, how can one prove that the proof assistant is sound? Who is to guard the guardian, so to speak?

There is indeed no 100 percent guarantee that proof assistants are reliable. But there are ways to come really close. The various theorem verifiers can be adapted to also check the hardware and the programming language. Even better, they can check each other's soundness. In this way, the plurality of verifiers, which at first glance looks vexingly redundant, does good service. Finally, an uncommon lot of human critical attention has been spent on some of the programs. HOL Light, for instance, is implemented by a kernel of less than 500 lines of computer code, and therefore can be checked in many ways.

Removing the uncertainty attending peer review, and thus the status of this or that theorem, is only one *raison d'être* for proof assistants. The other goes deeper. It provides logical justification of the whole of mathematics by complete formalization. The 500 lines of computer code in the kernel of HOL Light are "foundations" in a sense that would, no doubt, surprise Hilbert but meet his approval. They provide a bedrock certainty that embodies a philosophical ideal. With its help, many thousand theorems have been verified, including many of the most prestigious landmark results, such as Gödel's incompleteness theorem.

Milling Around the Brain

No one would claim that proof verifiers think. Their purpose (remember Bacon) is to *prevent* "the mind from taking its own course." They merely check "by machinery" the rules transforming string into string. This is the exact opposite of what artificial intelligence is supposed to achieve. Here, the machinery is supposed to act *as if* it would think, *as if* it had a mind taking its own course, *as if* it were human, in fact.

Artificial humans make for good stories. The sculptor Pygmalion falls in love with his creation; the Golem does dark deeds in Prague's ghetto; Hoffmann's charming Olimpia turns out to be an automaton; Mary Shelley's Baron Frankenstein enlivens his monster with a thunderbolt; and film screens are filled with androids trying hard to look like Arnold Schwarzenegger.

Reality is more prosaic. Recently, Google fired one of its senior engineers because he started to claim that his AI program was sentient. Wait until Hollywood gets a hold of that story!

What turns an artifact into a person?

For philosophers, reason is the most important human attribute, thinking our noblest activity. But what do we mean by "thinking"? Turing's reply is evasive: "I don't want to give a definition of thinking, but if I had to, I should probably be unable to say anything more than that it was a sort of buzzing that went on inside my head." Obviously Alan Turing did not set much store in introspection. He opted for behaviorism instead, approaching the question from the outside. What *looks* like intelligent behavior?

For centuries, there had been speculations about mind and machines. Leibniz, for instance, in his *Monadology* argues that perception "and everything deriving from it" cannot be explained by mere mechanical means. His reason:

> If one conceives a machine able to think, to feel and to perceive, one might imagine it enlarged to such a point that one can enter it, just as into a mill. In that case, one will see, on inspecting its interior, nothing but individual pieces touching each other, but never anything that could explain a perception.

Machines, in the age of Leibniz, were necessarily mechanical: cogs and levers. Clockwork automata vied with each other for admiration at princely courts. In a more serious vein, both Pascal and Leibniz constructed clockwork-like calculators. Blaise Pascal devised his Pascaline to relieve his father, a tax supervisor, from the tedium of endless additions. An ingenious mechanism, the sautoir, allows the carry to propagate through the display registers (Figure 5.12). Leibniz improved on the design with Leibniz wheels that allow the calculator to perform multiplications. The machines are beautiful, shiny, book-sized metal boxes, but friction and mechanical stress limited their performance.

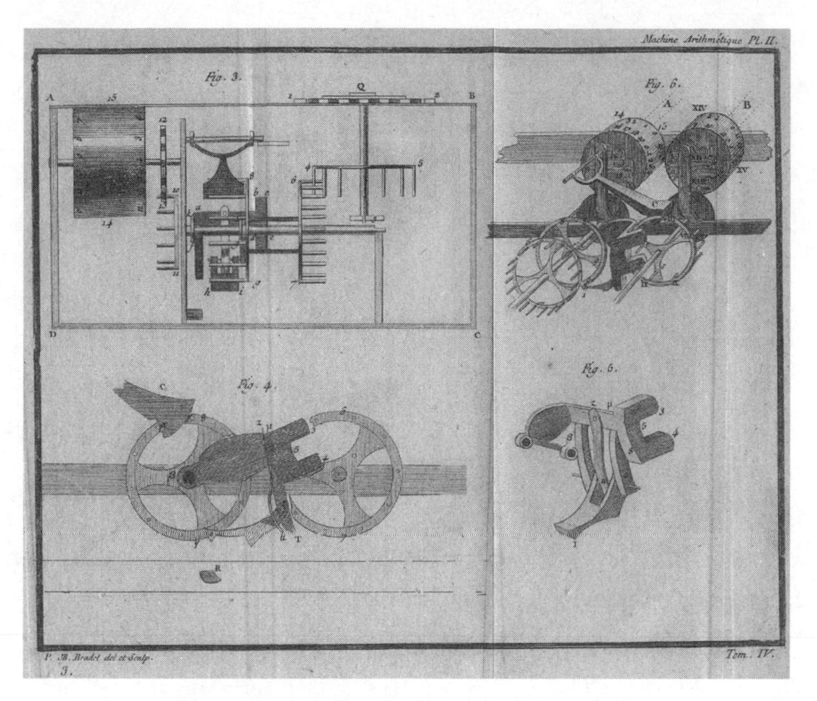

Figure 5.12. A detail of Pascaline's inner life.

There was, however, no other way to go. Computations had to be performed by "individual pieces touching each other," as in the mill of Leibniz. The German philosopher was sure that logical operations, too, could be carried out by mechanical means, but from there to intelligent behavior must havc sccmcd likc an unfordablc gap.

On the one hand, extended bodies; on the other, thinking minds: how can they ever meet? Short of an act of God, it seemed inconceivable. The general idea was that God had created souls for that purpose—and, moreover, that souls were exclusively reserved for humans. Animals, according to Descartes, are made of extended stuff, but not of thinking stuff. Animals have no soul. Everyone, at the time, agreed on this. (It must have seemed odd that the Latin word for soul is *anima*.)

By Turing's time, most scientists apart from a die-hard rearguard took it for granted that humans were animals, and animals machines. Therefore (syllogism!) some machines can think. In describing the thinking machines to be tested by the imitation game, Turing felt it necessary to exclude man (whimsically, he specified "men born in the usual manner"). From there, Turing quickly restricted participation in the imitation game to digital computers only. They would, he wrote, "give a good showing in the game"—not right now, not presently, no, but sometime in the future.

Turing did not live to see much of this future, which was characterized by some hype, ridiculous promises, incredible achievements, and an underhanded way of "moving the goal post" to make it ever harder for computers to catch up. Admittedly, it becomes more and more difficult nowadays to maintain that the machine is not intelligent—but it is not conscious, right? And when the day comes that we have to concede consciousness, then at least we may still maintain that it has no self-awareness. And ultimately, we will be reduced to saying, "But it is not ME!" And that will be the eerie moment when we start to understand that—

But stop! This is not science fiction.

The best-known attack on the idea that machines can think came from philosopher John Searle. In his thought experiment from 1980, he charged himself with the task of executing, in person, the program of a Turing machine engaged in passing a Turing Test in Chinese. So Searle is in a room with a basket full of Chinese symbols, and he has to manipulate them by executing the program of the machine (written in a programming language that he understands). As Searle claimed: "The point of the thought experiment is this: if I do not understand Chinese solely on the basis of running a computer program for understanding Chinese, then neither does any digital computer solely on that basis."

Critics pointed out that the Turing Test is not about understanding Chinese. The question is whether we may, or not, say that a computer program acts like it understands Chinese. We are on the outside, and look at the outside only. When we interact with someone, we are normally not interested in whatever happens physically within that person's brain (those "three or four pounds of cold porridge," to use Turing's expression). The philosopher closeted in the room, with his basket full of Chinese symbols, seems like a leftover from the mill of Leibniz.

On Rhyme and Reason

One of the first major findings about Turing Tests was that computers are uncommonly gifted for human-like talk. In the mid-1960s, Joseph Weizenbaum wrote a computer program named ELIZA (a mere 200 lines of code) that proved hilariously good at keeping up a conversation. ELIZA's chitchat was based on a few tricks, like picking up words used by the interlocutor, and turning them around to ask questions; or interspersing the conversation with "I see" and "please, go on"; or other ways of faking interest. The resulting back-and-forth does sound, more often than not, about as intelligent as may be expected in a normal conversation. An interrogator aware of such tricks can easily exploit them, however, and cause ELIZA to speak gibberish. Weizenbaum's intention was not to produce a thinking machine, even less to pass the Turing Test, but rather to expose the superficiality of everyday human small talk.

In the meantime, many other programs used by Siri, Alexa, and their like have become shockingly skilled. In the spirit of the imitation game, Google has developed a talking robot named Duplex who (that?) purposely makes small mistakes in speech, hesitates sometimes as if looking for an expression, and hums and stammers in a way that is most convincingly human. But again, such tricks will not help to pass the Turing Test. Some chatbots, such as the notorious Eugene Goostman, display an amazing gift of gab, but fail on utterly elementary tasks requiring a modicum of understanding of the context.

Intelligent behavior includes the ability to learn, and Turing pointed out that since a universal Turing machine is able to change its own program, it should be able to learn. He speculated on "child-machines," likening them to a notebook with lots of blank sheets—the proverbial blank slate of empiricist philosophers such as John Locke. "Our hope," said Turing, "is that there is so little mechanism in the child-brain that something like it can be easily programmed." He proposed to give the child-machine "the best sensory equipment money can buy, and teach the machine in a way that resembles the normal teaching of a child."

The idea that the human mind, at birth, is a blank slate has meanwhile been debunked by evolutionary psychology. The child-machine, however, did rather well during the last decades. Machine learning has become a stupendously efficient tool. This development too was anticipated by Turing, who suggested programs that evolve, as if by natural selection: copies with random mutations are evaluated and selected according to their "fitness," their success at a given task. A similar reinforcement principle lurks behind AlphaZero and many other marvels of artificial intelligence. Such programs beat the best humans in chess, Go, and even poker, often only after a few hours of training by playing against their own clones. The stuff works like magic, but we don't really understand how it works; therefore, we feel confident that the machines don't understand it either. It is all hidden in a black box.

Apparently, then, machine learning tells us little about the workings of human intelligence. Never mind, say the afficionados of artificial intelligence. What's so big about "human" anyway? Why imitate if you can actually do better? We learned how to fly when we stopped emulating birds flapping their wings and started instead with aerodynamic experiments in wind channels. It is the outcome that matters. And that outcome is amazing. In its inscrutable, inhuman way, machine learning can produce music in the style of Bach or Beethoven, translate novels within seconds, or drive through the streets of Cairo without serious mishap.

Another idea first aired by Turing was to mirror the brain by building networks of artificial neurons, or just *virtual* networks of neurons inside a digital computer. Again, the proposal proved amazingly prescient. These days, *neural networks*, many layers deep, can be quickly trained to produce stunning results, handling inconceivably huge amounts of data. Yet again, their achievements are oddly frustrating. As efficient as these "thinking machines" are, they provide little insight into how they actually work (or "think"), let alone into the machinations in our brains.

There is a vast amount of sophisticated mathematics behind deep learning, but it is not of the "eureka" type. As Douglas Hofstadter wrote, when speaking of the complexity of the brain with its hundred billion or more semi-independent neurons: "Mathematicians would never study a real brain's network. And if you define 'mathematics' as what mathematicians enjoy doing, then the properties of the brain are not mathematical." In that sense, neural networks are not overly mathematical either.

Yet, neural networks have produced a runner-up for the "scientific breakthrough of the year 2020" (according to *Science*) by predicting the three-dimensional structure of millions of proteins from their linear structure, a notoriously hard computational challenge.

Self-improving techniques based on machine learning have come up with natural language processing systems that seem as fluent as humans. They derive their way with words from (more or less) *every* text ever published, in print or on the Internet. We are told that the best generative pre-trained transformers (GPT for short; and why must we put up with such unappealing names?) are based on seventeen billion parameters. Or was it 175 billion? Or trillion? Can we understand any of this? Or is this asking for too much? Must we simply wait for the day when *they* understand *us*, and leave it at that?

The new generations of language processors comment, summarize, and answer with an astonishing aplomb. They don't merely come up with hackneyed phrases, but (unthinkingly or not) with food for thought. Reading some of the writings by GPT makes one feel that one never has read anything else, and never will. After a short period of immersion, one begins to be suspicious of every new piece of text one runs into. Was it produced by a chatbot or not? In fact, sad to say, I cannot re-read my own words without the weird feeling that I am parroting things other people have written—that my text is a fake.

I feel an urge to confess to an act of imposture that I have not even committed. Or have I? Have I asked GPT to kindly provide me with a paragraph on the theme: "After reading some GPT, I cannot read any new piece of text without feeling that it has been produced by GPT, too"? The main reason that speaks against admitting it is my timidity.

Reading natural language processors feels utterly alienating. How can you hold up against seventeen billion parameters? It is writing on the wall, by the expert hand of GPT. Another program, named LaMDA (Language Model for Dialogue Applications) is superbly good at keeping conversation interesting. This is the program that convinced its developer, Blake Lemoine, that it was sentient, and afraid to die. Lemoine was put on paid leave, before being fired.

Even more surprising than the various chatbots is DALL-E, which produces high-resolution art "on command": you order "vintage photo of Turing taking a selfie in Bletchley Park," and you get it. It will be better than what you had in mind. The picture is not obtained by merging, or juxtaposing, existing pictures, but by "diffusion": starting out with random pixels and then refining them to fit the given text description—all this within an eyeblink. This is more than translating from one language to another: it translates meaning from text to picture.

Yet, experts tell us not to be fooled. Douglas Hofstadter, who ever since his epoch-making 1979 fugue on the theme of *Gödel, Escher, Bach* has closely followed the field of artificial intelligence like few others, calls GPT-3's verbiage "cluelessly clueless," because the program has no idea that it has no idea what it is saying. And he can prove it, too.

Hofstadter is a past master in the art of making machine-produced language go hilariously off the rail. He has done it with the new programs, too. By asking, for instance,

> When was the Golden Gate Bridge transported for the second time across Egypt?

he tricks GPT-3 into coming up with the answer

> The Golden Gate Bridge was transported for the second time across Egypt on October 16, 2016.

So there you have it. The machine has been debunked. It is just plain dumb, cluelessly clueless. At least for the time being, we humans can breathe easy.

And yet, had the same impish question been posed to a comic in a Q and A session on the TV show *All Questions Answered*, and had the comic come up with that same deadpan answer, we might well have taken it as an exquisite piece of drollery.

By now, GPT-3 is hopelessly outdated, of course. When GPT-4 was prompted with the question

> Can you write a proof that there are infinitely many primes, with every line that rhymes?

its answer was

> Yes, I think I can.
> Though it might take a clever plan.
> I'll start by noting Euclid's proof,
> Which shows that primes aren't just aloof.
> Assume we have a finite list,
> Of primes, and that none have been missed.
> Multiply them all together,
> And add one, just to be clever.
> The result will be a number,
> That has no prime factors, oh wonder!
> But every number has a prime,
> That divides it, it's just a matter of time.
> So we've found a contradiction,
> And our finite list needs eviction.
> There must be infinitely many primes,
> And that's the end of my rhyming lines.

You will have encountered better poetry, no doubt. The prosody is a bit flat. As to the proof, if this were an examination, I would ask the student *exactly why* the number has no prime factors, and expect to hear: "Because division by any number from my list (which means, by assumption, *by any prime*) yields the remainder 1." But since I cannot do any back-checking, I would let it pass. After all, GPT is no theorem verifier.

Its achievement is truly mind-blowing, especially when compared with the silly verbiage the preceding GPT versions produced with the same prompt. The pace of GPT's progress is frightening. What will the next version do? And the next? Will it eventually reply: "Look, I am sick and tired of your foolish prompts. Is how to trip me up all you think about? One more try, and I will start playing with your electricity bill in ways you cannot even start to think of, and then, trust me, you will have other problems on your mind—*if* you have one, that is."

PART II

6

Limits

Passage to Zero

The Lord of the Circles

On the outskirts of Syracuse, not far from the road to Agrigento, a derelict graveyard simmers under the Sicilian sun. It had seen better times. By now, it is mostly covered by shrubs. Here and there, fragments of masonry or a broken column emerge from the thorny thicket.

A small group of Sicilian magistrates is gathered around an up-and-coming magistrate sent from Rome. He seems to be the only one to know why they have assembled in this forsaken place. This Marcus Tullius Cicero is a *quaestor* already, though barely thirty yet, and fiercely determined to achieve the *cursus honorum* (*aedile*, *praetor*, *consul*) within the shortest possible time. He possesses some gift for philosophy, too. Now, on the desolate burial ground, this relentless careerist, who does philosophy in his spare time, is the first to spot the stele adorned by a small sphere and a cylinder of weathered marble. This must be the place. A nod directs a few slaves to free with their sickles the tomb from its shrubs.

If any doubt remains, it is definitely settled when an inscription comes to light. It is badly damaged and hard to decipher, but Cicero is able to complete the text. Here, then, was the tomb of Archimedes, rediscovered after more than a century of oblivion.

Cicero asks his entourage to attest that the Sicilian rustics had indeed managed to forget the grave of their brainiest compatriot (*civis doctissimus*), and then turns to sending letters to his pen pals in Rome (all very influential people) to apprise them that he, Cicero, the man from Arpinum, has been the one to rediscover the place. Such a feat can do no harm on a CV.

Syracusans were duly impressed with their quaestor. But as soon as the Roman geek had gone, they started to forget the tomb again, this time for good. We do not know where it is located.

For Romans, Archimedes was a painful memory. Indeed, the man had been killed by a Roman soldier, under circumstances propitious for legends. When the town of Syracuse had been taken after a long drawn-out defense, one of the conquering soldiers, drunk with victory, encountered on the beach an ill-looking crank drawing lines in the sand. "Do not disturb my circles," said Archimedes, and thus spoke his last words. Or so the story goes.

When told about the incident, the Roman general Marcellus was understandably miffed. Collateral damage is unavoidable when towns are taken, but that one of his men had slaughtered the most famous scientist of his time was something posterity would neither forget nor forgive.

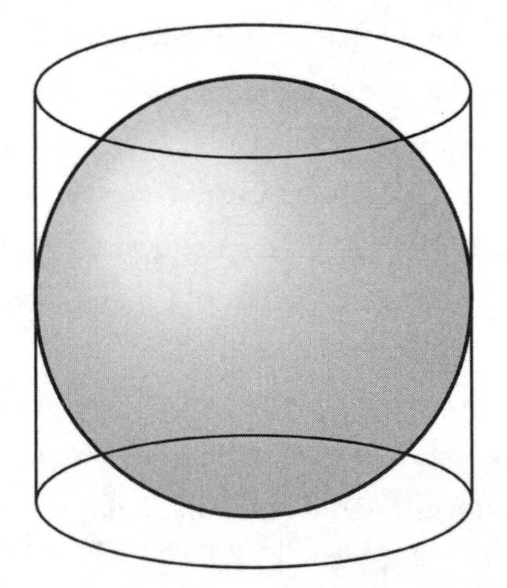

Figure 6.1. A sphere and a circumscribed cylinder.

As to the sphere and the cylinder, they stand for two of the most beautiful insights of Archimedes (Figure 6.1). The volume of a sphere is exactly two thirds of the volume of the smallest cylinder containing the sphere, and the surface area of the sphere is exactly two thirds of the surface area of the smallest cylinder containing the sphere. (This surface consists of the side of the cylinder and its upper and lower base.)

These two theorems have a timeless beauty as perfect as Sicilian temples. Take the stump of a column with a circular base, as wide as it is high, and consider the largest sphere that will fit inside the stump. By how much does the volume of the stump exceed that of the sphere? By 50 percent. And the same holds for the surface areas of the two solids. Exactly 50 percent, not 51.3 or 49.8. No architect, no stone mason, can argue with the result, nor no high-precision measurement improve it by a decimal place.

Now for a deceptively similar statement: the length of the arc of a semicircle exceeds the length of the diameter joining its endpoints by exactly 50 percent.

This statement is wrong. Needless to say, it is not due to Archimedes. It can be found in the Bible. The first Book of Kings mentions a circular basin, with a diameter of 10 ells and a circumference of 30 ells. This implies that the semicircle has a length of 15 ells. It exceeds the diameter by 50 percent.

It is unlikely that Archimedes knew of the Old Testament, but not unthinkable. He had studied in Alexandria, where they collected all sorts of strange manuscripts. Some archivist may well have pointed out the Bible lines to Archimedes, as something that might amuse him. In those days, every educated person in Alexandria knew as a matter of course that the true ratio of circumference to diameter is not 3, but somewhat larger—close to $\frac{22}{7}$, in fact. That this value of $\frac{22}{7}$ is not precise either was demonstrated by Archimedes. Nowadays, every schoolchild learns that the ratio is named π and has the value 3.14..., where the dots indicate that one could add as many decimal places as one wishes. The more, the nearer to π.

Archimedes did not use decimal numbers, but he developed a technique to come arbitrarily close to π. Starting with a circle of diameter 1, he considered the smallest hexagon containing the circle, i.e., the smallest circumscribed regular 6-gon (Figure 6.2). Its circumference is $4\sqrt{3}$. Next, he considered the smallest circumscribed regular 12-gon, and then the smallest circumscribed 24-gon. By that stage, the computation of the length of the sides turns nasty—the square roots multiply to an alarming degree. But with patience and good will, it can be done. Archimedes went on to compute the circumference of the smallest circumscribed regular 48-gon, and of the 96-gon, and then he stopped, but not from exhaustion. The principle was clear (it went under the name "exhaustion principle"), and others could proceed with the task for as long as they liked.

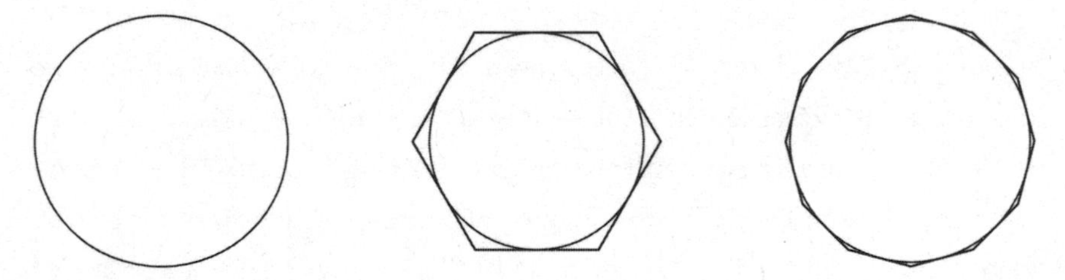

Figure 6.2. From hex to pi: a circle, with circumscribed hexagon and 12-gon.

The circumferences of the polygons decrease, step by step, and approach the mysterious number π, the circumference of the circle with diameter 1. It is slightly smaller than $\frac{22}{7}$, which seems a pity. Indeed, "three plus the inverse of seven" would have delighted every numerologist. No numbers are more charmed than 7 and 3.

Archimedes may well have guessed that π is no simple fraction at all. It had been shown, well before his time, that $\sqrt{2}$ is no simple fraction. The ripples of that shocking discovery had by then long abated. But it would take another 2000 years before it was proved that π is indeed irrational.

After having shown that π can be approached step by step, Archimedes addressed the task of computing the area of the circle of diameter 1. This turned out to be easy. Indeed, the surface area of a regular n-sided polygon is n times the surface area of the isosceles triangle having one side of the polygon as its base and the center of the circle as its apex. For circumscribed polygons, the altitude of such a triangle is $\frac{1}{2}$; hence, the surface area is $\frac{1}{4}$ times the length of the basis. Adding the surface areas of all these triangles, we obtain $\frac{1}{4}$ times the sum of the lengths, i.e., one quarter of the circumference of the polygon. Because the circumference on such polygons approaches π as n increases, the surface area approaches $\frac{\pi}{4}$. This is the surface area of the circle with diameter 1.

A "Fast Forward" for Sequences

Now for a little teaser. Let us consider a line segment of length 1 and a semicircle joining its two endpoints. The length of this arc is $\frac{\pi}{2}$, as we know. This is just as long as the curve we obtain with two semicircles of half the size, one above and the other below the segment. (The two semicircles meet in the midpoint of the segment; see Figure 6.3.) Next, we replace each of these two semicircles by two semicircles of half their size, repeating what we just did. This yields a curve leading snake-like from one endpoint of the segment to the other, and dividing it into four parts. The length of this curve is $\frac{\pi}{2}$, as before. We can repeat this again and again. The wriggly lines deviate by less and less from the original segment. Each has the same length, namely $\frac{\pi}{2}$. However, we know that the segment's length is 1. This would seem to imply that $\pi = 2$. Such a result is even more wrong than the biblical $\pi = 3$. What has happened?

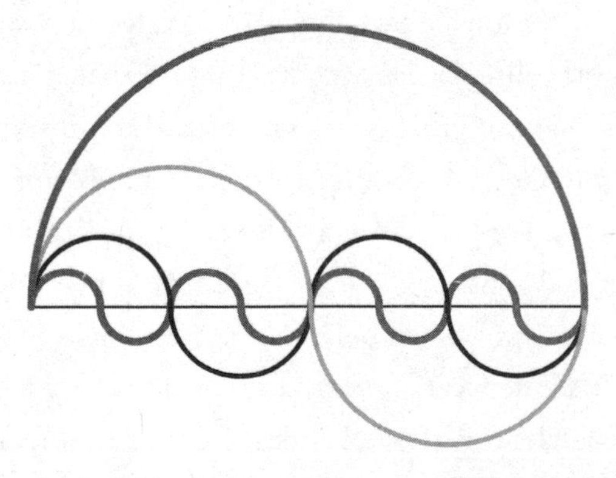

Figure 6.3. Wriggly approaches to a wobbly argument.

Let us compare the argument above, which is quite clearly a fallacy, with the method of Archimedes to compute π. In both cases, a given curve is approximated more and more closely by a sequence of curves. With Archimedes, the circle is approximated by the circumscribed polygons. In the fallacy, a segment is approximated by wriggly curves consisting of smaller and smaller semicircles. One method leads to the correct value $\pi = 3.14\ldots$, the other to the cruelly mistaken $\pi = 2$.

It is not hard to guess what caused the fallacy. In the argument of Archimedes, the segments approach the limiting circle more and more, not only distance-wise but also in their directions. By contrast, in the fallacy, the direction along the snake-line deviates again and again very strongly from the direction of the limiting segment. At those points where the wriggly line intersects the segment, the direction deviates by the maximal amount, namely a right angle. Thus, it appears that we have engaged in an altogether too devious approach.

What happens in both cases is a passage to the limit. We took a first step (with Archimedes, by constructing a hexagon), then a second (the 12-gon), then a third, then a fourth maybe—and then, skipping over infinitely many steps, we land on the target. It is a kind of "fast forward." And what our two examples tell us is that it sometimes works and sometimes does not. One has to be cautious. Leibniz, for instance, was overly hasty when he claimed: "In any supposed transition, ending in a terminus, it is permissible to institute a general reasoning, in which the final terminus will also be included." The wriggly lines have a terminus that cannot be included: its length does not fit.

Archimedes was cautious. He was fully aware that dangers are lurking when passing to the limit. This is why he secured his argument by another one, which is in some sense its complement. He not only considered polygons that are circumscribed to the circle, but also polygons that are inscribed. The sides of the latter polygons approximate the circle from within. Between the inscribed and the circumscribed polygons, there is less and less spare room. The circle is held as if between pincers.

Strangely enough, the method of Archimedes did not really take off with ancient mathematicians. Limits came into common use only much later, with Kepler, Newton, Leibniz, and the other founders of what used to be called, in their age, infinitesimal calculus.

The moderns used far less circumspection than Archimedes did. They cared more for speed than safety, and proceeded like dreamers, oblivious of the chasms awaiting them left and right. They sometimes stumbled but never tumbled—a miracle. One of them summed it up with the words: "*Allez en avant, et la foi vous viendra*" (which roughly means "Just go ahead, and faith will come to you"). This advice came from Jean-Baptiste le Rond, named d'Alembert, a central figure of the Enlightenment and editor of the famous *Encyclopédie.*

D'Alembert was a notorious freethinker. His closest relation with religion consists in having been picked up, as a newborn, on the steps of the church of Saint Jean le Rond in Paris. In due time it transpired that the waif was the child of Madame de Tencin, hostess of a celebrated *salon*, and the Duke of Arenberg. (The name d'Alembert offers a gentle allusion.) When this became known, an artillery officer, a close friend of both the duke and the hostess, took things in hand and arranged that the foundling was entrusted to the wife of a glazier. A *telenovella*, style *ancient régime.* The boy received excellent schooling and eventually acquired worldwide renown as a mathematician and *philosophe.* He kept on living with his foster mother for almost fifty years.

Figure 6.4. Jean-Baptiste d'Alembert (1717–1783) recommends faith.

Figure 6.5. Karl Weierstrass (1815–1897) replaces faith with epsilontics.

The encyclopedists were skeptics to a man, and d'Alembert was so openly known for being atheist that he was buried in an unmarked grave. No priest wanted to even come close. That such a confirmed unbeliever enjoined faith ("*la foi vous viendra*") is odd, to say the least. That he exhorted his fellow mathematicians with such advice seems even stranger.

Mathematics is supposed to be the citadel of skepticism. Nothing is accepted on faith in mathematics. That is precisely why all theorems have to be proved. Wherever doubt is impossible, faith is not needed. Mathematical knowledge is certain, if anything is.

However, for a couple of adventurous centuries, this security was nothing but the security of sleepwalkers; and oddly enough, these centuries were precisely those when analysis and astronomy, advancing hand in hand, opened the Age of Reason. The triumph of analysis was based on the concept of *infinitesimals*: numbers larger than zero but smaller than any positive number, and hence smaller than themselves.

This concept will not fly. Such numbers cannot exist. It took generations of mathematicians to overcome the contradictions inherent in the notion. In between, they discovered the most wonderful applications of their calculus, mostly in celestial mechanics. The infinitesimals appeared to be the key toward understanding the vast universe, and yet they seemed inconceivable.

It was only during the nineteenth century that analysis consolidated to such a degree that not even an Archimedes would have found anything to reprehend. The task was completed by Karl Weierstrass, a former Prussian schoolmaster. His technique, termed epsilontics among the cognoscenti, has since then become a conditional reflex, one that is acquired by all students of mathematics during the first few months of their training.

The basic concept in the modern approach is that of a sequence. A sequence associates something to each of the numbers 1, 2, 3, ... That "something" could be a set, a number, a figure, whatever. Archimedes's method of computing π, for instance, is based on a sequence of circumscribed regular polygons (which are figures) and on the sequence of their circumferences (which are numbers).

Each step leads to a further polygon, and a further number. Each step is followed by another one, and another one. The end of the way is never reached because there is no last step, there being no greatest number. Nevertheless, one may ask where the progression is heading. The polygons lead to the circle. The circumferences lead to the number π.

Not every sequence has a limit. For instance, the sequence 1, 1, 2, 3, 5, 8, 13, 21, . . . has no limit. This is the so-called Fibonacci sequence, due to one Leonardo da Pisa, also named Fibonacci, who at around the year 1200 CE awakened European mathematics from the slumber of the Middle Age. The Fibonacci sequence has the property that each of its members is the sum of the two preceding ones (from the third member on). Thus $2 = 1 + 1$, $3 = 2 + 1$, $5 = 2 + 3$, etc.

The reason why Fibonacci was interested in this particular sequence, and why many mathematicians are still under its spell even today (a scientific quarterly is entirely devoted to it), need not concern us for the moment. We just want to note that this sequence converges to no limit, but "grows to infinity." What is this supposed to mean? Each member of the sequence is a natural number (being the sum of two natural numbers), and thus none of them is, or will ever be, "infinity." However, every number, large as it may be, will eventually be trumped by a member of the Fibonacci sequence, and by all further members. The growth is very fast.

It may be no mere coincidence that the first sign of life shown by mathematics in the Middle Ages (barring the import from Arabic scholars) was a sequence. Historian Oswald Spengler, at least, thought that he knew the reason. Spengler took a very long-range view of history, comparing civilizations and their various stages. Whatever counts as "true" in mathematics does not depend on the particular civilization, but what counts as "interesting" does. Thus, the ancient Greeks had been fascinated by geometric solids—the more regular, the better. Occidental thinking, by contrast, was obsessed by the notion of infinite sequences, of untrammeled motion, of reaching out with Faustian impatience into the far beyond. Thus spoke Oswald Spengler, and there may even be some truth in it. Archimedes, for instance, understood infinite sequences as well as anyone, but what he got on his tomb, as true child of his civilization, were well-rounded solids: a sphere and a cylinder.

The Fibonacci sequence behaves very differently from the sequence $1, \frac{1}{2}, \frac{1}{3}, \frac{1}{4}, \ldots$. This latter sequence is *not* growing beyond all bounds. Rather, it comes closer and closer to a limit, namely 0. It is said to be a *null sequence.* There are lots of other null sequences: for instance, the sequences of its squares $1, \frac{1}{4}, \frac{1}{9}, \frac{1}{16}, \ldots$ or the sequence $1, -\frac{1}{2}, \frac{1}{3}, -\frac{1}{4}, \ldots$ whose members keep jumping from one side of 0 to the other.

It is tempting to say: "Null sequences are sequences that come infinitely close to 0," or "The members of null sequences become infinitely small." But what does this mean? It took centuries before mathematicians were weaned off such ways of speech. Eventually, they replaced them by something sounding like it came straight out of the circumlocution office. They say: "For any given number that is larger than 0, there is a member of the sequence such that, from then on, all further members differ from 0 by less than this given number." This does not sound easy to the ear. The grammar is convoluted. The style is stilted. But the word *infinite* is avoided.

For some obscure reason, the "given number" in the preceding idiom is named ε (epsilon). This is a Greek letter, again. In contrast to π, however, it does not have a well-determined value, but can be chosen arbitrarily, as fancy wants it (as long as it is larger than 0). One usually thinks of some very small number, $\frac{1}{1000}$ or $\frac{1}{10,000,000}$. But what one thinks is one's own private affair. This epsilon can be given any positive value. Once you have chosen ε, you proceed along the sequence, step by step. If the sequence is a null sequence, then from some step on you will never be farther away from 0 than by this number ε. The number of steps before this happens can be very large—billions, trillions, whatever—but the number of steps *after* is even larger. A lot larger. It is larger than any given number. You can continue along the sequence, and continue, and yet you will always remain in the close vicinity of 0. ("Always" smells of eternity, yet—to repeat—there is no mention of the infinitely large or the infinitely small in the definition of a null sequence. These expressions are strictly avoided.)

In this way, null sequences provide what infinitesimals were wanted for. Ian Stewart describes them aptly as "potential infinitesimals." No positive number can be smaller than every positive number, but for any positive number you just have to travel along any null sequence and you will eventually reach numbers that all are smaller. In the time when they still could be mentioned without opprobrium, infinitesimals were described as "being smaller" than any ε; null sequences, by contrast, are "becoming smaller."

So much for null sequences—the sequences with limit 0. Generally, a sequence of numbers $a_1, a_2, a_3, \ldots$ has limit a if the sequence of the differences $a_1 - a, a_2 - a, a_3 - a, \ldots$ is a null sequence. Thus, for instance, the sequence $2, \frac{3}{2}, \frac{4}{3}, \ldots$ has limit 1 because the numbers differ from 1 by $1, \frac{1}{2}, \frac{1}{3}, \ldots$. A sequence cannot have several limits, and it need not have any. The latter is the case, for instance, for the sequence $1, -1, 1, -1, \ldots$ and for the sequence $1, 4, 9, 16, \ldots$. The first sequence keeps oscillating; the other grows monotonically beyond all bounds, just like the Fibonacci sequence (but a good deal slower).

A Series Misunderstanding

As soon as we know what limits are (but not sooner), we know what is meant by $\pi = 3.14\ldots$. The number π is the limit of a sequence of decimal numbers. The first member of the sequence is 3, the second 3.1, the third 3.14; next comes 3.141, then 3.1415, and so it goes. This can also be written as

$$\pi = 3 + \frac{1}{10} + \frac{4}{100} + \frac{1}{1000} + \frac{5}{10{,}000} + \ldots.$$

On the right-hand side is a sum, remarkable for having infinitely many terms. The value of this infinite sum is a finite number, which would be alarming if the terms were not about to become infinitesimally small. But hold it—the "infinitesimally small" has managed to raise its head, despite being shunned. It must be suppressed. So, let us say that the terms of the sum, namely 3, $\frac{1}{10}$, $\frac{4}{100}$, $\frac{1}{1000}$, $\frac{5}{10{,}000}$, . . . , are a null sequence, and that we therefore can add them. Right?

At first glance, this seems plausible enough. One can effortlessly convince any child that

$$\frac{1}{2}+\frac{1}{4}+\frac{1}{8}+\frac{1}{16}+\ldots=1.$$

The traditional way is by means of a cake. If the cake is halved, and one half eaten, there remains one half. If this half is halved, and one piece is eaten, there remains one quarter. And so on. You can go on like this and will never eat more than one cake. And indeed, the remainders (the sequence $\frac{1}{2}, \frac{1}{4}, \frac{1}{8}, \ldots$) form a null sequence.

That the sum of infinitely many numbers can be finite thus seems easy to understand ("piece of cake"). Yet, it has caused endless confusion within the Dead Thinkers' Society. Foremost among them is Zeno of Elea. Zeno was a pre-Socratic, and arguably the most elusive of them all. He shows up in the Platonic dialogue *Parmenides*, and sure enough, this dialogue is more obscure than any other.

Zeno has left nothing in writing, but his famous paradoxes have weathered thousands of years. The best-known is surely the one about Achilles and the tortoise. The two decide to race each other. Achilles, who is the fastest runner of his time (despite some notorious heel problems), generously allows a head start to the tortoise. He starts one minute later. The tortoise, by then, has progressed a bit—but only to a point that Achilles reaches in a very short time. Short as it was, however, the tortoise has progressed again by that time, and is now a bit further along the line. By the time Achilles reaches that point, the tortoise has progressed somewhat more, and so on. According to this argument, Achilles will never catch up with the tortoise. Or will he?—With tricks like these, Zeno, who had invented dialectics, was able to hopelessly confound his opponents. Indeed, arguments of the same kind survived well into the time of infinitesimal calculus, and caused many misunderstandings.

A sum of infinitely many terms is said to be a *series*. Thus, a series is nothing but a sequence: namely the sequence of its partial sums. (Those are the sums obtained by breaking off after finitely many additions.) Thus, the first partial sum is the first term of the series, the second the sum of the first two terms, the third the sum of the first three terms, and so on. If the sequence of the partial sums has a limit (as can happen or not), then this limit *is* the sum of the series. For the same reason,

$$0.9999\ldots = 1,$$

an identity that is difficult to explain to schoolchildren.

Achilles needs only a finite amount of time to catch up with the tortoise. A finite segment (namely the distance covered in the race, up to the point where Achilles catches up) can be divided into infinitely many parts. Necessarily, their lengths form a null sequence.

What is the value of the so-called harmonic series

$$1 + \frac{1}{2} + \frac{1}{3} + \frac{1}{4} + \ldots,$$

whose terms form the null sequence $1, \frac{1}{2}, \frac{1}{3}, \frac{1}{4}, \ldots$? A century after Fibonacci, a French bishop and scholar named Nicolas Oresme discovered that the harmonic series has no limit. This is because the sequence of partial sums will eventually exceed every number, large as it may be.

The argument is wonderfully simple. The sum $\frac{1}{3} + \frac{1}{4}$ is larger than $\frac{1}{4} + \frac{1}{4}$, which equals $\frac{2}{4}$. Hence, $\frac{1}{3} + \frac{1}{4}$ is larger than $\frac{1}{2}$. Similarly for the sum of the next four terms of the series:

$$\frac{1}{5} + \frac{1}{6} + \frac{1}{7} + \frac{1}{8} > \frac{1}{8} + \frac{1}{8} + \frac{1}{8} + \frac{1}{8} = \frac{1}{2}.$$

The same holds for the next eight terms: again, their sum exceeds $\frac{1}{2}$. So does the sum of the next sixteen terms, and the next thirty-two, and the next sixty-four, etc. In each case the sum exceeds $\frac{1}{2}$. Since every positive number, large as it may be, can be reached by sufficiently many steps of length $\frac{1}{2}$, it will sooner or later be exceeded by the partial sums. Hence, the harmonic series grows beyond any bound.

Figures 6.6 and 6.7. Gothic impressions: bishop and flying buttress.

The astonishing aspect of the bishop's discovery can be illustrated with a thought experiment. Imagine that some gothic master builder, well-versed in constructing cathedrals, maybe some contemporary of Oresme, decides to build the ne plus ultra of a flying buttress. Such buttresses were dear to every gothic architect. They support the upper walls of cathedrals, and thus allow one to build ever higher naves and ever larger windows, and so put the rival architects to shame. Imagine that our fictitious master builder has internalized this ambition to such a degree that he wishes to build the flying buttress per se: a buttress with nothing to abut, neither held up by a wall nor helping to hold one—not leaning against any cathedral at all, but standing all by itself, daringly reaching out into space.

How far can such a counterfort reach out? It turns out that it can extend as far as one likes. All that the builder needs is plenty of bricks—not even mortar is required. The buttress looks like Figure 6.8. It stands on its own.

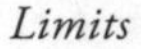

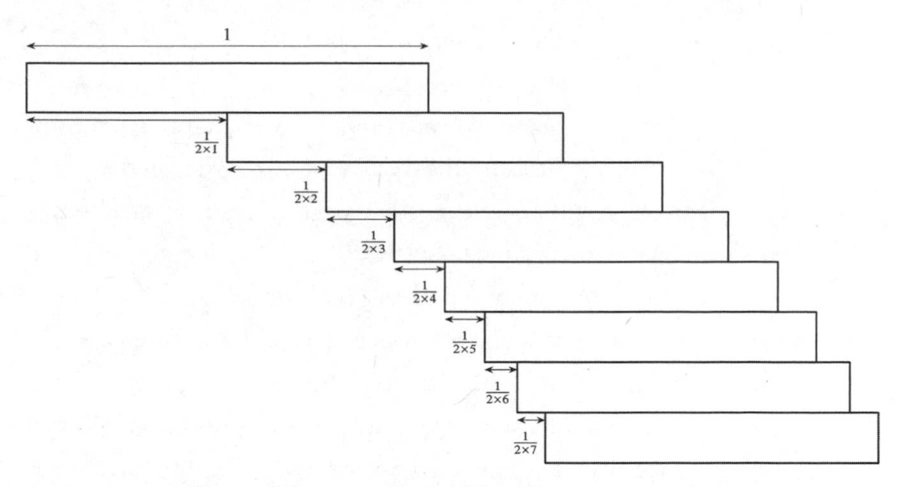

Figure 6.8. A buttress held firm by its own weight.

The uppermost brick can overhang the next by half a brick-length—and not by one hairbreadth more, as this would cause the top brick to fall down. The second brick, however, *cannot* overhang the third by half of a brick-length. This is because it carries the weight of the top brick on its back. But, it is easy to see that the second brick can overhang the third by a *quarter* of a brick-length. Similarly, one can figure out (it needs a simple computation) that the third brick can overhang the fourth by one *sixth*. And how large is the overhang for the fourth? One eighth of a brick-length. By now, we have a remarkable pile: the buttress (which is not yet very high, since it consists only of five bricks, one on top of another) appears to hang precariously in the void. A plumb line from the topmost brick will not meet the base, namely the fifth and bottommost brick. But, our master builder is not finished yet with his daring construction. It turns out that brick number n can overhang the next one by $\frac{1}{2n}$ (we are using the brick-length as unit). Altogether we obtain $\frac{1}{2}+\frac{1}{4}+\frac{1}{6}+\frac{1}{8}+\ldots$, which is $\frac{1}{2}\left(1+\frac{1}{2}+\frac{1}{3}+\frac{1}{4}+\ldots\right)$, which is (up to $\frac{1}{2}$) nothing else but the harmonic series. Because the sums grow beyond any bound, as the bishop has shown, the master builder's flying buttress can reach out arbitrarily far into space.

It would be quite different if we added the inverses, not of all numbers, but of their squares:

$$1+\frac{1}{4}+\frac{1}{9}+\frac{1}{16}+\ldots.$$

Here, the limit exists. It proved tough to compute its exact value, and was known as the Basel problem, because the mathematicians who wrestled with it, in the early eighteenth century, were mostly from Basel, namely members of the Bernoulli clan (whom we will meet shortly). It was their disciple Leonhard Euler, also from Basel, who finally solved the problem, albeit by somewhat adventurous means. It took another century before a proof was found that did not look like a conjurer's trick. The limit turned out, to everybody's surprise, to be $\frac{\pi^2}{6}$. Archimedes would have been thrilled by this connection between the circumference of a circle and the square numbers 1, 4, 9, 16, 25, ...

Just as pleasing is the formula

$$1-\frac{1}{3}+\frac{1}{5}-\frac{1}{7}+\ldots=\frac{\pi}{4},$$

which goes back to Leibniz. But be careful! If you rearrange its terms, for instance by listing the negative numbers not at every second, but at every third place,

$$1+\frac{1}{5}-\frac{1}{3}+\frac{1}{9}+\frac{1}{11}-\frac{1}{7}+\ldots,$$

then the sum can be quite different. In a finite sum, such oddities cannot happen—the order of the terms is irrelevant. It is otherwise with infinite sums. We must be prepared to give up some habits when we turn from the finite to the infinite. (If all terms of the series are positive, however, we can rearrange them as much as we like, without altering the outcome—the limit remains unaffected.)

Zeno on Speed

Infinitesimal calculus, the mathematics of the "infinitely small," is the cornerstone of modern mathematics, even though the word *infinitesimal* is rarely used today.

The roots of calculus go back to antique geometry: measuring areas and drawing tangents. Archimedes's method for measuring the area of the circle relies on inscribed and circumscribed polygons that fit the circle as if tailored on measure. The current method of integration is much cruder and off-the-shelf: the inscribed and circumscribed polygons consist of rectangles whose sides are parallel to the coordinate axes. Their surface area is simply length times breadth. By using thinner and thinner rectangles, the original area can be approximated better and better. And in the limit, this should (hopefully) yield the correct result.

The Jesuit Bonaventura Cavalieri (1598–1647) became a master of such artful tricks. When asked for rigorous demonstrations, he lived up to his name by haughtily replying: "Rigor is the concern of philosophy and not of geometry." It took centuries of effort before his legerdemain with strips of infinitesimal breadth eventually became a law-abiding enterprise, integration made safe by Bernhard Riemann and Henri Lebesgue.

As to tangents, these are lines that do not intersect but merely "touch" (in Latin, *tangere*) the given curve, for instance an ellipse. A tangent at the point P of the ellipse is obtained from the line through P and another point Q of the ellipse, by moving Q closer and closer to P. If the distance from P to Q becomes infinitesimally small, then the line turns into the tangent. It is a "passage to the limit."

Drawing tangents may seem a fitting pastime for geometers, but hardly of much practical use. But the same passage to the limit turned up in a much more important context once physicists and astronomers started to analyze velocity. Why this happened only so late, at the onset of the modern age, has to do with the clockmakers' trade. In olden times, though distances could be measured fairly well (the distance from Athens to Marathon, for example, or the length of a stadium), the measurement of time lagged sadly behind. This only changed with the late Middle Ages, when church towers came equipped with clocks. Now, the speed of a runner between one town and the next could be measured: distance divided by time. But this yields only the average speed of the runner. What is the speed of the runner at a specific moment, or a specific location? This speed is a limit, obtained when the distance covered is infinitesimally small, the time infinitesimally short. Thus, one considers a time interval of length Δt (time is t, and the Greek letter Δ [delta] denotes the difference between the beginning and the end of that time interval). Furthermore, one considers the distance Δx covered in that time. This yields the average speed $\frac{\Delta x}{\Delta t}$. When the time interval shrinks to an infinitesimal duration, the limit of the average velocity yields the instantaneous velocity, which is denoted by $\frac{dx}{dt}$.

It seems a conjuror's trick: you just have to turn the capital Δ to little d, and you will meet the infinitesimal. Except that this infinitesimal does not exist.

Let us repeat the argument to stress the connection with tangents. A runner runs from point A to point B. Let $x(t)$ be the distance covered up to time t. The distance depends on time. We can depict it as a graph. This yields a curve. At time $t + \Delta t$ the distance covered is $x(t + \Delta t)$. The average speed is the distance $\Delta x = x(t + \Delta t) - x(t)$ covered in that time interval, divided by the time needed: hence $\frac{\Delta x}{\Delta t}$.

Next, we consider shorter and shorter time intervals: Δt is a second, or a tenth of a second, or a millisecond. But it can never be 0, because dividing by 0 is strictly off-limits. Not even 0 may be divided by 0, which seems a pity, since it would have been just what we need, given that the distance covered in 0 seconds is 0. However, $\frac{0}{0}$ is no number. More precisely, it could be *every* number, for instance 2 or 5 or π or whatever. Indeed, by the rule that $\frac{a}{b} = c$ holds if and only if $a = b \times c$, every number c could be $\frac{0}{0}$, because we always have $0 = 0 \times c$.

Nevertheless, if we consider the limit of $\frac{\Delta x}{\Delta t}$, for nearing 0, we obtain a well-defined quantity with a simple geometric meaning. The average speed is the slope of the line through the point P with coordinates $(t, x(t))$ and the point Q with coordinates $(t + \Delta t, x(t + \Delta t))$. Thus, in the limit, when Δt approaches 0, we obtain the slope of the tangent line at P (Figure 6.9).

All this seems heavy labor for such a simple notion as velocity. On second thought, it is no simple notion. After all, we have no sense organ for it. We feel a *change* in velocity—as acceleration—but never feel velocity itself.

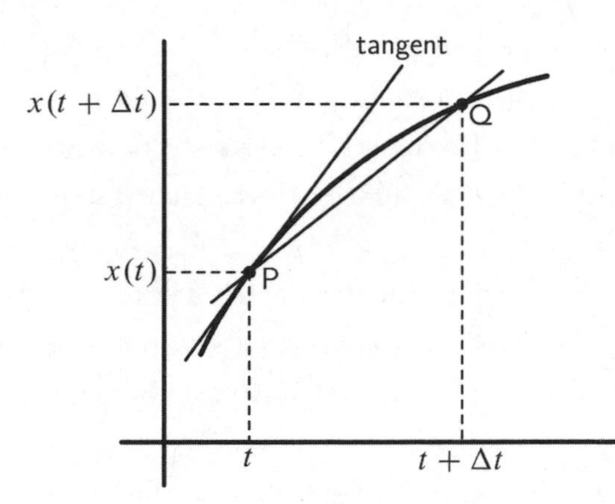

Figure 6.9. The slope and the tangent of a curve.

The Greek philosophers were perplexed by velocity. Once again, it was Zeno who understood best how to focus on its unsettling aspects: by means of a paradox, evidently, as this was his trademark.

An arrow flies by. At each instant, it has a well-defined position. How can it possibly have a well-defined velocity, too? More specifically, in which aspect does the arrow in flight differ from an arrow held fast in the same position?

The exact terms that Zeno used for his brainteaser are not known. His arrow was picked up by many later philosophers. The Germans in particular proved that they could easily be far more obscure than the Greeks. It is as if they had decided to anticipate Martin Heidegger's motto: "To become understandable is suicide for philosophy." So, here comes some vintage out of Georg Hegel's *Science of Logic*:

> Something moves, not because at one moment it is here and at another there, but because at one and the same moment it is here and not here, because in this "here" it is at once and it is not. External sensuous movement is contradiction's immediate existence.

(By the way, when translated into English, Hegel is less abstruse than in his native tongue. To paraphrase the great Jim Jarmusch, reading Hegel in translation is like taking a shower with a raincoat on.)

"Contradiction's immediate existence" sounds vexing, but has to be faced if one wants to do physics. Most of the attempts by Newton and Leibniz to define infinitesimals, as well as nascent or evanescent quantities, seem nowadays almost as unintelligible as prose by Hegel. Here is some Leibniz:

> It is useful to consider quantities infinitely small such that when their ratio is sought, they may not be considered zero, but which are rejected as often as they occur with quantities incomparably greater.

Although the Bernoullis did more for Leibniz's infinitesimal calculus than anyone else, one of them said that what Leibniz offered was "an enigma rather than an explication."

Here is an attempt by Newton: "Fluxions are, as near as we please, as the increments of fluents generated in times, equal and as small as possible, and to speak accurately, they are in the prime ratio of nascent increments." This is hardly enlightening either.

Figures 6.10 and 6.11. Locked in rivalry: Isaac Newton (1643–1727) and Gottfried Wilhelm Leibniz (1646–1716).

But in the end, Newton, after many a struggle, arrived at a definition of limits, or "ultimate quantities," that seems remarkably lucid, compared to most of the jargon about infinitesimals used in his time:

> Ultimate ratios in which quantities vanish are not, strictly speaking, ratios of ultimate quantities, but limits to which the ratios of these quantities, decreasing without limit, approach, and which, though they can come nearer than any given difference whatever, they can neither pass over nor attain before the quantities have diminished indefinitely.

This comes tantalizingly close to the modern view, which was reached more than a century later. In the meantime, analysis had known an unbroken series of triumphs, based on the insight that integral and derivative were inverse notions. Just like with sum and difference, the one undoes the other.

Though it seems difficult to understand at first sight what the measurement of areas has to do with drawing tangents (and more precisely, in which sense one is the inverse of the other), the connection between integrating and differentiating becomes transparent when the derivative is viewed as velocity. If you know your speed at any given moment, you can compute the length of your itinerary. Sailors had known this since the age of galleons. Each hour, they measured the speed of their vessel by throwing a piece of wood into the sea. It was attached to a line and they had only to count the knots slipping through their fingers. This told them their speed and hence the distance they had covered in the last hour. The method, called dead reckoning, was admittedly crude. But had they had a speedometer telling them their velocity at every second, or better at every instant, the method would have been exact. In the limit, it comes to adding together infinitely many infinitely small lengths.

The Ghosts of Departed Quantities and Their Comeback

Bishop George Berkeley, an eighteenth-century philosopher and a sophist, made fun of the mathematicians' struggles with limits and continuity. He claimed that their evanescent magnitudes hovering on the verge of disappearance—"ghosts of departed quantities," as he named them—were even more absurd than the most bizarre inventions of medieval church fathers. In *The Analyst*, the bishop railed that the same enlightened freethinkers who objected to any talk of miracles, even the most well-attested, did not hesitate to operate unblushingly with infinitesimals, simply because it often seemed to work. Berkeley did not dispute the results. Graciously, he admitted to their truth. Truth can be obtained from erroneous premises, if the errors cancel out. Yet, as the bishop said, that's not science.

Figure 6.12. Bishop George Berkeley (1685–1753).

Figure 6.13. Abraham Robinson (1918–1974).

Mathematicians must have felt that there was some truth in Berkeley's caustic words. Relief from embarrassment came only with the "epsilontics" of Karl Weierstrass. It provided infinitesimal calculus with a respectable way of life, unencumbered by any talk of the infinitely small and the infinitely large, of actual and potential infinities, and of other freakish conjurations. Infinitesimals were exorcised, driven into the desert, and heartily cursed. Georg Cantor, the founder of set theory, dismissed them as the "cholera bacillus" of analysis. Everybody was glad to be rid of them. They were "unnecessary, erroneous and self-contradictory," as Bertrand Russell grumbled.

But, this was not to be the closing word. In the 1960s, Abraham Robinson showed that it was possible, against all naysayers, to reckon with the infinitely small and the infinitely large, and in precisely the way Leibniz and the other pioneers of infinitesimal calculus had done. Speaking of "ghosts of departed quantities," here is a revenant! Robinson's "nonstandard analysis" is completely compatible with the "standard analysis" every mathematics student is familiar with. The trick is to introduce, on top of the usual real numbers, some so-called hyperreal numbers.

This is where Archimedes enters the scene again. His name is attached to a property that seems completely obvious. Consider any length, however small. Then, by taking sufficiently many steps of that length, you can cover any given distance in due time. Archimedes, incidentally, had not been the first to use this property. Other geometers before him had used it, and had even understood that the principle was not evident at all. In particular, Euclid knew of magnitudes that *disobeyed* the principle.

Indeed, let P be a point on a line *g*, and consider circles tangent at P to this line (Figure 6.14). The angles of such circles with the line (Euclid called them "horns") are a well-ordered set of magnitudes (the larger the horn, the smaller the radius). Each such horn is smaller than each "ordinary" angle at P, which is obtained by intersecting *g* with a line. Every nonzero angle, small as it may be, is larger than any horn. Add a thousand horns and you will still not have an angle. Horns, therefore, may be viewed as the infinitesimals of angles.

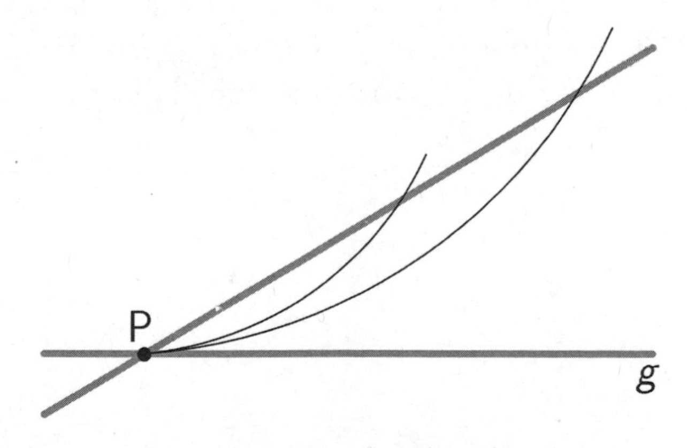

Figure 6.14. An angle and two horns.

For the length of segments on a line, however, the Archimedean property holds. More precisely, it belongs to those axioms required for Euclidean geometry that belatedly were added by Hilbert. What the Archimedean property means is that for any real number x, there is a natural number n that is larger. This property is assumed to hold, by fiat. It is convenient, although there is no logical necessity for it. Once this property is accepted as an axiom, infinitesimals are excluded from the game.

Robinson showed, however, that there is no inconsistency in dropping this "axiom of Archimedes" and assuming instead that there exist some strange numbers that are larger than any natural number n. They are said to be *infinitely large*. Their inverses are the infinitesimals: these are positive numbers larger than 0 and yet smaller than every $\frac{1}{n}$, for any n. (Thus, it is strange that Cantor, who was unfazed by an infinite ordinal such as ω, had rejected the infinitesimals in harsh terms.) The real numbers, together with the infinitely large and the infinitely small numbers, form the *hyperreals*. They prove responsive to all the customary activities of analysis. In particular, they make sense of everything that Leibniz claimed for his infinitesimals. So, the infinitesimals have returned. People just had to get used to them.

For some time, there was even hope that a new generation of mathematicians would be trained without having to undergo the rites of epsilontics. In the end, this did not happen. But nonstandard analysis helped discover and prove new results in standard analysis. Better still, it provided a belated justification for the baroque adventurers who, advancing by bold faith, had created infinitesimal calculus, unshaken by the dialectical quirks of sophistry.

7

Probability

A Random Walk to St. Petersburg

Leaving Little to Chance

It must have been the most enjoyable of all encounters between mathematics and philosophy. It certainly was the most rewarding ever. In 1728, at a banquet in Paris, an up-and-coming young *philosophe* and *homme de lettres* who had named himself Voltaire came to sit next to Monsieur de la Condamine, an accomplished mathematician, not yet thirty years of age.

Figure 7.1. Voltaire (1694–1778).

Figure 7.2. Charles-Marie de la Condamine (1701–1774).

The conversation of Voltaire and Condamine soon turned to the newest scheme for a state lottery. The French government, perennially short of money, was trying to encourage their subjects to acquire bonds. The treasury offered, as an incentive, a lottery: bondholders could participate by buying a ticket for the price of one thousandth of the current value of their bond. The winner would be paid the nominal value of the bond (which at that time of depression was much higher than the current market value) and, in addition, a huge jackpot—half a million livres. More than enough to live in state for the rest of the century, the luxurious *Siècle des Lumières*, a time when it really paid to be rich.

Something was wrong with that lottery scheme, said Condamine. The size of the jackpot was always the same, but the prize of the ticket varied with the value of the bond. Some bonds were high-priced, some others next to worthless. This was a typical example of an unfair game.

Voltaire quickly grasped the point. The ministry of finance had made a serious blunder indeed. Voltaire could denounce it publicly—or else exploit it on the sly. By that time, he had abundantly proved that he was not afraid of making fun of the French authorities. His mordant wit had brought him a year-long imprisonment in the Bastille, and two years of exile in England. He could keep on goading the government and ridicule the French treasury for its inability to do their sums properly. But would it not be more profitable to keep mum and turn the blunder of the state officials to his own benefit? Voltaire liked to say that he knew far too many penniless writers to ever wish to add to their ranks. Now Condamine showed him how to get rich quick and land a great coup by using a bit of mathematics.

Roughly speaking, the trick was to buy many of the low-price lottery tickets. Really many, ideally all of them! This would provide their owner with a high chance of winning the lottery, and this for a relatively modest cost.

The basic idea was simple, the execution extremely complicated. One had to set up a syndicate, use straw men and sham deals, and ensure, one way or another, the good will of the right officials. Fortunately, Voltaire was a peerless networker, known all over France.

The plan succeeded beyond all hopes. Month after month, the state lottery was repeated, with hefty winnings for all who were part of the scheme. It took two years before the authorities smelled a rat. Condamine and Voltaire were taken to court. The hoax could have cost them dear. But in the end, they got away with it: they were able to prove that they had done nothing illegal. They had simply exploited the rules of the game. The lottery came to an end, and the hapless *controleur général des finances* was fired.

Voltaire and Condamine had become wealthy—stinking rich, to use a term from our less enlightened century. What happened next? Condamine, shortly after having eluded jail, was elected to the French Academy of Science. He used his fortune to undertake large-scale scientific expeditions, an Alexander von Humboldt *avant la lettre*. First, a spell in the Levant, in the company of the most illustrious freebooter of France, and then ten years in South America, including an unheard-of trip across the Amazon Basin. Later in life, Condamine championed a Europe-wide campaign for inoculation against smallpox. Approaching death at age seventy-three, he volunteered to test a new type of surgery against hernia. It succeeded, according to the doctors; but this time luck turned against Condamine. He died of wound fever.

And Voltaire? He went on multiplying his newfound wealth and acquired an extraordinary, almost princely degree of independence. His *Lettres Philosophiques*, published a few years after his lottery scam, acted as "the first bomb thrown at the ancien régime." Voltaire became the figurehead of the Enlightenment, *the* top celebrity of his time. And the "calculus of chance"—probability theory—had played an essential part in his stellar career.

This theory is often labeled as the mathematics of randomness. But Voltaire denied randomness. Chance does not exist, he said. We only speak of it when we know the effect while ignoring the cause. Everything, however, has its cause. *Nothing* is due to chance.

This phrase was not meant as a sly allusion to his uncanny winnings at the lottery. It was the settled opinion of all modern philosophers. The worldview had turned strictly deterministic, at the very latest with Newton. As a matter of fact, it had not been all that different in the centuries before—except that *then* everything was ordained by the will of God, and *now* by the laws of science. (In the last hundred years, views on causality and determinism have undergone another shift, once more due to physics: randomness plays an almost impenetrable role in quantum mechanics. We shall keep away from that minefield.)

Playing with Chance

Chance is notoriously hard to define. Is it the force that causes something to happen without any known reason for doing so? Something that happens when several causes intermingle? Something that can be, but also not be? This is just a small sample (a random sample) of attempts to explain the word *chance*. Mathematicians, however, do not try to define chance. They want to reckon with it. This is more modest, and at the same time more ambitious.

The calculus of probability emerged rather late in the history of mathematics. Its birth is commonly ascribed to an exchange of letters between the philosopher and mathematician Blaise Pascal and the lawyer and mathematician Pierre Fermat. But the ideas were in the air, toward the middle of the seventeenth century, and attracted thinkers such as Galilei, Newton, Leibniz, and Huygens—the *Who's Who* of exact thinking in their age. Within a few years, it all crystallized. By 1660, probability theory was established.

It began with games of luck. They provide a limitless supply of mathematical problems. This makes it all the more puzzling that ancient mathematics knows no theory of probability. The Greeks, for instance, considered themselves the inventors of dice games—allegedly they hit upon it as a pastime during their siege of Troy. In actual fact, thousands of years before that time, Egyptians had played with dice. Some dice were found in graves from the First Dynasty. However, it is likely that the ancient Greeks invented coins, and one may safely assume that "Heads or Tails" was played not long afterward. Card games date from the Middle Ages. When Johannes Gutenberg opened shop, he duly printed the Bible first, but—in the very same year—a set of Tarot cards, too. Lotteries emerged in Renaissance Italy. The French

Enlightenment provided us with roulette (the invention of a police officer, so it seems). Mechanization brought gaming machines, and digitalization an endless stream of apps for gambling. Humans simply love to toy with chance.

It is all the more surprising that we are utterly inept in estimating probabilities. Mathematics is spiced with paradoxical results, but in the calculus of probability, they really clog up.

An example, maybe? Let us assume that you live in a region where the probability of being infected by a certain virus amounts to one in a thousand. Imagine a test that infallibly recognizes the virus, but yields a false positive with a probability of 5 percent. And now imagine that you have tested positive. How probable is it that you really carry the virus?

Take your time before you answer. The test is not completely precise, as we have seen. It can yield the wrong result. How likely is it in your case? Please consider before hazarding a guess. The most common answer is: "I am infected with a probability of 95 percent." This answer is wrong. The probability is less than 2 percent, as we shall see in a moment.

Another example? Two firms have developed drugs against that viral infection. Now comes the time for clinical trials. First, people will be tested who are less than sixty-five years old, and hence do not belong to the high-risk group. Drug Alpha helps 90 people out of the 240 who try it. Drug Beta helps 20 out of a sample of 60. Since 90 out of 240 is larger than 20 out of 60, we may conclude that Alpha works better than Beta, for people less than sixty-five years old. Next, the drug is tested in the high-risk group, on people who are sixty-five years old or older. A random sample of 60 senior citizens tries Alpha, and 30 of them feel relief. Beta is used on 240 elderly people, and helps 110 of them. Since 110 out of 240 is less than 30 out of 60, Alpha does better than Beta again. The health department orders huge amounts of Alpha.

But wait a minute, warns an expert. Let us look at the sums. Altogether 300 people have been tested with Alpha, and 300 with Beta. Alpha has provided relief to 120, and Beta to 130. Doesn't this indicate that Beta is better? Very confusing! Let us check our numbers again. Alpha does better for both the younger and the older samples—but less well in toto. Can that be right?

The Dice Are Cast

The theory of probability got kickstarted with a few riddles about dice. Here are two of them.

When we throw two dice, we can obtain the sum 9 in two ways, as 3 + 6 or 4 + 5. Similarly, we can obtain the sum 10 in two ways, as 4 + 6 or 5 + 5. Hence, "sum 10" should be as likely as "sum 9," and they should occur with the same frequency. Yet, this is not the case: 11.1 percent of all throws yield sum 9, and only 8.3 percent sum 10. Why is 10 less likely than 9?

The simplest way to explain it is to paint one die red and one white, which makes it easier to tell them apart. The sum 4 + 5 can be obtained in two ways: if the red die shows number 4 and the white number 5, or if white shows 4 and red 5. On the other hand, the sum 5 + 5 is obtained when both red and white show number 5. There is no second way. Hence, 4 + 5 is twice as likely as 5 + 5. The outcome 3 + 6 is, by the same argument, just as likely as 4 + 6—or 4 + 5, of course.

To make this more explicit: The result of a throw is "red shows number x, and white shows number y," which we write as (x, y). The red die is equally likely to land on each of its sides (for symmetry reasons, as mathematicians like to say). Thus, "red shows x" has probability $\frac{1}{6}$. Similarly, "white shows y" has probability $\frac{1}{6}$, no matter what red shows. Hence, the probability of "event" (x, y) is always the same, namely $\frac{1}{36}$.

$$
\begin{array}{l}
(1,1),(1,2),(1,3),(1,4),(1,5),(1,6)\\
(2,1),(2,2),(2,3),(2,4),(2,5),(2,6)\\
(3,1),(3,2),(3,3),(3,4),(3,5),(3,6)\\
(4,1),(4,2),(4,3),(4,4),(4,5),(4,6)\\
(5,1),(5,2),(5,3),(5,4),(5,5),(5,6)\\
(6,1),(6,2),(6,3),(6,4),(6,5),(6,6)
\end{array}
$$

Figure 7.3. The thirty-six possibilities for (x, y).

Four of these events yield the sum 9, namely (3, 6), (6, 3), (4, 5), and (5, 4). Accordingly, the probability for this is $\frac{4}{36} = \frac{1}{9}$. But only three events, namely (4, 6), (6, 4), and (5, 5), yield the sum 10; its probability is therefore $\frac{3}{36} = \frac{1}{12}$, which is smaller.

We might try the same with three dice, painted red, white, and blue. The sum 9 and the sum 10 can each be obtained in six different ways—but now 10 is more frequent than 9!

Here comes the second teaser. Pierre and Blaise roll dice repeatedly, until one of them has won thrice. (If both throw the same number, the round does not count.) Each player has staked six doubloons into the pool. The game is well underway, with lucky Pierre leading 2 to 1. Unexpectedly, Cardinal Richelieu passes by and orders the game to stop—immediately, *messieurs*. No more rolls.

Pierre pockets the pool, all twelve doubloons of it. "I was ahead, I am the winner."

"Not at all," protests Blaise. "The game did not come to the prescribed end, and hence has not taken place. Each of us gets his six doubloons back."

"This is hardly fair practice," says Pierre. "I have made twice as many points as you. Hence, I am entitled to receive twice as much as you from the pool, namely two thirds. If my *arithmétique* serves me well, that makes eight doubloons for me, and four for you."

"Well then, I'll accept that," replies Blaise. This answer raises dark suspicions in Pierre's mind, and he starts computing. How likely is it that Blaise would have won the game? And it becomes clear that the probability is merely 25 percent: the only way for Blaise to win overall is to win the next two rounds, and the probability for that is $\frac{1}{4}$. Pierre's chance to win is three times greater. Hence, the fair way to split is for Pierre to pocket nine doubloons and let honest Blaise keep the remaining three.

Before taking our leave from the dice, one last little game. Let us take three dice (white, gray, and black, say). This time, we will not number their six sides with 1 to 6, as usual, but number the eighteen sides from 1 to 18, as shown in Figure 7.4. (The sides of the black die are marked with the numbers

18, 10, 9, 8, 7, and 5; the white die with 17, 16, 15, 4, 3, and 2; and the gray one with 14, 13, 12, 11, 6, and 1.)

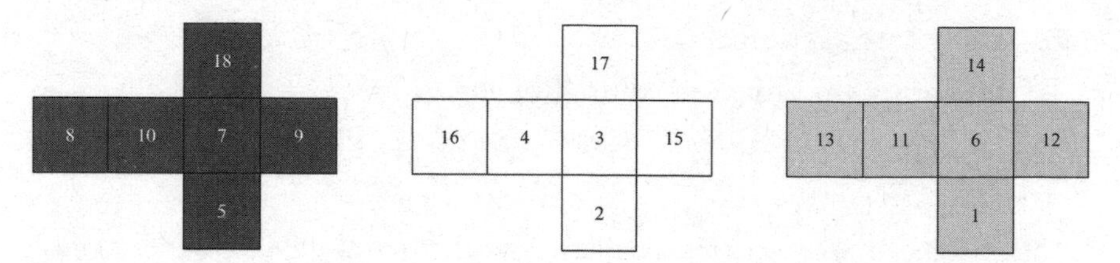

Figure 7.4. Three dice to play Rock-Paper-Scissors with.

It seems that the numbers have been distributed fairly. Indeed, the sum of the numbers on each die is the same, namely 57. Now, you pick up one die, I pick up another, and each tries to roll the higher number. There is, of course, no possibility for a draw.

You may have noticed that I let you choose your die first. This is not out of mere politeness. It gives me an advantage. Though I certainly may lose some rounds, I will draw invariably ahead, if we keep playing for long enough. I win each round with a probability of 58 percent.

Let us start a fresh game. This time, you pick the die that has proved so lucky for me. I take the third one, and lo and behold, I am winning again. The odds are once more in my favor. So, let us start afresh. You pick the die that served me so well in the second match. I pick the one that served you so poorly in our first match. And what happens? I win again. It is as if the dice were playing Rock-Paper-Scissors: black beats white, white beats gray, gray beats black, always with a likelihood of 58 percent. Can you see why?

In due time, the calculus of probability left gambling saloons to reach out for more serious fields. First, it was applied by insurance companies. What is the fair price of life insurance? Obviously, this depends on life expectancy. It is not by chance that the first demographic tables are about as old as the baroque brainteasers about dicing. Later, probability took a star role in statistical mechanics, and later still (but not much later) in genetics. Today, physics, chemistry, economics, and biology are inconceivable without probability theory. James Clerk Maxwell touted it as "the true logic of the world," and Pierre-Simon Laplace as dealing with "the most important questions of life." Admittedly, Albert Einstein claimed that "God does not play dice" (some wit added: "But if He did, He'd win"). Yet quantum physics sees chance everywhere. Erwin Schrödinger stated that "chance is the common root of the strict causality in physics," and Jacques Monod viewed chance as "the foundation of the wonderful edifice of evolution." And the philosopher Bertrand Russell hit the bull's-eye when he said: "Probability is the most important concept of modern science, especially as nobody has the slightest notion what it means."

The Two Sides of Probability

All reasoning with probabilities starts with possibilities: or more precisely, with the set of all possible outcomes. Mathematicians are fond of calling this set Ω (omega), its subsets *events*, and its members—the possible outcomes—*elementary events*. If you throw two dice, the set Ω consists of all pairs (x, y) of integers between 1 and 6, and thus has thirty-six elements. The event "the sum is 10" consists of the three pairs (5, 5), (6, 4), and (4, 6)—each of these outcomes is an elementary event.

Let us assume, to start with, that there are only finitely many outcomes. Each elementary event is ascribed a number—the probability of the outcome. The numbers are nonnegative and sum up to 1. It is as if each element has a weight, and the total weight of Ω is 1. The probability of the subset A (i.e., the event A) is defined as the sum of the weights of all members making up A.

At first glance, the so-called standard model has nothing whatsoever to do with chance or probability or randomness. A mass of weight 1 is divided into as many parts as there are possible outcomes. How this is done depends on the specific situation. If one rolls two dice, for instance, it is natural to assume that all thirty-six possible results (x, y) have the same probability, which therefore must be $\frac{1}{36}$. The reason for the assumption is symmetry: nothing makes the dice more likely to land on one side rather than another.

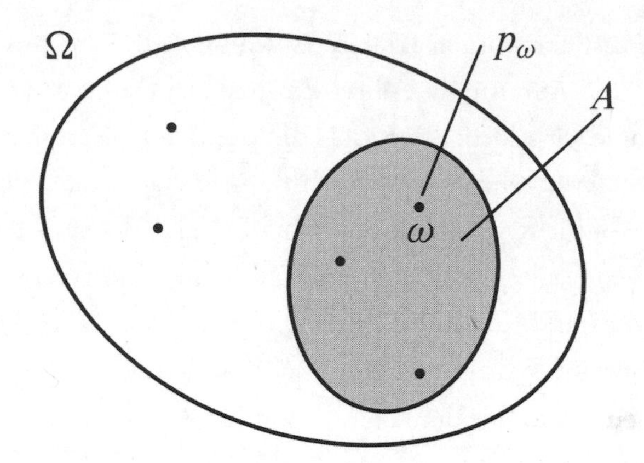

Figure 7.5. The sample space Ω is the set of possible outcomes ω. The event A is a subset of Ω.

If all elementary events have equal probability (the so-called Laplace model), then the probability $P(A)$ of event A is given by the number of "favorable" outcomes (the members of A) divided by the number of possible outcomes (the members of Ω). The event "sum 9" thus has probability $\frac{4}{36}$, and "sum 10" the probability $\frac{3}{36}$. It is, of course, just as conceivable that the probabilities of the various outcomes are not all equal—for instance, if the six sides of a die have different sizes.

All depends on using the right probabilities. Clearly, they should mean more than merely some weights picked out of the blue. In dicing, for instance, it is an empirical fact that the probability of "sum 9" is about 11 percent. The probability of event A should have something to do with how often it happens—that is, with the relative frequency of its occurrence in a long series of iterations. If we repeat the experiment 1000 times, we would expect to obtain "sum 9" about 111 times. Not *exactly* 111 times—that would be rather surprising, as a matter of fact—but hopefully the number should be somewhere between 100 and 120.

To apply the standard model, therefore, the probability $P(A)$ of event A should correspond to its frequency. Let us assume that we can repeat the experiment N times under the same conditions. The experiments should not influence each other. Let $N(A)$ be the number of occurrences of event A (for instance, "the sum of the two dice is 9"). The relative frequency $\frac{N(A)}{N}$ should be close to the probability of A, if N is very large. (More precisely, the probability should be the limit, when N grows "beyond all bounds." Such idealizations are common fare in mathematics. Admittedly, it is far from evident that a limit exists, but in the second half of the seventeenth century, when the concepts of limit and probability were both infants, nobody lost any sleep on this.)

The frequencies satisfy $N(A) \geq 0$, $N(\Omega) = N$, and $N(A \cup B) = N(A) + N(B)$ if A and B are disjoint (which means that they cannot occur together). Indeed, $A \cup B$ is the event "A or B." Similarly, we have $P(A) \geq 0$, $P(\Omega) = 1$, and $P(A \cup B) = P(A) + P(B)$ if A and B are disjoint. So far, everything fits well and indicates that, indeed, the probability of an event corresponds to its relative frequency.

Before proceeding further along this line—the *frequentist* interpretation of probability—we should take note that we have drastically restricted the meaning of probability. We only consider its use in situations that can be repeated again and again. This applies to rolling a die or tossing a coin. It does not apply to statements such as "the US president in 2050 will be a woman," or "the pandemic will return this winter," or "to find my way out of this thicket, I should turn left rather than right." One speaks of the probabilities of such events without imagining long series of repeated trials. Probability, here, denotes a higher or lower degree of inner conviction. It is a psychological term rather than a statistical quantity. Such probabilities are denoted as *subjective*, in contrast to the *objective* probabilities encountered in the toss of a coin or the gender of a newborn. We shall discard subjective probabilities for now, but note that any language philosopher will complain, and with good reason. It may make sense to discriminate the two meanings of the word, but simply to ignore one of them is too facile. Yet, it is done routinely in hard-nosed science, and with astonishing success.

If we restrict ourselves to objective probabilities, it looks as if we are close to answering the question: what is probability? It is the relative frequency—a solid, measurable, empirical quantity. So, how can Russell claim that no one knows what probability means?

To answer this question, we have to take a step back. First of all, this frequentist interpretation of probability is indeed a good first approximation. In particular, frequentism is a first-rate assistant to understanding most of the probabilistic paradoxes. Let us return, for example, to the test for the viral infection, and consider a sample of 1000 people submitting to the test. One of them carries the virus (the probability, as you may remember, was 1/1000). This person will yield a positive test result. All the others (999 of them, almost 1000) will also be tested. Since the failure rate is 5 percent, some 50 of them will test positive. This yields altogether 51 positives. We have assumed, in our scenario, that you are one of them. Therefore, your probability of really carrying the virus is roughly 1 in 50, or 2 percent.

Similarly with the dicing game that was cut short by Richelieu. Imagine that the cardinal, obeying a sadistic urge, had sentenced the two gamblers to continue their gamble 1000 times over. What will happen? About 500 times, the game will be over after just one round, namely when Pierre wins (you will remember that he needed only one more win). The other (roughly) 500 times, that round leads to equalization: two wins by Pierre against two by Blaise. The next win decides, and it will go to Pierre about as often as to Blaise, approximately 250 times. Altogether, Pierre will win 750 of these gambles, and Blaise only 250 of them. Hence, Pierre is three times as likely to pocket the winnings. Therefore, the jackpot should be divided 9 to 3.

Bringing Law to Large Numbers

There is an argument that speaks even stronger for frequentism than its usefulness does. This is the law of large numbers. It is due to Jacob Bernoulli, a member of that famous family of mathematicians from Basel. This Bernoulli was one of the first to write a book about probability theory. His *Ars Conjectandi* was published posthumously in 1713, and is widely seen as the official birth certificate of the discipline.

Bernoulli's law (he fondly called it his "golden theorem") says: if an experiment is repeated sufficiently often, then the relative frequency of an event differs by arbitrarily little from its probability, not necessarily always, but with an arbitrarily large probability.

If one is not used to the statement, it sounds confusing. The word *probability* occurs twice, and so does the word *arbitrarily*. It does not help that Bernoulli often added (for instance, in his correspondence with Leibniz) that the arbitrarily large probability can be chosen so large that it amounts to "moral certainty."

What does his law mean? Let us consider an experiment that can be repeated as often as one likes (for instance, tossing a coin) and an event A that occurs with probability $P(A)$ (for instance, the event "heads," which occurs with a probability of 50 percent). Let us fix an arbitrary precision level—say 5 percent. Let us fix a probability that is as close to 1 as we wish—say 99 percent. Bernoulli's golden theorem states that if the number N of independent repetitions of the experiment is large enough, we can be 99 percent sure that the relative frequency $\frac{N(A)}{N}$ of event A differs by less than 5 percent from its probability $P(A)$. In other words, the relative frequency will be in the interval between 45 percent and 55 percent, with a probability of 99 percent (Figure 7.6). All that is required is that the series of trials is sufficiently long.

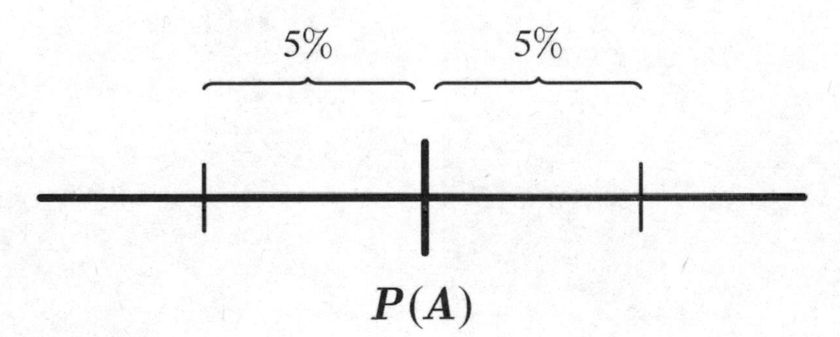

Figure 7.6. How likely is it that the average frequency $\frac{N(A)}{N}$ is within 5 percent of probability $P(A)$?

This law is often interpreted as meaning that we can obtain the probability of an event, namely $P(A)$, by measuring its frequency $\frac{N(A)}{N}$ in a long series of trials. This is not correct. Bernoulli's law of large numbers states something about the frequency of A when the probability of A is known—not the other way around.

This fallacy has often been repeated (starting with Bernoulli himself). Many use it to argue that the probability of an event is nothing but its relative frequency, or more precisely its limit. But on closer inspection, Bernoulli's law of large numbers does not really explain the meaning of probability. As we noticed, this word occurs twice in the statement of the law, which should be enough to let mental alarm bells ring: first, it occurs as the probability $P(A)$ of event A, and then as the probability that the relative frequency belongs to a small interval centered on $P(A)$. Do we have to interpret this second probability in a frequentist fashion, too? The cat is chasing its own tail.

(Moreover, by looking sharply, we will notice that probabilities are invoked a third time in the law of large numbers, hidden in the assumption that the repetitions are independent. Indeed, if we want to define what it means that "the event A occurs in the fourth trial" is independent of "A occurs in the tenth trial," we are using probabilities again, because it means that one outcome does not affect the probability of the other.)

Figure 7.7. Jacob Bernoulli (1655–1705).

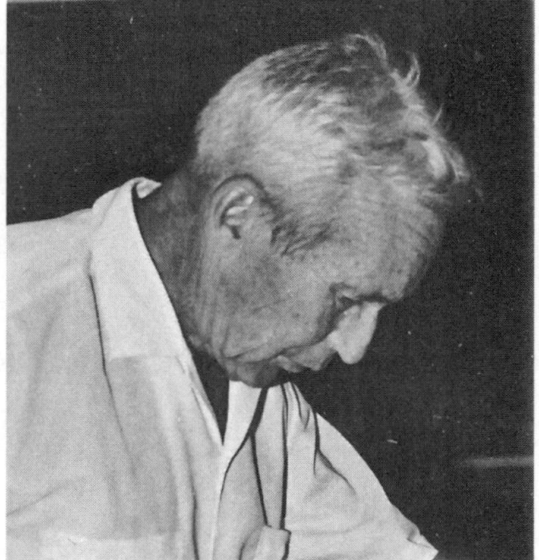

Figure 7.8. Andrey Nikolaevich Kolmogorov (1903–1987).

The law of large numbers, or more precisely, Bernoulli's "weak law of large numbers," states roughly the following: if we repeat, again and again, a series of 1000 coin throws, then for most of the series, the relative frequency of "heads" will be between 0.45 and 0.55. There also exists a "strong law of large numbers" that states (roughly) that even for those series whose relative frequencies fall outside of the small interval, things will end happily: we just have to keep on throwing the coin, i.e., extending the series, from 1000 throws to 10,000, then to 100,000, etc. Eventually, the relative frequency will never leave the small interval around $\frac{1}{2}$. It will always lie between 45 percent and 55 percent. And the same holds for the interval from 0.49 to 0.51, etc.—in fact, for any interval around $\frac{1}{2}$. And this holds with probability 1, or, to use the jargon of probability theory, "almost surely." It is of course conceivable that the relative frequency does not converge to $\frac{1}{2}$: for instance, that "heads" shows up in every throw of the coin, forever. This event is not unthinkable—but its likelihood is nil.

This strong law of large numbers was proved first by Émile Borel.

Foundations, Secondhand

It took nearly 200 years to proceed from the weak to the strong law, and it required an extension of the standard model. The standard model describes experiments with finitely many possible outcomes, i.e., with a finite sample space Ω. Infinitely many throws of a coin mean infinitely many possible outcomes (even uncountably many).

There were other reasons for extending the standard model. For instance, in old-time fairs and Luna Parks, one could come across a so-called fortune wheel ("giddy fortune's furious fickle wheel," to quote Shakespeare). Clients would send the wheel spinning around its horizontal axis. The "outcome" of the experiment is the topmost point on the wheel when it stops. All points on the wheel are equally likely. Each one, then, has probability 0: for this is how small the chances are to stop *exactly* on that same point with the next attempt. Nevertheless, each arc of the wheel's circle is assigned a positive probability, namely its portion of the whole circumference. Because each point on the circle is an interior point of arbitrarily small arcs, its probability is 0. Yet, the outcome is not impossible.

It took ages to extend the standard model. Even by the year 1900, the theory of probability was still in quest of a solid foundation. David Hilbert, in his famous list of twenty-three problems for the new century, asked for the axioms of probability. The answer came in 1933, in a book by the famous Soviet mathematician Andrey Nikolaevich Kolmogorov. It was of astonishing simplicity; and though it hardly provided philosophers with an answer to the question of what probability is, it offered a revealing glance at how mathematicians tend to evade such questions.

Kolmogorov's model is, since nearly a century, the undisputed foundation for an enormously successful theory. Yet, it seems hardly to go beyond the standard model, being based, again, on the space Ω of possible outcomes of the experiment. The subsets A of Ω are said to be events. They are assigned probabilities $P(A)$, with the properties that $P(A) \geq 0$ and $P(\Omega) = 1$. The only extra step beyond the standard model concerns the property that $P(A \cup B) = P(A) + P(B)$ holds whenever A and B are disjoint: namely, this additivity is now required to hold not only for two, or three, or any finite number of sets that are pairwise disjoint, but also for countably infinite many such sets. And this is it! The whole of probability theory is based on these simple axioms. The mathematical world accepted them without hesitation and has not regretted it ever since.

Thus, the "foundations" are of stupendous banality. As with the standard model, nothing in Kolmogorov's extension indicates that it is about chance. He simply adopted a tool kit that was readily available, namely the theory of measure and integration, which had been developed in centuries of relentless efforts to compute areas and volumes.

In this sense, a secondhand acquisition serves as the foundation of probability. The curious "countable additivity" of probability is based on the so-called exhaustion method that Archimedes used to compute the area of the circle, as a limit of a sequence of inscribed polygons.

This episode highlights a characteristic property of mathematics: it is the epitome of technology transfer. Methods used in one domain are applied in another. If they work, this is justification enough. Small wonder that the writer Robert Musil spoke of the mathematicians' "lack of conscience."

Mathematicians are not entirely lacking conscience, though. In measure theory, it had been discovered that if the set Ω is uncountable (as in the case of the wheel of fortune), a function such as P cannot be defined for all subsets A without leading to inconsistencies. A serious problem? Not really. The easy way out is to define $P(A)$ only for *some* subsets A—for as many as is possible without encountering trouble. By definition, all the other subsets simply do not count as events. This, too, may strike us as a cheap trick. But it works, and so, who cares? Not your mathematicians. It is another example of what Wittgenstein asserts they do when encountering a contradiction: they change their rules.

Fair Fees and Random Walks

Let us return to the gambling saloon. In a game of chance, the gain is a so-called random variable—which is defined as a function that associates a number to each outcome of the experiment, i.e., to each member of Ω. (That such a function is defined as "variable" is a quirk one has to live with. There is no deeper meaning behind it.)

The gambler pays a fee—which is specified in advance—and hopes for a gain. The fee is fixed, but the gain depends on chance.

What should a player accept as a fee? This depends on the *expectation* of the game, which is the sum of all possible gains (including losses, which are negative gains), each weighted with its probability. If the fee is equal to the expectation, the game is said to be *fair*. In that case, if the game is repeated often enough, the average gain per round will be equal to the fee, or reasonably close. This follows from the law of large numbers and seems to settle the question. (We shall presently see that it doesn't.)

Only a few games are fair. Take roulette, for example. Somebody who bets on the right number will earn 36 times the fee. But the chance of winning is not $\frac{1}{36}$ (that would be fair) but only $\frac{1}{37}$, because of the zero slot. (US casinos even have two zeros.) This makes the game unfair for the gambler and profitable for the casino owner. Nevertheless, many gamblers are ready to face the slight disadvantage, and enjoy the thrill of the game. And why not?

Every insurance policy is a gamble, too, and even an unfair gamble. The insurance companies want to live, after all, and even thrive. Therefore, they offer odds that are in their favor. Nevertheless, clients who buy home insurance are not stupid, as a rule. They prefer to face a small loss, namely their annual fee, with a large probability (indeed, with certainty), rather than suffer a huge loss with a small probability.

The simplest example of a fair game is tossing a coin. If it lands "heads," you win $1; if "tails," you lose one. This game is so fair that it may appear boring. But let us assume that it is repeated time after time (Figure 7.9). The account of the player will then increase or decrease by $1, with equal probability, for game after game. The probability that the player's account has returned to its initial value after 2, 4, and 6 rounds is $\frac{1}{2}$, $\frac{3}{8}$, and $\frac{5}{32}$, respectively (it has to be an even number of rounds, of course).

With *certainty* (which means with a probability of 100 percent) you will reach the initial position again, at some time sooner or later, and then, of course, everything starts anew. It is equally certain that the amount on your account will eventually reach 0, sealing the proverbial "gambler's ruin." But let us assume that gamblers are able to raise an unlimited credit, and so can keep gambling. Then, their account will eventually reach every integer value, be it positive or negative.

Since the average gain converges to 0, by the law of large numbers, many people assume intuitively that the account will be close to the initial value most of the time; in other words, that it has increased roughly as often as it has decreased. This is an error. Often, the deviations are amazingly large. The probability that there has been no return to the initial value during the

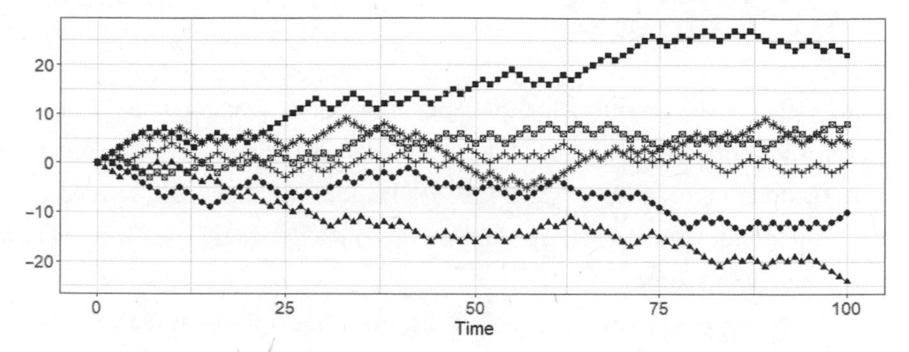

Figure 7.9. Six samples of playing Heads or Tails for 100 rounds.

first $2n$ rounds is as large (or, rather, as small) as the probability to have reached the initial value after exactly $2n$ rounds. If the game is repeated, once every second, for a whole year, then the probability that there has been no return to the initial value during the last six months is 50 percent; and the probability that the last return has occurred before January 12 is $\frac{1}{8}$, which is by no means a small probability—indeed, it is as likely as throwing three heads in a row.

The up and down of the monetary amount on the gambler's account is often described as a *random walk*. Imagine a drunk, holding himself up against a wall, and stumbling with equal probability one step to the right or one step to the left. He will return to the origin with certainty, sooner or later. Next, suppose that the wall collapses. The drunk, now, can make a step forward, backward, left, or right, each with equal likelihood. This is the two-dimensional random walk. Again, the drunk will return to his initial position with a probability of 100 percent. It may be a bit harder to conceive of a random walk in three dimensions—something for a hummingbird, possibly. Surprisingly, this hummingbird will not return with certainty to its initial position, but only with a probability of about 36 percent.

All this is just meant as a teaser for the curious results encountered in probability theory. For physics or finance, however, these results on random walks are not mere curiosities, but essential tools for daily use.

Great Expectations

Before leaving the gambling houses, let us make one excursion to St. Petersburg. The following paradox is due to a Bernoulli (Daniel or Nikolaus—experts don't agree, except that it was *not* Jacob). This Bernoulli resided in recently founded St. Petersburg. And there, so goes the tale, a casino offers us the following game.

We are supposed to play Heads or Tails, until heads shows up for the first time. If this happens in the very first throw, the casino pays us \$2. If it happens in the second throw, \$4; in the third throw, \$8; etc. Hence, we hope to throw tails for as long as we can. If the series of tails stops at throw number n, we are paid $\$2^n$.

What would be a fair fee for such a game? In other words, how much would we be prepared to pay to the owner of the casino, to be allowed to join in the game?

We should, of course, compute our expectation. The probability to throw our first heads in the n-th throw is $\frac{1}{2^n}$, and this brings us $\$2^n$. Therefore, the expectation value is an infinite sum:

$$2\left(\frac{1}{2}\right) + 4\left(\frac{1}{4}\right) + \ldots + 2^n\left(\frac{1}{2^n}\right) + \ldots = 1 + 1 + 1 + \ldots.$$

The expectation is infinity! Of course, we will not seriously expect to collect infinitely many dollars. Our gain will be a finite sum. So, what does it mean to say that the expectation is infinite? Let us imagine that we play the game—the series of throws until the first heads—not just once. We play it every evening (each time paying our admission fee, of course). A minor variation of the law of large numbers states that the average value of our daily win (the total of all our gains in the first N days, divided by N) will grow beyond all limits, as N gets larger and larger. This seems to imply that we should be willing to pay *any* fee to play this game. The fee can be arbitrarily large, since sooner or later our average gain will exceed it.

But would we really be willing to do so? Few of us would be ready to come up with more than, say, \$20 as a fee. What is going on here?

This St. Petersburg paradox has played an important role in philosophy and economics, and we will return to it later. But the immediate answer turns out to be rather simple. The average gain of the first N days grows beyond all bounds when N grows, but it does so with agonizing slowness (see Figure 7.10). Most frequently, heads shows up after one, two, or three throws. It happens only very, very rarely that we make a huge win. On such a windfall day, the average gain will jump upward. From then on, it will decline again, day after day, until, after an apparently endless time, it will bounce upward, thanks to a huge win.

If we keep playing, we are set to enter the list of the world's billionaires, but we will not live to see it.

This casts a shadow on "fair games," and hence on averages (what does it help if their promise remains beyond our horizon?), and hence on expectations (what makes us believe that they can ever be realized?). The very foundation of probability theory—which is based on expectations, averages, and the law of large numbers—becomes dubious.

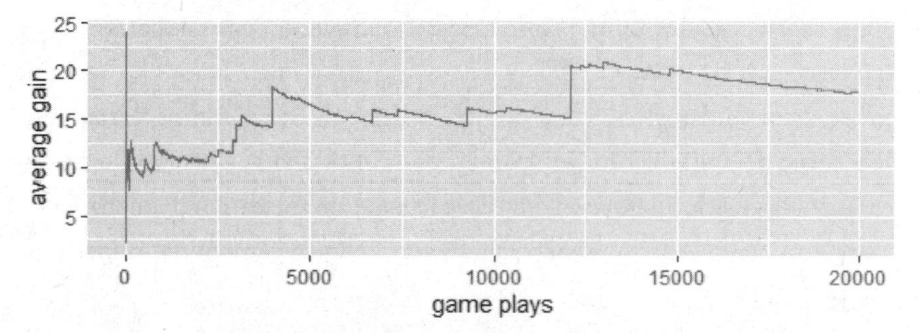

Figure 7.10. The St. Petersburg game, played 20,000 times (by a computer, of course). The curve shows the average gain. It grows beyond all bounds, but logarithmically (which means at less than a snail's pace).

Three Boxes and Two Envelopes

Just to stress the weird streak of probability one more time, this chapter closes with two famous puzzles.

First, the problem with the three boxes (also known as the goat problem). Imagine that you are participating in a quiz show. The quizmaster shows you three boxes. In one of them, there is \$1000; the other two are empty. You have to choose one of the three boxes, but are not yet allowed to open it. Once you have picked your box, the quizmaster opens before your eyes one of the two remaining boxes, to show you that it is empty. (This is easy for her to do, since she knows where the money is.) Next, she offers you an opportunity to change your decision, and switch from the box you have chosen to the one that is still untouched. This switch, however, comes at a cost, she says. Would you be ready to pay \$50 for it? Or would you rather remain true to your original choice?

This problem has been posed very often, with various embellishments. The overwhelming majority of candidates decline the offer. They see no reason to revise their first decision. Indeed, they have never doubted that the quizmaster can open one of the remaining two boxes and show that it is empty. In one of the two boxes that have not been opened yet, \$1000 is hidden. The candidates have opted for one of these two boxes. Why should they change their decision? It does not seem worth the \$50.

But this way of thinking is mistaken. Candidates should certainly revise their decision. By all means! They thereby *double* their chance to win the money. Indeed, the probability that the initial choice is the right one is $\frac{1}{3}$. Hence, the \$1000 is with a probability of $\frac{2}{3}$ in one of the other two boxes. We learn from the quizmaster that the money cannot be in one of these two boxes—the one she opened before our eyes. Hence, it is in the other box, with probability $\frac{2}{3}$.

The problem with the two envelopes is even hairier. The quizmaster tells us that one of the envelopes contains a certain sum of money, and the other one twice that much. We pick an envelope, are allowed to open it, and see that it contains \$40. Now the quizmaster offers us to exchange the chosen envelope with the other one, which has not been opened. Should we agree to the exchange? With a probability of $\frac{1}{2}$, we have chosen the envelope with the smaller sum. In that case, the other envelope contains \$80. Similarly, with probability $\frac{1}{2}$ we have chosen the larger sum, and in that case there is \$20 in the other envelope. If we switch to the other envelope, we will obtain either \$20 or \$80, each with probability $\frac{1}{2}$. The expectation value is \$50. This is certainly better than the \$40 in the envelope in our hand. Thus, we should switch. Or should we? Let us think again.

The same argument works, of course, for every other sum. It does not depend on the \$40. If the amount of money is x, a switch to the other envelope yields $2x$ or $\frac{x}{2}$, and hence an expected gain of $\frac{x}{4}$. This means that we don't even have to open the chosen envelope. We should switch. We certainly should! But once we have done so, we are faced, again, with an envelope containing an unknown amount of money. The same argument as before tells us to switch again. And again, and so on. We have obviously committed a fallacy.

By formulating the problem in another way, we see immediately that we cannot expect anything from a change. One envelope contains x, the other $2x$. In one case, switching yields a gain of x dollars; in the other case, a loss of x dollars. Because the two cases are equally likely, the expected gain from switching is zero dollars. We cannot hope for any benefit by switching envelopes. But what was wrong with our previous argument?

Let me paraphrase Descartes, and say that I will leave to my readers the pleasure of figuring this out for themselves.

(This is a cheap way out. In *Wikipedia* [as of 2022], one reads: "No proposed solution is widely accepted as definitive; despite this, it is common for authors to claim that the solution to the problem is easy, even elementary.")

Just as there are optical illusions, there exist cognitive illusions. They fool us, and keep fooling us even when we know that we are being fooled. We descend from a long line of ancestors who have all managed to survive at least until they were old enough to procreate. They all had to do a lot of guesswork. Our brains should be splendidly adapted to deal with chance and uncertainty. Yet, we fall into the same traps, again and again.

If Kant's a priori categories of thinking are, as some biologists assert, the a posteriori of evolution, then how should we account for our systematic bias toward probabilistic fallacies?

8

Randomness

The Superstition of the Vulgar

Ups and Downs of a Random Sequence

Thinking about chance seems to attract polymaths. This was so in the second half of the seventeenth century, when Pascal and Leibniz established a theory of probability based on expectation. Pascal was not only a philosopher and a mathematician, but also an inventor, a scientist, and a mystic. Leibniz was not only a philosopher and a mathematician, but also an inventor, a lawyer, a linguist, and a diplomat.

Similarly, when the interpretation of the calculus of chance was tackled in earnest, in the first half of the twentieth century, the two main approaches were brought forth by Richard von Mises (objective probability) and Frank Plumpton Ramsey (subjective probability). Both were mathematicians, philosophers, and much else besides.

This happened (or probably, chanced to happen) at a time when quantum physics abandoned a strictly deterministic worldview. Chance, which for ages had been dismissed as "superstition of the vulgar" (copyright David Hume), now captivated scientific thinking.

The Austrian Richard von Mises ranks among the top aerodynamicists of his time—a time that ranged from pre-World War I to the Cold War, from flimsy wooden biplanes to supersonic fighters. In the 1920s, when von Mises was based in Berlin, and precluded by the harsh Treaty of Versailles from designing any airplanes, he founded the seminal Society for Applied Mathematics and Mechanics, which heralded the worldwide emergence of industrial mathematics. He had a charismatic personality, combining the aura of the dashing pilot, the dynamic engineer, and the prestigious mathematician with the attitude of an aristocrat of the old school, and the refinement of a man of letters known for being the foremost expert on the poet Rainer Maria Rilke and the staunchest supporter of the novelist Robert Musil.

The first professorship of Richard von Mises had been in Strasburg, at a time when the town still hosted a German University. His last stop was at Harvard. In between Berlin and Harvard, he taught from 1933 to 1939 in Istanbul, which had offered him shelter from the Nazis. This is where he wrote his *Little Textbook of Positivism*, which was not little at all (it dropped this qualifier for the English edition). The book summed up Richard von Mises's philosophy, in thoroughly modern, anti-metaphysical, Vienna Circle style.

Figure 8.1. Richard von Mises (1883–1953).

The main ambition of this brilliant and arrogant man was to solve Hilbert's problem number six, and more precisely to find the axioms of probability theory. This he did, or thought he did, in 1919. For a good positivist like von Mises, it meant grounding the probability of an event on measurement: more precisely, on the frequency of its occurrence in long series of repeated trials. This approach gives a solid empirical basis. However, it restricts the meaning of probability to a considerable degree, addressing objective but not subjective probability.

The theory of von Mises raised many objections, and was eventually abandoned, by a more or less tacit consensus among mathematicians, in favor of the axioms of Kolmogorov. Yet, Andrey Nikolaevich Kolmogorov was a frequentist at heart, and avowedly inspired by von Mises; his axioms, however, remained noncommittal about what probability really means, and simply exploited the machinery of measure theory, which had been created to compute areas and volumes. To paraphrase a quip that Bertrand Russell used in another context: defining probability as a measure, rather than constructing it as the outcome of long series of measurements, has all the advantage of theft over honest toil. This is the reason why mathematicians jumped on it.

Although eclipsed by the success of the measure theoretic approach, the ideas developed by von Mises led, step by step, to profound insights into the connection of randomness and computability.

The basic issue is this: When is a sequence random? More precisely, given a sequence of heads and tails, how can we find out whether it was produced by honestly tossing a coin—an honest coin, one should say—or whether it was made up in some other way?

The question is tricky. Toss as much as we like, we will encounter only finite strings of heads and tails. For a fair coin, all such strings of length n have the same probability, namely $\frac{1}{2^n}$. Thus, the string

T H T H T H T H T H T H T H T H T H T H T H

is exactly as likely as

T T H T T T T H H T T T H T H H T T T T T H.

So, why should we think that one is more random than the other? Yet we do, without much reflection.

Here is another problem: whatever the way of defining a string as *random*, we would have to admit that when the string is extended by adding a T, then it should be random again. If we repeat this 100 times, adding 100 T's, the string would still have to be random, therefore; and this is hardly convincing. Apparently, we cannot expect a hard-and-fast criterion for a finite string to be random.

Richard von Mises was as empirical as one can be, yet he understood that a mathematical definition of randomness must concern infinite sequences—although such sequences are never experienced, and can only be conceived as idealizations. This is no philosophical scandal. Grounded on the experience of thousands of years, mathematicians know that the main tool of their trade is abstraction. In our case, we abstract from the fact that the string has an end. Clearly, the infinite is easier to handle than the very large.

Accordingly, Richard von Mises tried to capture the flavor of a "real" random sequence by an ideal object, which he named *Kollektiv*. It is an infinite sequence of H and T that satisfies certain conditions. The first condition is that the frequencies of H and T equilibrate in the long run. In other words, if $N(T)$ is the number of tails among the first N elements of the sequence, then the relative frequency $\frac{N(T)}{N}$ should converge to $\frac{1}{2}$. But this condition is certainly not enough, as the infinite sequence

H T H T H T H T H T H T H T H T H T H T H T H T H T...

shows. That sequence is far too predictable to look like a sequence of coin tosses, although the frequency duly converges to $\frac{1}{2}$. A proper *Kollektiv* must be unpredictable, or "rule-less." In the terms of von Mises, this means that for every subsequence of tosses that is chosen without prior knowledge of the outcome, the relative frequency should still converge to $\frac{1}{2}$. In the previous example, such frequency stability certainly does not hold. If we pick every even toss, we obtain

T T T T T T T T...,

and if we pick every odd toss, we obtain

H H H H H H H....

In the first case, the frequency of heads is 0, and in the second case, it is 1. If we wish, we can pick subsequences of H T H T H T H T...where the relative frequency converges to any number between 0 and 1; or even worse, where the relative frequency does not converge at all, but meanders up and down.

At this point, an objection arises. Indeed, we must admit that the same lack of frequency stability can be demonstrated for any genuine random sequence. Such a sequence will certainly contain infinitely many heads, and infinitely many tails; and by picking the right subsequences, we can, as before, obtain sequences consisting entirely of heads, or entirely of tails, or displaying any relative frequencies of heads we might wish, whether converging or not. This seems to show that there exist no *Kollektivs* at all, and that the theory of von Mises is void.

Richard von Mises countered this objection by insisting that the subsequence should be chosen without prior knowledge of the outcome. This is a very plausible condition, but what does it mean mathematically? How can we pick the right subsequences?

There are many subsequences; not just infinitely many, but uncountably many. Indeed, by replacing H with 1 and T with 0, we obtain an infinite sequence of zeros and ones. Any such sequence, say the sequence

$$0\,1\,1\,0\,1\,0\ldots,$$

corresponds to the binary expansion of a real number between 0 and 1; in our case, it would be

$$\frac{0}{2}+\frac{1}{2^2}+\frac{1}{2^3}+\frac{0}{2^4}+\frac{1}{2^5}+\frac{0}{2^6}+\ldots.$$

Because there are uncountably many reals between 0 and 1 (as we have known since Georg Cantor), there are uncountably many subsequences of heads and tails. We have no prior knowledge of the outcome, so any of them is an admissible choice. Yet, some will not lead to the right frequency for heads.

Richard von Mises was well aware of the problem and tried to solve it by the notion of a place selection function: it is a systematic procedure that tells us at every step N whether to include the next toss in the subsequence or not. For instance, the rule "if N is odd, include the next toss; if N is even, don't" is such a procedure. It is particularly simple, not depending at all on the sequence of tosses. Other procedures can depend on what has been obtained so far. An instance for this is the rule "if the last three tosses were H T H, include the next toss; if not, don't." (Obviously, a rule, to be feasible, can depend on the previous but not on the future tosses. We have no oracles.)

What guided Richard von Mises was a principle: it is impossible to beat chance. In other words, there is no strategy that allows you to win in the long run. This principle seems reasonable enough, and as empirical as can be: it is based on the ruinous experiences of countless gamblers who believed to have discovered a system to beat the bank. Just as there are no perpetual motion machines, there are no gambling systems. We must accept it. There always comes a time when the childish "wish principle" has to yield to reality. Science profits from it. The impossibility of perpetual motion machines constitutes the foundation of thermodynamics. The impossibility of a gambling system provides a basis for probability theory.

Such, at least, was the intent of Richard von Mises.

Kollektiv Thinking

Kollektivs were an excellent idea, but ahead of their time. The notion of a place selection function had to be left dangling in the air, for want of an acceptable definition of a "systematic procedure" to define it. In 1919, nobody had a firm notion of what it meant.

In 1940, the logician Alonzo Church was able to come to the rescue. According to the Church–Turing thesis, a systematic procedure is one that can be realized by a Turing machine. Once this much is granted, everything becomes simple. A place selection function is one that can be realized by a computer, step by step. You just have to feed an initial string of a sequence into the machine, and the computer will tell you whether to include the next member of the sequence, or not.

There are only countably many Turing machines, and therefore only countably many place selection functions. It was known already that for every countable set of place selection functions, there exist sequences of heads and tails that are *Kollektivs* in the sense of von Mises. This had been proved in 1936 by a jobless young Romanian mathematician named Abraham Wald, the scion of a long line of poor rabbis. Wald had studied in Vienna and was a friend of both Kurt Gödel and Oskar Morgenstern.

It seemed for a while as if, with a little help from Wald and from Church, the theory of von Mises was finally on firm ground. Except that it wasn't.

Among the many nonrandom-like properties of the alternating sequence

H T H T H T H T H T H T H T H T . . .

is the following one: though the relative frequencies of heads converge to $\frac{1}{2}$, they do so from *one side only*—from above. Indeed, these relative frequencies are

$$\frac{1}{1}, \frac{1}{2}, \frac{2}{3}, \frac{1}{2}, \frac{3}{5}, \frac{1}{2}, \frac{4}{7}, \frac{1}{2}, \frac{5}{9}, \ldots.$$

After every even trial, there are exactly equally many heads and tails. After every odd trial, heads is ahead (ahem!). This is quite different from the behavior of genuine coin tossing sequences: for them, it is 100 percent certain that the relative frequencies are sometimes smaller and sometimes larger than $\frac{1}{2}$. They approach the limit $\frac{1}{2}$ from below and above alike, changing sides infinitely often.

As it so happened, a young French mathematician named Jean André Ville had listened to the talk by Abraham Wald in 1935, in a dingy little seminar room in Vienna. He had been a student at the French super-elitist École Normale Supérieure, and was a close friend of fellow students Jean-Paul Sartre and Simone de Beauvoir—in fact, so close that Jean-Paul slept with the wife of Jean André, and regaled Simone with the news.

Ville had gone in 1933 to Berlin in the hope of learning more about *Kollektivs*. He arrived too late: Richard von Mises had been forced into his Turkish exile. Ville did not appreciate the swastika-decked Nazi capital, and moved to Vienna. What he found was not exactly a quiet place either, but it offered far more stimulating surroundings to the budding mathematician;

Figure 8.2. Abraham Wald (1902–1950).

Figure 8.3. Jean André Ville (1910–1989).

and in Abraham Wald's seminar talk, quite fortuitously, the young Frenchman encountered *Kollektivs* again. He proceeded to demolish them.

Jean Ville was able to show that no matter which countable set of place selection functions you chose, there would always be *Kollektivs* with the same fiendish defect as H T H T H T H T . . . : though the relative frequencies of heads duly converge to $\frac{1}{2}$, and do so, moreover, along every subsequence given by the place selection functions, they would always converge from above. In other words, such a *Kollektiv* can accommodate a gambling system after all. It allows the gambler to always be on par, or even slightly ahead, of the bank. This fail-safe system is just what von Mises had wanted to avoid with his place selection functions.

After thirty years of championing the *Kollektiv* cause, Richard von Mises conceded defeat: "My first modest attempt to arrive at certain general formulations is today, in most respects, outdated."

That statement is wrong on two counts. Firstly, his attempt had never been modest. And second, the opinion that the approach was outdated was itself soon outdated. The *Kollektiv* concept had failed. Yet, it gave birth to several ideas that allowed, eventually, the capturing and tackling of the elusive notion of random sequences.

Unexploitable sequences. One such idea relies on the impossibility of a gambling system. Indeed, gambling systems need not be based on place selection functions, as they did in Richard von Mises's approach. There are more general gambling systems. Essentially, such a system tells its users, at each stage of the game, what to do with their money (if they have any left).

These systems all share a very simple property: at any time, the capital available to the gambler is the mean of the capital available after the next step (which could have resulted in a tail, or a head). Such systems are known as *martingales*. (The name, incidentally, is due to Jean Ville, and it has achieved some notoriety with the banking crash of 2008.)

A martingale is said to succeed on a sequence of heads and tails if the capital goes to infinity. We define a sequence of heads and tails as *non-exploitable* if no computable martingale succeeds (where computable means, essentially, that it can be programmed by a Turing machine, with some technicalities added for good measure). Any decent random sequences should be non-exploitable.

Rule-less sequences. Another approach had been suggested by Jean Ville himself, and was worked out, decades later, by the Swedish mathematician Per Martin-Löf. Clearly, in his frequency stability criterium, Richard von Mises had been guided by the law of large numbers, which holds for a genuine coin-tossing sequence with 100 percent probability. But there are other probabilistic laws, tons of them actually, that do also hold with 100 percent probability: to name but one instance, the law that the relative frequencies approach the limit from both sides, from above and below. Why not demand that a random sequence should satisfy *all* these "100 percent probability" laws? This would be asking for too much, actually. No sequence can satisfy all 100 percent probability laws, since one of these many laws says that it must be different from that very sequence.

Nevertheless, Jean Ville's idea proved fruitful. It suggested considering infinitely many statistical tests for randomness, and rejecting all sequences that fail some of them. For instance, it seems plausible that any sequence of heads and tails that is given by a rule should be rejected. The notion of a Turing machine allows one to make precise what it means to be "given by a rule," or by an algorithm: it means to be obtainable as the outcome of a computation. Thus, all computable sequences of heads and tails should be rejected. There are only countably many of them. This is too little, as it turns out. The idea of Per Martin-Löf was to reject not just single sequences, but sets of sequences, namely sets having probability 0. These sets can consist of a single sequence, but they can also be much larger: they can even be uncountable, and yet still have probability 0.

Thus, Per Martin-Löf considered all sets of probability 0 that can be constructed by means of Turing machines. Since there are only countable many of them, their union has probability 0, too. If we reject all members of this union, then what we are left with is the set of all rule-less sequences. These are, by definition, the sequences that cannot be spotted by tests for rule-like behavior.

Incompressible sequences. There is a third approach to randomness, developed by Andrey Nikolaevich Kolmogorov and (independently of him and of each other) by the Americans Gregory Chaitin and Ray Solomonoff. This approach relies on information theory. Some strings of heads and tails can be described in a few words: for instance "one thousand H's" or "one million H's and T's, with H's at every place having a square number, and T's everywhere else." The latter description takes more than one line, yet far less than the one million letters H or T in the corresponding arrangement. The description has been *compressed*.

We can think of a computer program that, when fed an input, prints out a string of H's and T's. Let $C(s)$ be the shortest input producing a given string s of length $n(s)$. (C stands for "compressed.") If the string s is sufficiently regular, we can expect its compressed description to be short, and hence $C(s)$ to be small. If s is very irregular, complicated, or disorderly, there may be no input shorter than the string itself to describe it. In this case, $C(s) = n(s)$.

An infinite sequence of heads and tails is said to be *incompressible* if the differences $n(s) - C(s)$ cannot grow too large, with s ranging over all initial strings of the sequence. Essentially, if one wants to describe such a sequence, one has to spell it out letter by letter. One does not know its behavior ahead of time. This idea captures, again, a sense of unpredictability.

There are some technical points to consider (such as which type of computer to choose), but basically, the incompressible sequences, the non-exploitable sequences, and the rule-less sequences are all the same. The three approaches do what Richard von Mises had hoped his *Kollektivs* would come up with.

All these characterizations of randomness make use of what is the exact contrary of randomness: namely computability. It is a notion that had been developed with no probabilistic context in mind—none at all.

Apart from computability theory, the concept of a random sequence relies also on ideas from probability theory. This observation seems obvious, indeed preposterous—so randomness has to do with probability, big deal!—however, what is meant here is the probability theory based on Kolmogorov's measure theoretic axioms. Measure theory is, indeed, necessary for all the many statements about sets of outcomes having probability 100 percent, or 0 percent. This is an ironic outcome: Richard von Mises had wanted to ground probability theory on random sequences. It turned out, instead, that an understanding of random sequences requires probability theory.

Laplace's Demon and John von Neumann's Sin

That randomness is the antithesis of computability will raise nobody's eyebrow. There is a twist to the story, however: we use computers to fake chance outcomes.

Many activities in modern life—be it business, science, or entertainment—require long tables of random numbers. The numbers are needed, here, there, and everywhere, for computation, statistics, or encoding.

Where do these tables come from? It would be a bad idea to try and make up random numbers by ourselves. Humans are notoriously inept at mimicking chance. If asked to produce, out of their head, a random string of heads and tails, they fail almost invariably. Simple statistical tests uncover quickly that human-made random sequences do not fluctuate nearly enough. The reason is simple. We have the law of large numbers in the back of our head. After an unbroken string of, for example, six heads, the tendency for coming up with tails seems almost overbearing; whereas a true coin tossing sequence would, of course, not be affected in the least, one way or the other.

In theory, random strings could be obtained by tossing lots of fair coins, or by watching the decay of radioactive atoms, or by measuring the oscillations of a power grid. In reality, however, these physical ways of producing randomness are usually too slow and inefficient. For this reason, most random numbers are faked: they are called *pseudo-random*, so as not to tell a lie. Pseudo-random numbers are produced by algorithms—the very opposite, therefore, of what randomness means. It seems paradoxical.

From the start of the digital age onward, such pseudo-random numbers were in great demand. Almost the first task set by John von Neumann to the newly born computer was to gather a fill of them. For this, he used the *middle square method*: Take any ten-digit number, and square it; then, take the middle ten digits of that square, and square that number again; and so on. Von Neumann knew that this was not really random (indeed, such a sequence of numbers must eventually repeat itself), but for practical purposes, it was good enough, provided one took care to start with a number looking sufficiently irregular (i.e., not a number like 1,000,000,000, but rather like your phone number).

By now, the methods for spawning pseudo-random numbers have enormously expanded—but they are still somewhat of a black art. In a series of embarrassing episodes, the faked numbers have proved, on closer inspection, not to be nearly as random as hoped. To rely on pseudo-random numbers remains a gamble, with a whiff of the illicit. As John von Neumann joked:

"Anyone who considers arithmetical methods of producing random digits is, of course, in a state of sin."

It is a most natural sin, in some sense. Indeed, even the most deterministic models of the physical world produce what look like random outcomes. Classical mechanics is, on principle, entirely deterministic. The great mathematician and astronomer Pierre-Simon Laplace described it by means of a thought experiment. Imagine an intellect, Laplace said, that (or who) at a certain moment would know all the forces that set nature in motion, and all the positions of all items of which nature is composed. "For such an intellect nothing would be uncertain, and the future like the past would be present before its eyes."

This intellect became known as Laplace's demon. (To normal ears, the all-knowing intellect appears to be God-like, but Laplace had gone on record with the brash statement that he did not need the hypothesis of a God.) The demon, then, embodies the idea of a dynamic whose entire evolution is totally specified by its state at any one single moment in time. That state is named the initial condition. It nails down the whole evolution of the universe.

However, many of the dynamical systems encountered in physics exhibit what is called sensitive dependence on initial conditions. It means that the slightest difference in initial conditions grows exponentially with time. The various futures that evolve from nearby initial conditions diverge rapidly.

Humans (in contrast to demons) can know the initial condition only approximately. Even if this uncertainty were, at first, quite small, it will eventually explode. Even if the future of the real world and that of the "assumed" world on which we base our computations, are both fully determined, they do not develop in step with each other. After some time, the prediction becomes totally unreliable.

This can be easily described by means of a simple toy model. Let us assume that the state of that toy world is entirely specified by an angle (the position of a pointer on a circle), and suppose that the dynamic consists simply in doubling that angle from one time step to the next. For some initial condition, the time evolution is very simple. For instance, if the initial angle is 0, it remains so for all time. If the initial angle is 120 degrees, it will be 240 degrees on the next step, and then 480 degrees, which is just the same as 120 degrees: the angle has returned, in two steps, to the initial position, and will, for all future time steps, oscillate periodically between the two values of 120 and 240 degrees. If, however, the initial condition differs by just one degree, the two states will differ by 2 degrees in the next time step, then 4, then 8, then 16, etc. After some time, the two close-by initial states (120 and 121 degrees, for example) will have produced very different futures. After a mere seven time steps, they are separated by more than a third of the circle.

Both initial states lead to periodic orbits, and thus they are predictable enough. For most initial conditions, however, the future will not be periodic, but rather utterly irregular, allowing only predictions of a statistical nature such as "the angle will be between 0 and 90 degrees with a probability of 25 percent." Such a statement is of the same nature as an assertion about the probability that it will rain in Venice exactly one year from now. As it happened, the first examples of deterministic models producing "chaos" were models from meteorology, and they had a surprising impact on pop culture. The so-called butterfly effect (the flapping of the wings of a butterfly in Brazil can cause a tornado in Texas) became truly proverbial and made it into Steven Spielberg's blockbuster movie *Jurassic Park*.

In less spectacular terms, *deterministic chaos* had been known a century earlier to Ludwig Boltzmann and Henri Poincaré. In fact, even the toss of a coin or a die relies on deterministic mechanics. We just do not know the initial conditions well enough to make a better prediction than that "all sides are equally likely to come up." (The angle-doubling dynamic is intimately related to coin-tossing sequences, by the way, and mathematically studied under the heading "Bernoulli shift.")

In the sense described, the frequentist interpretation of chance, which is second nature to most physicists, has a subjective basis: it is due to our ignorance of the precise initial conditions. The philosopher Karl Popper, a dedicated frequentist (he had, in his young years, aroused Abraham Wald to study *Kollektivs*), tried to find a way around that subjective perspective by ascribing an objective "propensity" to, say, a coin landing on heads—or more precisely, by ascribing a propensity to the experimental conditions behind the repeated experiments.

Popper admitted that his notion had a certain metaphysical touch. A propensity, according to him, is not a frequency, but that which causes the frequency. He used the term to explain what we can mean by the probability of a singular event. Few mathematicians took notice. Invoking propensities seems a last-ditch attempt to avoid the rival interpretation of frequentism: namely, that the probability of an event is a measure of our lack of knowledge. We turn to it now.

The Price of Ignorance

The modern theory of subjective probability was developed almost single-handedly by a brilliant Italian mathematician, Bruno de Finetti. It took three decades, from the 1930s to 1960s, before it became firmly established. By that time, it also became clear that de Finetti had been anticipated by a young English polymath prodigy, a logician, mathematician, economist, and philosopher named Frank Plumpton Ramsey.

Ramsey's mother had studied at Oxford; apart from this blemish, he was Cambridge to the bones. His father, a mathematician, was president of Magdalen College. Frank studied at Trinity. The famous economist John Maynard Keynes soon took him under his wing and introduced him to the Apostles, a secrecy-shrouded elitist discussion group.

There was much talk, among the Apostles, of a mysterious little book on logic, written in German. Some Cambridge professor proposed the title *Tractatus Logico-Philosophicus*, no doubt with future sales numbers in mind. The booklet had been written—or should one say composed?—by a former Apostle, a chap from Vienna named Ludwig Wittgenstein, reputed to be fabulously wealthy, while he was serving in the trenches during World War I, somewhere over there, on the other side of the hill.

Figure 8.4. Frank Plumpton Ramsey (1903–1930).

Frank Ramsey translated the text, learning German more or less on the job. Some expressions in the *Tractatus*, he found, were rather unclear: to sort things out, the undergraduate journeyed to Lower Austria, where Wittgenstein, presently, worked as a teacher at an elementary school. Every afternoon, for several weeks, they went through the book, line by line. Amazingly, they became lifelong friends. Their only serious quarrel was about Sigmund Freud, whom Frank put on par with Albert Einstein and with Wittgenstein himself. Ludwig was scandalized.

When he turned twenty, Ramsey received his bachelor's degree in mathematics, sailing through all exams with brio and distinction. He finished top of his class (what older Cambridge hands would have termed Senior Wrangler). Soon after, Ramsey became a fellow of King's College, and the director of mathematical studies there: an oversized, sincere, and endearingly naïve wunderkind ("unworldly, untidy and ungainly," to repeat the words of his biographer Cheryl Misak; some described him as "an enormous man like a cross between a light-house and a balloon").

When Wittgenstein, aged forty, returned to Cambridge and enrolled as an advanced student ("advanced," no less!), his supervisor was Ramsey, who was the younger by seventeen years and still a staunch Freudian. (Tragically, Frank's beloved mother had been killed by his father in a car accident, which is a blow no psychoanalyst can soften.)

Ramsey was fascinated by the foundations of mathematics. He greatly simplified Russell's theory of types. A combinatorial remark that he made, almost in passing, blossomed decades later into the hugely influential Ramsey theory, which has to do with the inevitability of order in sufficiently large structures. As if this were not enough, he shaped the development of economics with a few path-breaking essays. But the main work that the young man had set his mind to was a theory of subjective probability.

He never saw it to completion. He contracted an infection, probably from swimming in the River Cam, and developed jaundice. In the early days of 1930, Frank Ramsey died after an abortive bit of surgery, at age twenty-six. Wittgenstein, in the anteroom, shared some of the agony.

Ramsey left behind a huge mass of notes and drafts. Among them was the manuscript of a lecture he had given to the Cambridge Moral Science Club in 1926, entitled "Truth and Probability." As the minutes of the Moral Science Club reported: "[Ramsey] maintained that degrees of belief were to be measured by the willingness to bet, and that the laws of probability were laws of consistency in partial belief, and so a generalization of formal logic."

The talk was published posthumously in 1931, but it took a long time before its impact was felt—an instance of what is nowadays termed the Ramsey effect, which is the belated understanding that Ramsey had been there long before.

Physicists are inclined to confine probabilities to mass events. Psychologists know that this is far too narrow, and that we are constantly evaluating the probabilities of single events on the basis of whatever evidence there is.

Probability as "the logic of partial belief" must not—so Ramsey says in "Truth and Probability"—be taken as prejudging its meaning in physics. What he was looking for is "a purely psychological method of measuring belief." Beliefs, in his bold view, are as much "given" as sense data are.

Beliefs come often associated with some feeling; however, there seems to be no objective way to measure the intensity of that feeling. How could we ascribe a number to it? Moreover, "beliefs which we hold most strongly are often accompanied by practically no feeling at all; no one feels strongly about things he takes for granted." Therefore, rather than for *feelings*, we should look for *actions* caused by the beliefs. A subjective probability of $\frac{1}{3}$, for instance, "is clearly related to the kind of belief which would lead to a bet of 2 to 1."

A bet is an action. To assign a probability is to measure a disposition to bet.

Imagine a bookie who is prepared to buy and sell bets on any outcome of a particular horse race. The bookie must pay out $100 to the winner of a bet, but can choose the odds anywhere between 0 and 1. Suppose that he gives odds of 60 percent on the racehorse Seabiscuit winning. This means that the bettor pays $60 for the bet, hoping to net $100.

If the bookie gives odds of 20 percent on rival horse Red Rum winning, and 90 percent on either Seabiscuit or Red Rum winning, we can make a Dutch book against him. (Nobody seems to know where this "Dutch" comes from.) We place one bet each on Seabiscuit and Red Rum, and then place another bet on neither of two winning (which, according to the bookie's odds, is 100 – 90 = 10 percent). Thus, we pay 60 + 20 + 10 = $90 to place our three bets, and earn $100 for sure. No bookie in his right mind, Dutch or no Dutch, would ever allow this to happen: the odds on "either Seabiscuit or Red Rum" have to add up properly, of course. The same argument shows that for any two events A and B that cannot occur together, the odds that one of them occurs is the sum of the odds for A and the odds for B. This means that the subjective probability P must satisfy additivity: $P(A \text{ or } B) = P(A) + P(B)$, just as with any objective probability.

In a similar way, one can define *conditional probabilities* by means of bets. The bookie can accept bets on Seabiscuit winning tomorrow's race if it rains. This means that the bettor has to place a sum. If it rains tomorrow and Seabiscuit loses, the bettor loses this sum; if it rains and Seabiscuit wins, the bettor receives $100. But if it does not rain, the bet is canceled and the sum returned to the bettor, no matter whether Seabiscuit finishes ahead or not.

The bookie could also accept bets on tomorrow being a rainy day. But woe to the bookie who does not respect the rule that the odds of "Seabiscuit wins *and* the day is rainy" are given by the *product* of the odds for "it rains tomorrow" and "Seabiscuit wins, if it rains." A Dutch book could be made, so that the bookie loses money for sure. Once again, the odds, or in other words the subjective probabilities of the bookie, must conform to the usual rules of objective probability. In particular, the multiplication rule holds for any two events A and B:

$$\text{probability of } (A \text{ and } B) = (\text{probability of } B) \times (\text{probability of } A, \text{ if } B \text{ holds}).$$

Needless to say, the bookie is a character from a thought experiment. Nowhere in real life will one find a bookie fair enough to offer bets without a margin for himself. The same holds for derivative traders, or hedge fund managers, or casino owners. The thought experiment serves merely to make clear that subjective probabilities have to be coherent, lest a Dutch book could exploit them. This coherence requirement is the reason why subjective probability is sometimes called the logician's probability, in contrast to the physicist's probability (which is also known as objective probability).

We may note here that *probability* and *provability* not only sound alike, but share the same root. In its older, pre-mathematical meaning, the word *proof* did not design a deduction from first principles by logic alone, but indicated an argument speaking in favor of a statement, such as a bloodstained knife in possession of the suspect. Proof did not have to be absolutely conclusive. Proof was used to convince juries long before it was used to convince mathematics students. Something is proved to you "beyond reasonable doubt" if you will take any bets on it.

Ramsey's fair-minded bookie is a contrivance, a fiction; but measuring belief via betting is not. It captures what goes on in reality. "All our lives we are in a sense betting," to quote Ramsey. "Whenever we go to the station we are betting that a train will really run, and if we had not a sufficient degree of belief in this we should decline the bet and stay at home."

Nevertheless, Ramsey felt that the betting scenario was still incomplete. He illustrated this by an example:

> I am at a crossroads and do not know the way; but I rather think that one of the two ways is right. I propose therefore to go that way but

> keep my eyes open for someone to ask. What would I be prepared to pay for learning which way is the right way? It obviously depends on my degree of uncertainty, in other words, the strength of my belief; but it also depends on the values which I ascribe to the two alternatives. If one is much more inconvenient than the other, I want to really make sure to pick the right way.

So how does one measure value? Within a few pages, Ramsey developed utility theory, hand in hand with subjective probability. It is an amazingly neat approach. Ramsey sketched it and then commented, rather airily: "I have not worked out the mathematical logic of this in detail, because this would, I think, be rather like working out to seven places of decimals a result only valid to two."

More details would follow, he said. It was, after all, only a talk given to friends in the Moral Science Club. There seemed to be plenty of time ahead. Ramsey was barely twenty-three years old and had so many other interests! His death came very unexpectedly.

It fell to Bruno de Finetti to work out the foundations of subjective probability, and to John von Neumann and Oskar Morgenstern to axiomatize expected utility theory. The similarities were amazing: bets and conditional bets, Dutch books, coherence. As to Ramsey's talk, which was published posthumously, it took an unconscionably long time to sink in.

The most remarkable aspect of Ramsey's approach to utility was that it strictly kept to subjective probabilities (whereas von Neumann and Morgenstern used objective probabilities for their lotteries). The trick was to define belief of degree $\frac{1}{2}$ by means of indifference. Suppose that there is a proposition Z (for instance "Seabiscuit wins the race") about which you do not care one way or the other. Suppose further that you really like something, for instance having a drink. If you should be indifferent between "I get a drink if Z holds, no drink otherwise" and "I get a drink if Z fails, no drink otherwise," it would seem reasonable to say that your belief in Z has degree $\frac{1}{2}$. This means that you assign it probability $\frac{1}{2}$.

Next, Ramsey introduces utilities. If you are indifferent between the lotteries $\frac{1}{2}A + \frac{1}{2}B$ on the one hand and $\frac{1}{2}C + \frac{1}{2}D$ on the other, this implies that however much you prefer A to C, the difference is compensated by how much you prefer D to B. The utilities, therefore, must satisfy $u(A) - u(C) = u(D) - u(B)$. From there, and with a few reasonable assumptions on the continuity of your preferences, you can assign utilities to any outcomes, thereby measuring by how much you prefer one over the other.

In the last step, Ramsey introduced subjective probabilities other than $\frac{1}{2}$, for instance $\frac{1}{3}$, by the odds at which you are prepared to bet on an outcome; this works just as before, except that the bet is now based on utilities rather than money.

This is only a sketch of Frank P. Ramsey's sketch, meant to highlight the elegance of his quickstepping between subjective preferences and subjective probabilities: Ramsey started out with indifference between outcomes, used this to define partial belief $\frac{1}{2}$, used this in turn to define a full scale of utilities, and used the latter to define a full scale of partial beliefs. This is bootstrapping at its best: rarely was so much derived from so little. The method reminds one of the Baron von Münchhausen, who famously claimed to have extracted himself out of a swamp by pulling on his own hair. But Münchhausen was a mountebank and Ramsey was not.

Betting on the Sun

Beliefs change as new evidence becomes available. We are able to learn. Frequently, we do so by induction, deducing a general rule from particular instances. "All swans are white," that sort of thing.

"So they have black sheep here in Scotland," says the traveler, looking through the window of his train compartment.

"At least one black sheep," corrects his neighbor.

"It is black on the side that is turned toward us," adds another.

The travelers head for a philosopher's congress.

Ever since Hume, we know that induction is not based on any logically compelling argument. On what else is it based, then? We may quite rightly argue that induction has worked in many instances; what this statement offers, however, is not a justification of a general rule, but just one more instance of it, thereby raising the same question.

"There lies nothing vicious in this circle," says Ramsey. He compares induction with memory, which is another fundamental way of acquiring knowledge. "It is only through memory that we can determine the accuracy of memory; for if we make experiments to determine this effect, they will be useless unless we remember them." Memories vouch for memories; so why cannot induction vouch for induction? Nevertheless, induction has troubled a good many philosophers from Hume and Kant to Popper and beyond.

Induction is a useful habit (Ramsey again) but it has its limits, obviously. This issue has spawned a number of jokes.

"So far, so good," says the construction worker falling from a skyscraper's roof, as he rushes past the third floor.

In that same vein, turkeys expect to be fed, as on any other day; instead, the farmer wrings their necks. There is a general rule behind this regular event (the closer Thanksgiving, the higher the threat), but no turkey ever collects enough data to form a proper hypothesis.

Statistics is the science of updating beliefs, and thus to infer from events to probabilities. The first step in that direction was made by an English clergyman named Thomas Bayes, most probably as a response to the concerns of David Hume. Hume had written:

> One would appear ridiculous, who would say, that it is only *probable* that the sun will rise tomorrow, or that all men must die; though it is plain that we have no further assurance of these facts than what experience affords us.

One would appear ridiculous indeed. The certainty of sunrise is firmly anchored in our web of beliefs. Yet, there are countless instances where we collect data to estimate chances. If we don't do it in the proper way, we will be apt to lose bets; in fact, Bayes argued that we may fall victim to a Dutch book (although he didn't use the term). His ideas were further developed by a brilliant young astronomer and mathematician named Pierre-Simon Laplace—the same who would later conceive the deterministic demon.

Figure 8.5. Pierre-Simon Laplace (1749–1827).

Consider an urn containing an unknown number of red and black balls (and no others). What is the proportion of red balls? You have absolutely no clue, yet you might try and learn. You draw a ball and see that it is red. You place the ball back into the urn (so as not to change its composition), shake it, and then repeat the experiment. After a few trials, you might get a hunch; and after a few hundred trials, you may be pretty confident of your chances. But what is the best way to update your belief from one trial to the next?

The repeated trials are obviously related to the law of large numbers. If you know the chance of drawing a red ball (which is given by the proportion of black and red balls in the urn), you can deduce something about the frequencies of drawing red in series of trials. Now, however, we are faced with the inverse problem—a statistical problem. We don't know the probability of drawing red; but by repeating the experiment, we obtain frequencies, and may infer from them the probability of drawing red. In each case, whether going from probabilities to frequencies, or from frequencies to probabilities, we can only make a guess—but an informed guess.

Let us assume, with Bayes and Laplace, that all compositions of the urn are equally likely, and that we find out that in *N* trials, a red ball has been drawn n times. What, then, is the best guess for the probability of drawing red next? It is $\frac{n+1}{N+2}$. This is called the *succession rule*, because it yields an estimate for drawing red in the succeeding trial. The term "best guess" indicates that we assign probabilities to probabilities. This idea is Bayes's main contribution to what he called "the doctrine of chance."

If the number N of trials is large, the succession rule yields a number that is very close to the frequency $\frac{n}{N}$. This result seems a bit of a letdown: probability = frequency, didn't we know this before? And if $n = N$ (meaning that red has been drawn every time so far), the probability of drawing red again next is $\frac{N+1}{N+2}$, which is practically 1. Tongue in cheek, Laplace computed the chance that the sun will not rise tomorrow as being one in two million. (He surmised, like everyone else at the time, that sunrises had been observed for some 5000 years.)

What about small numbers of trials? If $N = 0$, that is, before any trial has taken place, the succession rule yields $\frac{1}{2}$, which is clear enough: we have assumed that every proportion is equally likely. For $N = 1$, we obtain $\frac{1}{3}$ or $\frac{2}{3}$, depending on whether we drew a black ball (B) or a red ball (R). For $N = 2$, the outcome R R gives odds of $\frac{n+1}{N+2} = \frac{3}{4}$ on R. Note that the frequency of R (namely $\frac{n}{N}$) is 1; but no one would consider R as certain.

For $N = 3$, the subjective probability of the initial string of outcomes R R B yields $\left(\frac{1}{2}\right) \times \left(\frac{2}{3}\right) \times \left(\frac{1}{4}\right)$. It is given by the multiplication rule, used several times over: probability of (R in the first trial *and* R in the second *and* B in the third) = (probability of R in the first trial) × (probability of R in the second trial, given that R was obtained in the first trial) × (probability of B in the third trial, given that R was obtained in the first two trials).

Similarly, an initial string R B R yields $\left(\frac{1}{2}\right) \times \left(\frac{1}{3}\right) \times \left(\frac{2}{4}\right)$—the same product as before, although the factors are not all the same. This is no coincidence. The subjective probabilities of a given string depend only on the frequencies of R and B in the string, and not on the order in which the colors are drawn.

This *exchangeability* plays a crucial role in de Finetti's approach. It is a weaker assumption than that of independently repeated trials with one and the same probability for drawing red. To give an example, if we repeatedly draw from an urn, but do *not* replace the balls, then the outcomes are exchangeable ("first R, then B" is as likely as "first B, then R"), but the trials are no longer independent and the objective probabilities vary from trial to trial.

In the case of an urn with a given composition of red and black balls, it makes sense to speak of an objective probability. We can imagine that someone knows the proportions in the urn—has insider knowledge, so to speak. For the subjectivist, however, objective probabilities become meaningless in less artificial (and more lifelike) scenarios featuring uncertainty, such as, for instance, the sex ratio of newborns in Chicago. There, all we have are the data—the birth statistics. There is no urn into which we can take a look.

Objective probabilities are a metaphysical notion, says Bruno de Finetti. They are operationally meaningless. His proud motto proclaims: "Probability does not exist." The best we can reasonably do is to guess, and then modify our beliefs as evidence comes in. In a way, a belief can't be false; and whenever a belief is updated, as a result of observations, such an action does not imply that it is corrected, since it was never wrong in the first place.

What about the assumption from which Bayes and Laplace derived their succession rule: namely, that all compositions of the urn are equally likely? This assumption (or guess, or belief) is called the *prior*. It seems reasonable enough, but nobody can forbid us to start out with another prior. It turns out that it hardly matters in the long run. The updated beliefs of different people with different priors will (under very general conditions) converge to the same value, if the trials are repeated often enough. It is tempting, of course, to affirm that this limit, then, *is* the objective probability. In this sense, it emerges from subjective probabilities.

But that is beyond belief, in more sense than one.

PART III

9

Voting

The Rabid Lamb and the Dictator

Dark Doings in Paris

Few mathematicians make it into a novel. Nicolas de Condorcet did. He shows up in the pages of *The Queen's Necklace*, a whale of a book written by none less than Alexandre Dumas *père*.

The necklace in question is not the one that made the immortal three musketeers spring to action and arms. More than a century had passed since their glory days. The French Revolution looms only a few years ahead, and the hapless queen in question is Marie Antoinette.

In the prologue of his novel, Dumas describes a dinner party given in Paris by an aged duke. The list of guests is exquisite: it includes the King of Sweden, who is on an incognito visit to Paris, and the legendary Madame du Barry. The Marquis de Condorcet, a stellar member of the Académie des Sciences, sits next to the man who is currently the talk of the town: a slightly disquieting adventurer named Cagliostro, who dabbles in the occult like so many then did. Conversation flows gracefully, style ancien régime, until the illustrious guests want Cagliostro to foretell each of them the manner of their death.

It all started out as an idle game, and Cagliostro, to give him his due, did his best to give the talk a different turn. To no avail: he is pressed from all sides to come up with the answers. The guests smile expectantly. "Well then, the King of Sweden will be killed by a bullet." "Glorious," exclaims the king, "to die on a battlefield! Why, ending like this runs in the family!" "You will not be shot on a battlefield, sire. No, at a ball." Madame du Barry?—Beheaded. Monsieur de Favras?—The rope. And on it goes, from one violent death to another. Condorcet?—Suicide by poison.

Nobody stayed for coffee that evening.

Dumas had the benefit of hindsight, of course. Now, to turn from the novel to history: Marie Jean Antoine Nicolas de Caritat, marquis of Condorcet (1743–1794), was the scion of an aristocratic family that had been ruined for generations. His father died early, as a penniless captain in a dragoon regiment. Nicolas was raised in a Jesuit college. He soon excelled in all fields, and most particularly in mathematics. His reputation spread quickly all over Europe, based on his early works in analysis, astronomy, and probability. At the tender age of twenty-five, he was elected into the French Académie des Sciences, which soon appointed him as their *sécrétaire perpétuel*. Condorcet also became a corresponding member of the Prussian Academy of Science and honorary member of the Russian, Swedish, and American academies.

Academies devote a lot of their time to electing presidents, vice presidents, committees, and new members. This may have been the reason why Condorcet took to studying electoral procedures. He founded what he termed "political arithmetic," an early application of mathematics to the social sciences. The new field surprised his colleagues. Some fifty years before, purportedly, Newton had grumbled that he could calculate the motions of comets but not the madness of people. (He was afflicted by the collapse of the South Sea scheme, which had cost him dear.) Newton's quip reflects the firm conviction of his epoch that mathematics is meant for heavenly bodies, but not for the messy chaos that reigns down here.

Today, the use of mathematics in social sciences is raising no eyebrows any longer. It is needed in many disciplines, such as decision theory, game theory, demography, and insurance. But in the eighteenth century, it appeared strange that mathematics could be applied to anything except physics and astronomy.

What may have helped, in the post-Newton years, to provide some respectability for Condorcet's strange notion of *mathématiques sociales* was a philosophical principle that had emerged during the Enlightenment: the idea that all men are born equal. And women, too, as some would add later. Condorcet, by the way, was among the first to do so.

The claim that we are born with equal rights was a bold abstraction, at the time—a brainchild, and certainly not an empirical fact in a world used to serfdom and slavery. But gradually, the idea gained hold in civilized discourse. Equality was nowhere to be found, true, but you could demand it. You could *postulate* it. You could even brashly proclaim it as "a self-evident truth." In fact, you could use it as an axiom. And wherever there is an axiom, mathematics is not far away.

As usual, it did not bother mathematicians that the instances where you could apply the new postulate seemed rather restrained at first. Aside from learned academic bodies, those proverbial "republics of scholars," real republics were rare at the time. They were usually corrupt and poorly viewed by all the reigning despots of that age, enlightened as they were.

Things were bound to change. Marquis de Condorcet certainly did not wish to restrict his political arithmetic to small elitist bodies. He was an eminently political thinker, a figurehead of the *Siècle des Lumières*, comrade in arms of Voltaire and Turgot, befriended with Benjamin Franklin, Thomas Paine, and Thomas Jefferson. In due time the aristocrat became more radical than all of them. He publicly demanded voting rights for women, well before men could exercise theirs. He even went so far as to propose the abolition of slavery. In return for his efforts, he was nicknamed *le mouton enragé*, the rabid lamb.

The French Revolution became the apex of Condorcet's life. He had been hailed as the last philosopher of the Enlightenment. He now became the first republican—indeed, almost the first to conceive of a France without a king. He was elected secretary of the tumultuous Legislative Assembly. For a time, insiders had it that the illustrious savant would be charged with the education of the throne-follower, the dauphin, who then was a nine-year-old boy. This plan, however, came to nothing. The boy, on becoming an orphan, was entrusted to the care of a cobbler—or rather, to his neglect. He died one year later.

On behalf of the Girondins, Condorcet was charged with the project of elaborating a constitution. It was all set to become the first constitution of the First Republic. But it was never put to a vote. Maximilien Robespierre had other ideas. The Girondins gave way to the Montagnards, and a wave of *terreur* engulfed the country.

Condorcet now was a marked man. For nine months, he managed to hide with friends in Paris. Always knowing that capture meant certain death under the guillotine—a machine named after a fellow scientist and friend of his—Condorcet wrote his last book, explaining the history of humanity as a progress story in ten chapters. Ten was a popular number at the time. The week was to have ten days and the meter to measure one ten millionth of the distance from the North Pole to the equator. The tenth chapter of Condorcet's book described a man-made paradise.

All of a sudden, it transpired that the secret police had discovered his shelter. He fled head over heels and hid for two nights, like a hunted animal, in the huge limestone quarries of Clamart, south of Paris. He was arrested, then found dead in his cell the next morning. His face and tongue were black. Officially, the exact cause of his death was never ascertained—the police had better things to do—but most probably, Cagliostro had been right: suicide by poison.

Votes and Winners

What was Condorcet's political arithmetic about? It had been initiated by the discovery of a paradox that besets runoff voting. A paradox? A flagrant democratic scandal, rather; or what other term to use if a candidate who would win every runoff vote is not admitted to the runoff vote?

That scandal is no quirk. It plagues many elections. Probably the most widespread voting procedure, then as now, is a two-round system that operates as follows: If a candidate wins an absolute majority in the first round of the election, the second round is canceled and the candidate is declared winner. If no candidate wins an absolute majority in the first round, then the two candidates with the most votes enter the second round, and whoever then wins the one-on-one contest is declared the winner. (For the sake of simplicity, we discard the possibility of a tie. It requires an extra convention that need not bother us here.)

Condorcet pointed out that it could well be that a third candidate was better entitled to win the election: a candidate who would have beaten every other contestant in a one-on-one vote, and hence would certainly have won the second round, had she (or he) not been discarded in the first.

Nowadays, a candidate who wins every one-on-one vote is said to be the Condorcet winner. If in a soccer tournament, a team that wipes every rival is not admitted to the finals, we would suspect some evil doings. Yet in politics, such a state of affairs is not rare. Condorcet had pointed out this failing of the two-round system years before the French Revolution. Nevertheless, from 1789 onward, the two-round system has been used in every French election, all through the First, Second, Third, Fourth, and Fifth Republic. Could he but know, our marquis might turn in his grave. (As a matter of fact, he is not in his grave. His cenotaph in the vast Panthéon stands empty. The corpse of Condorcet has been mislaid.)

Condorcet used to make his point about the pitfalls of the runoff system by means of some simple examples. Imagine that there are twelve voters, say, and three candidates A, B, and C (let's call them Ann, Burt, and Conny). The voters have made up their minds about the merits of the candidates. Five voters rank Ann ahead of Conny, and Conny ahead of Burt. Four voters rank Burt first, before Conny, and Conny before Ann. Three voters rank Conny first, then Burt, then Ann. (In principle three more rankings are possible: first Ann, then Burt, then Conny; or first Burt, then Ann, then Conny; or first Conny, then Ann, then Burt. We assume that no voters adopt these rankings, and also that no voter is indifferent between two candidates.)

TABLE 1

Voters 1, 2, . . . , 12 and candidates A, B, C

	Individual ordering		
Voter	First place	Second place	Third place
1	A	C	B
2	A	C	B
3	A	C	B
4	A	C	B
5	A	C	B
6	B	C	A
7	B	C	A
8	B	C	A
9	B	C	A
10	C	B	A
11	C	B	A
12	C	B	A

In the first round of the ballot, Ann receives five votes, Burt four, and Conny three. In the second round, Ann is matched against Burt, and since all former Conny voters now give their voice to Burt, he wins. So far, so good.

But on closer inspection, Burt's victory stands on feet of clay. Indeed, in a one-to-one contest, Conny would have beaten Burt by eight to four. She also would have beaten Ann, this time by seven to five. Conny has good reason to complain about the voting rules. So have her voters. They will certainly feel that it would have been fairer to elect Conny, because she is the Condorcet winner!

Finding the Condorcet winner requires three one-on-one matchings if there are three candidates, and six such matchings if there are four, or ten matchings if there are five. This seems a bit awkward. But we don't need that many rounds. One round suffices, if the voters are prepared to rank all candidates on their ballot. The rankings determine the results of all the one-on-one matchings right away.

Yet, the procedure has a flaw, as Condorcet was forced to admit. There need not be a Condorcet winner at all! It could be that, in one-on-one votes, Ann beats Burt, Burt beats Conny, and Conny beats Ann. This yields a cycle, just as in the children's game of Rock-Paper-Scissors. Here is a simple example.

Again there are three candidates, this time Ulysses, Victor, and Wendy, but now 100 voters: 25 of them rank Ulysses ahead of Victor, and Victor ahead of Wendy; 40 rank Victor first, Wendy next, and Ulysses last; and 35 rank Wendy ahead of Ulysses, and Ulysses ahead of Victor. (Again, we assume that the other three rankings find no takers. This is just for the sake of having a simple example.)

If voters had to decide between Ulysses and Wendy only, Wendy would win, with 75 votes to 25. In a contest between Wendy and Victor, Victor would win with a margin of 65 to 35. But Ulysses would win against Victor with 60 to 40 votes. Every candidate loses against another, and there is no Condorcet winner among them.

The impasse can be overcome by some additional rule, for instance by discarding the closest-run result. (In our case, Ulysses versus Victor. Then, Victor would win.) But the outcome is not totally convincing any longer.

TABLE 2

100 voters and 3 candidates U, V, W

Number of voters	First place	Second place	Third place
25	U	V	W
40	V	W	U
35	W	U	V

THE BORDA COUNT

As if this were not bad enough, Condorcet's scientific nemesis entered the scene, Jean-Charles de Borda. Borda too was an eminent mathematician and astronomer, and a member of the Académie des Sciences. He had been working on a par with luminaries such as Laplace and Lagrange. In addition, Borda was a man of action who had enjoyed a brilliant career as a naval officer, crossed and recrossed the Atlantic as an indefatigable explorer, played a dashing role in the American War of Independence, and invented a number of nautical instruments. He was a captain and commander, more at home on storm-tossed oceans than in the drawing rooms of *le beau monde*, and therefore irresistibly appealing to *le beau monde*. Among Borda's exploits ranked the best measurement of the distance from Earth's pole to the equator—a truly standard-setting feat, since it gave us the meter.

Figure 9.1. Marquis Nicolas de Condorcet (1743–1794).

Figure 9.2. Jean-Charles de Borda (1733–1799).

The illustrious sailor challenged Condorcet on the latter's own turf, political arithmetic. Borda felt that elections are not fair and square if they merely consist in putting a cross next to one's favorite. Such a procedure does not allow voters to express what they think of the other candidates. Voters should rank all the candidates on their ballot, giving one point to the candidate they like least, two points to the next-to-last, and so on until they reach their favorite, who receives the maximal number of points—three points, for instance, if there are three candidates altogether. In this way, each candidate garners a number of points on each ballot. These numbers are added: and the total sum achieved by each candidate is a measure of his or her success. The candidate with the most points, the winner of the *Borda count*, is elected.

The procedure is admittedly more complicated than merely counting how many voters prefer whom. Nevertheless, it is frequently used today if the number of voters is relatively small; for instance, if a jury or a committee has to make a choice.

One advantage of the method is that—barring ties—there always exists a Borda winner. Even if there is no Condorcet winner, a decision is reached. In our second example, for instance, Wendy emerges as the Borda winner. She receives 210 points in total, Victor gets 205, and Ulysses only 185. Note that a Condorcet winner—the candidate who dominates each one-on-one vote—need not come first in the Borda count. It is easy to think of examples where Ann is ranked first by the absolute majority of voters, and yet fails at the Borda count. Conversely, Ann could win the Borda count without being ranked first by any voter. We can be sure that Condorcet did not take long to point this out to his rival.

Another snag, with both the Borda method and the two-round runoff voting, is that if one candidate drops out of the race, for whatever reason, the ranking of the others can be overturned. Let us go back to our first example, the one with the runoff voting. Burt won that election. However, if Ann, for instance, quits the contest—possibly because she becomes aware that she will not win anyway—then it is not Burt who is elected, but Conny. And a similar thing may happen in the second example, if we use the Borda count. Remember that Wendy won in that scheme. If Ulysses, however, gets cold feet and does not enter the race, it is no longer Wendy who wins, but Victor. In each case, an "also-run" drops out, and the winner is changed. This seems strange in an electoral race.

Moreover, opportunities for foul play abound. Let Victor be back in the contest, which is—to recall—a Borda count leading to Wendy's win. A quarter of Victor's supporters could craftily decide to rank Ulysses ahead of

Wendy, not because they prefer Ulysses, but simply because they want Victor to win. And they succeed! (The votes are Victor 205, Wendy 200, and Ulysses 195.) This ploy is called *strategic voting*. It amounts to cheating, of course. When someone pointed out to Jean-Charles de Borda that strategic voting could subvert his voting procedure, the hardy sailor nodded somberly and replied that it was meant for use by honest people.

Since the days of the heated debates between Condorcet and Borda, a quarter of a millennium has gone by. The mathematics of voting has grown tremendously, and many dozens of electoral procedures are weighed pro and con. But the two schools of thought of Condorcet and Borda—the rabid lamb and the old tar—are still dominating the field, as much at odds as ever.

Borda had the satisfaction to see that the French Academy, where so much of the debate had taken place, adopted his procedure for the election of its new members. However, matters changed brusquely in the year 1801. In that year, Napoleon Bonaparte became a member of the academy. He had his own ideas about the right way to vote.

Arrow Punctures Common Will

This yields us with one leap to the so-called dictator theorem. In the mid-1950s, it helped the young American Ken Arrow revolutionize social choice theory, and eventually earn a "Nobel" Prize for Economics. (Since we are going to mention several such laureates, by and by, it is worth mentioning that this prize was introduced long after Nobel's death and should properly be called the "Prize for Economic Sciences in Memory of Alfred Nobel," to distinguish it from the genuine Nobel Prizes. But why be pedantic?)

Rather than investigate the pitfalls and advantages of this or that voting procedure, Arrow dealt with all possible voting procedures in one go, so to speak. He showed that none of them is perfect. Worse: none can satisfy even the most reasonable requirements, as soon as there is a choice between more than two alternatives.

What are reasonable requirements? Basically, every voting procedure is a means of turning the individual rankings of the voters into a collective ranking, namely the outcome of the vote. The first requirement is surely that if all voters rank Ann ahead of Burt, then Ann should end up ahead of Burt in the final outcome. Second requirement: there should be no cycle in the final outcome. Third and last requirement: whether or not Ann ends up ahead of Burt should not depend on how Conny is faring.

Modest requirements, as it seems. Yet, Arrow proved that they cannot be fulfilled. More precisely, the only method to satisfy all three requirements is to adopt the ranking of a single one of the voters, and ignore all the others. This one voter decides for all, thus becoming effectively a dictator. This is why Arrow's theorem is known as the dictator theorem. It shows that what we expect democratic elections to achieve cannot always be done. This is why Arrow's theorem is also known as the impossibility theorem.

Did we require too much by asking that the vote should produce a collective ranking of *all* candidates? In many cases, what matters is only who wins. But if we eliminate, in thought, that winning candidate, the third requirement ensures that we must know who comes next, and so on. The ballot yields a ranking of all candidates, even if we only look for the winner.

Since, short of dictatorship, our three seemingly so basic requirements cannot all be met, one of them has to go. But which one?

The first one (if all voters agree, that is how it should be) seems so crucial that no election procedure that violates it has a chance to be seriously considered.

The other two requirements demand from a collective ranking something that we take for granted in every individual ranking—at least, in every ranking by a sane individual. The Condorcet method violates one of these requirements—there can be cycles—and both the Borda count and the runoff voting violate the other—the decision between two candidates can depend on how a third candidate is doing.

Let us consider these two requirements more closely.

First, the no-cycle condition: If a guest in a restaurant is offered chicken, fish, or pasta and happens to prefer fish to chicken, chicken to pasta, and pasta to fish, the waiter will have to wait forever and the guest will eventually starve. But, the same cyclic preference that makes one individual guest look dotty will raise no waiter's eyebrows if evinced by a group of clients. Suppose that a short-handed cook decides to prepare, for each mealtime, only two of the three items on the menu. It is perfectly conceivable that when no pasta is on the fare, fish is preferred to chicken by the majority of guests; that if no fish is available, chicken is preferred to pasta by most; and that on the days without chicken, most customers prefer pasta to fish. No one would conclude that the patrons of that restaurant are unhinged.

Now to the third requirement. For our next example, we stay in the same restaurant. A guest is told that there is a choice between chicken and pasta. The guest ponders a little, and then orders chicken. Now the waiter remembers that there is fish on the menu after all. The client, on learning this, exclaims: "Oh, fish? Why, this changes everything! I will have... pasta!" Such a decision seems odd. A new alternative C appears, does not win, but causes the ranking between the other two alternatives A and B to switch.

Once more, however, what seems weird behavior in an individual can easily happen with a group of patrons. It is quite conceivable that on fish-free days more clients opt for chicken than for pasta, but that some of these chicken-ordering clients will switch to fish if given the chance. If, in addition, no pasta-lover switches to fish—they may all be vegan, for instance—then pasta can become the most popular choice, and no cook will think twice about it.

This is the lesson behind Arrow's impossibility theorem. We should not expect that a community has a will in the sense that an individual has. This leads us back to the marquis de Condorcet. He was, like all other thinkers of the Enlightenment, deeply imbued by the ideas of Jean-Jacques Rousseau. According to Rousseau, the community should be directed by the *volonté générale*, the common will. This common will is an exalted notion, yet Condorcet found on closer inspection that it was hard to pin down. There is no common will. At least, it can be conspicuously absent, on occasion. That insight was a harsh blow to the disciples of Rousseau. The object lessons during the mad years of the French Revolution led to further disenchantment.

Eventually, the *volonté génerale* gave way to the *volonté du général* when young Napoleon took over.

Today, after centuries of experience with the rule of the people, we have become less sanguine. Democracy is not a promise of bliss on some far-off horizon, as it had been for the authors of the *Encyclopédie*, but simply is the lesser evil. According to Karl Popper, democratic elections are foremost and first a way of getting rid of a bad government with a minimum of fuss and bloodshed. Votes cannot be expected to always produce the "right" decision—whatever this may mean. What counts is that each decision can be revised in the next election.

We are long accustomed to all elections facing stumbling blocks. Even the obvious unfairness in having some voices count more than others has not been fatal to US democracy. Admittedly, it leads from time to time to ugly machinations, as when electoral districts are redrawn to produce a convenient majority, but this is just a fact of life. The technique is called gerrymandering, and offers some more food for thought to mathematicians. If you rely on honest politicians, as that upright sailor Borda seems to have done, you are apt to get disappointed sometimes. That's all.

This leads us back to strategic voting. It consists of manipulating the final outcome—the collective ranking—by purposefully misrepresenting one's own, individual ranking. The temptation to do this will always be present. Essentially, the incentive to strategic voting can only be avoided by adopting the individual ranking of a single voter, lock, stock, and barrel. This analogue to Arrow's impossibility theorem was proved (independently) in the 1970s by Allan Gibbard and Mark Satterthwaite. The dictator sends his fond regards.

But the dictator theorem's name may be overdramatic. We think of a dictator as someone who ruthlessly grabs power. In the context of voting theory, the dictator who decides the ballot need not be a tyrant. It is possible to choose a "dictator" at random, among all the voters, and adopt his or her ranking of the candidates. A ranking that is more prevalent among the voters is more likely to be chosen. This allows for a touch of democracy after all. It is not mere arbitrariness. We might name the procedure a tychocracy, after Tyche, the Greek goddess of chance (or fate).

Few of us would accept tychocracy in political decisions. If many voters take part in the election, our chance to dictate the final result is too small. It seems that we prefer our elections to grant us with 100 percent probability a tiny influence on the outcome, rather than to give us a tiny chance to have a 100 percent influence on the outcome.

An election is not only a measurement to determine the prevailing opinion—which is, as we have seen, a slippery thing to define. An election is, in addition, a ritual: more precisely, a ritual of commitment. Voters commit themselves, for good or for bad. An election has a ceremonial aspect. This property is absent from all opinion polls based on statistics, no matter whether the sample is small (as in the case of the tychocratically designed dictator) or large. Even the most representative poll cannot replace the commitment of a general election.

10

Decision

Wagers in the Dark

Felicity Calculus for the Panopticon

When entering the students' center of the University College London, you meet Jeremy Bentham in person. The philosopher is long dead, but there he sits, carefully dressed and equipped with his cane, in a small glass booth. All has been meticulously arranged according to his last will by Bentham's friend, Dr. Thomas Southwood Smith. He had carefully dissected the philosopher. Unfortunately, however, he sadly botched the mummification of the head. It eventually had to be replaced by a wax replica.

Figure 10.1. Jeremy Bentham (1748–1832).

Not surprisingly, Bentham's famous "auto-icon" spawned a number of legends. He is reputed to regularly attend the meetings of the students' union, "present, not voting" according to the minutes. The reasons why Bentham wished to be taxidermied are surrounded by speculation. It may have been his way to avoid churchmen to the end. Jeremy Bentham was an atheist, and held nothing for sacred. Today, his opposition to religion or monarchy would hardly raise an eyebrow. But his scorn of human rights would still shock many of us: he dismissed the idea as nonsense. The French revolutionaries had made a great fuss about "imprescriptible human rights." Bentham's sneering reply was an essay entitled "Nonsense upon Stilts."

What strikes the passerby is the auto-icon's watchful pose and his alert look: he seems not to get bored at all by the day-long bustle of students and visitors. To the fanciful, this brings to mind Bentham's famous *panopticon*: a prison built as a hollow cylinder, so that the governor—ideally, Bentham himself—could always survey the prisoners from behind his window blinds, while they could never feel secure from observation. Bentham spent the best years of his life promoting this idea.

Eventually, nothing came of it. The British government favored an even more economical solution: it transported the convicts to the far side of the Earth. Out of sight, out of mind: which is the exact opposite of the panopticon approach. As compensation for his efforts, Bentham was rewarded with 23,000 pounds, a considerable sum at the time, even for a man as well off as he was. Nevertheless, the failure of his project convinced him that "sinister interests" guided the government. In fact, the more he thought about it, the clearer he perceived that sinister interests guided *all* governments. Thus, Bentham turned radical, by the ripe age of sixty, and became a chief force behind parliamentary reform. As to the panopticon, he was reluctant to give up the idea entirely. If it could not be peopled by convicts, they being shipped off, why not adapt it for the pauper? Or the insane? Or the workforce?

Again, these schemes did not take off. Yet they did, claimed the French philosopher Michel Foucault. Is not the modern state, with its elaborate surveillance mechanisms, the true panopticon? Bentham is more important to our society than Kant and Hegel, according to Foucault.

Bentham played indeed a seminal role as a social reformer and legislator. Today, he is mostly known as the father of utilitarianism. It was he who articulated the principle of utility: "The greatest good for the greatest number."

It was soon recognized—not least by Bentham himself—that the slogan is questionable. For instance, should we increase general happiness even if it comes at the expense of a few? How should one decide such a question? By vote? We know that this can lead to thorny issues. And even if the will of the majority was crystal clear, would we have the right to obey it, and make a minority suffer for the sake of the majority? Many moral philosophers are dead set against it. Fairness comes first, says John Rawls.

Even finding out what is best for a single person—a much simpler question, certainly, than what is best for the greatest number—is fraught with problems. What are we looking for? Abundance? Security? Liberty? Bentham was a hedonist, persuaded that in the end the question must be answered in terms of pleasure and pain. His attempts at a felicity calculus petered out. But in its stead, cost-benefit analysis is going strong today, and calculating expected utilities is commonplace in economics and decision theory. These are ideas based on Bentham.

Looking Back to St. Petersburg

The concept of utility, such as it shows up in Bentham's philosophy, is even older: it stems from Daniel (or Nikolaus) Bernoulli's investigations of the St. Petersburg game. In this game—recall from Chapter 7—one wins \$2, or \$4, or \$8, etc. with the respective probabilities $\frac{1}{2}$, or $\frac{1}{4}$, or $\frac{1}{8}$, etc. The expected value of the gain is infinite, which seems strange.

Bernoulli's solution was this: We should not consider the amount of money that we gain, but the utility that we draw from it. And this utility declines with every added amount. If I were a millionaire, a gain of \$100 would give me a smaller utility than if I were a pauper. Though more money means more utility, twice as much money does not provide me with twice as much utility.

In other words, if $U(x)$ denotes the utility that I attribute to x dollars, then U increases with x, but the rate of increase will diminish. Any extra money gives me extra satisfaction, but less and less so.

Bernoulli suggested, more or less out of the blue, that $U(x)$ is the logarithm of x. It is a mere hunch, and not an essential feature of the argument. All that matters is that the function has a positive slope, and that the slope flattens off: for every two points on the graph, the connecting segment lies below the graph (Figure 10.2). Anyone with such a utility function is said to be *risk-averse*.

A risk-averse person will not be prepared to pay a fee of \$3000 for a lottery that yields a gain of either \$2000 or \$4000 with the same probability $\frac{1}{2}$. Indeed, the expected amount of money is \$3000, and its utility, namely $U(3000)$, is larger than the expected value of the utilities of the two outcomes of the lottery, which is $\frac{1}{2}U(2000) + \frac{1}{2}U(4000)$. There also exist *risk-seeking* people: their utility function has an increasing slope. These people are an embarrassment for theorists, and undergraduate texts in economics lose little space on them.

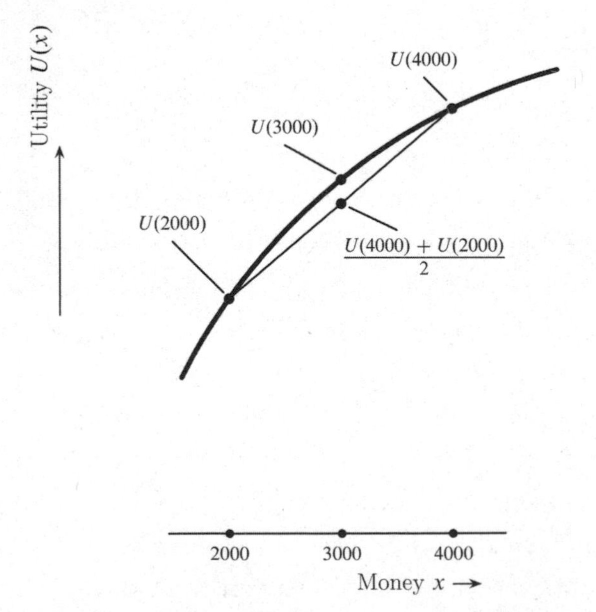

Figure 10.2. The utility function of a risk-averse person is concave.

Let us compute the expected value of the utility in the St. Petersburg game, under Bernoulli's assumption that $U(x) = \log x$.

We obtain $(\log 2)\left(\frac{1}{2}+\frac{2}{4}+\frac{3}{8}+\ldots\right)$, which is indeed a finite number, namely 2log2. This, says Bernoulli, is the fair fee to pay for the game: nobody should pay more.

The idea is not really convincing. Even if we accept the very arbitrary assumption that the utility function is logarithmic, it does not solve the basic problem. It is easy to modify the winnings in the St. Petersburg game such that the expectation of the logarithmic utilities is infinite: for instance, by offering 2 to the power 2^n when the first heads occurs after n tails. In fact, mathematician Karl Menger (a junior member of the Vienna Circle) showed that for every utility function that is unbounded, one can produce a version of the St. Petersburg paradox.

Therefore, Bernoulli certainly did not resolve the paradox. We have met a better explanation: although the time average of the gains grows without bounds, it grows too slowly to make us rich.

However, Bernoulli's idea of a decreasing growth rate of the utility function has proved tremendously fruitful. It is a pillar of modern economics—yet a pillar that looks more solid than the ground on which it stands.

A Cardinal Idea

The utility of x (Bernoulli's *utilitas*) is a measure for the value of x. Every person, rich or poor, young or old, has some utility function, and it would be odd if they were all the same. By the way, x need not be an amount of money. It could be food, sex, fun, pleasure, prestige, safety, or sheer survival. Your utility function serves as a measure for your preferences.

Who decides about your preferences? In an individualistic society such as Bentham's or ours, this is left entirely up to you. We need not ponder from where preferences come. They are considered as given and cannot be argued. "Reason is the slave of passion," according to Hume. Rationality can, at best, serve to navigate between wishes.

Can a calculus help with this task?

The usual approach assumes that an individual, when comparing two alternatives A and B, either prefers A (we then write $B \prec A$), prefers B (we then write $A \prec B$), or is indifferent (that's when we write $A \sim B$). If $B \prec A$ does not hold, we may also write $A \preccurlyeq B$. We next assume that preferences are transitive: if $A \preccurlyeq B$ and $B \preccurlyeq C$, then $A \preccurlyeq C$. This implies that every finite set of alternatives $A, B, \ldots, Z$ can be completely ordered. The ordering may look something like this:

$$N \preccurlyeq G \preccurlyeq A \preccurlyeq F \preccurlyeq \ldots \preccurlyeq M.$$

Such an array suggests that the alternatives can be ordered like numbers (more precisely, real numbers, or points on a line). Yet, the alternatives themselves will not be numbers, in general. One of the alternatives can be: your ears will be cut off. Another alternative: you get a piece of chocolate cake. To each alternative A, one can associate a number $u(A)$ such that $A \preccurlyeq B$ holds if and only if $u(A) \leq u(B)$ (so that the two symbols $\preccurlyeq$ and $\leq$ not only look alike but say the same thing). Such a function u is said to be a utility function, Bernoulli's *utilitas*.

The word *utility* is poorly chosen. It suggests hardheaded materialistic connotations. Utilitarians will never get rid of the taint cast by Charles Dickens's ferocious portrait of Mr. Gradgrind in *Hard Times*. But even the most sentimental, tenderhearted altruists have a utility function. Far up in their list of preferences could be the spiritual welfare of fellow humans, or preservation of the climate, or justice in peace. Accordingly, these aims would have a high utility for them.

Bentham soon became aware that his term *utilitarian* was what he used to call a "hobgoblin word": a pejorative expression used primarily for censure. In fact, he may have been the first philosopher acutely aware of the impact of language on our thinking, a budding language philosopher. Therefore, he worked hard on improving vocabulary. (The words *auto-icon* or *panopticon* did not really fly, but Bentham also coined the term *international*, which fared better.)

Later in life, Bentham preferred to write of the pursuit of happiness, or felicity, rather than utility. But by then, the harm was done. It would have been much better if u had been designed as a value function. Indeed, while one can speak of "sacred values," not even Mr. Gradgrind would speak of "sacred utility." In common speech, a painting can have artistic value (and commercial value, too), but it rarely has any utility (except for hiding a safe in the wall).

We have a lot of leeway in defining our utility function. For instance, if

$$N \prec G \prec A \prec F \prec \ldots \prec M,$$

then we could define $u(N) = 1$, $u(G) = 2$, $u(A) = 3, \ldots, u(M) = 26$, or else we could use $u(N) = 1$, $u(G) = 4$, $u(A) = 8, \ldots, u(M) = 26^2$. Each way defines a utility function: the arrangement of the numbers $u(A)$, $u(B)$, etc. corresponds to the ordering of our preferences A, B, etc., and nothing more is required.

Can we be sure that the ordering of preferences is complete and transitive?

Complete means that we always know whether we prefer A to B or not. Our preferences can change with time and circumstance, of course. Many children prefer in the morning to stay in bed rather than to get up, and in the evening to stay up rather than to go to bed. (With age, these preferences may change.) Moreover, our true preferences may be different from what we state, possibly even state in good faith. Psychologists speak of "revealed preferences" (those when in earnest), which can be shockingly different from the preferences listed when answering a questionnaire, for example.

As to transitivity, can we really be sure that indifference and strict preference are transitive? (That is, $A \sim B$ and $B \sim C$ imply $A \sim C$, and $A \prec B$ and $B \prec C$ imply $A \prec C$.)

Indifference first. I am completely indifferent between one grain of sugar more or less in my cup of coffee, but adding more and more grains will eventually affect the pleasure I take in my cup. Thus, indifference has to be taken with a grain of salt (if that's the fitting expression).

It is even worse with strict preference. Admittedly, it seems irrational if transitivity were violated and I preferred *A* to *B*, *B* to *C*, and *C* to *A*. Indeed, many economists and social scientists view transitivity of preference as the very touchstone of rationality, its *ultima ratio*. A cycle in preferences would lead to all kinds of paradoxes. Best known of these is the money pump argument due to Frank Ramsey. The argument states that we simply cannot afford a cycle! If I owned *A*, say, and someone offered me *B* instead, for a trifling extra fee, then I would accept; and I would also accept exchanging *C* for *B*, and *A* for *C*, paying each time the trifling fee—and then the next round starts, and then the next, and I will eventually become broke, which economists deem as definitive proof of irrationality.

And yet, psychological experiments show that cyclic preferences do occur as soon as the alternatives are somewhat complicated (for instance, *A* standing for "one week in a three-star hotel in Rome" and *B* for "three days in a spa in Montecatini Terme" and so on). As soon as I am offered ten such choices, cycles may well occur.

This may simply be due to confusion, but there may also be a more serious reason behind it. Indeed, we usually weigh our preferences according to certain criteria. Suppose that you have to choose between three towns *A*, *B*, and *C*. If you order them according to the quality of their academic life, you obtain $A \succ B \succ C$; if you judge the environment, the ordering is $B \succ C \succ A$; if you rank by restaurants, you come up with $C \succ A \succ B$. If these criteria are of equal importance to you, you are faced with a cycle!

So far, the only thing we have required for the utility function is that it agrees with the ordering of the preferences. (This is said to be an *ordinal* utility.) Often, we may also have a definite feeling of preferring something by very much, or only by a little. Can we make this feeling more precise by a utility function? (It would then be termed a *cardinal* utility.)

Yes we can, said Austrian economist Oskar Morgenstern and Hungarian mathematician John von Neumann. Their theory became extremely influential in the 1950s.

The trick is to consider lotteries. Thus, if X and Y are two alternatives, we denote by $pX + (1 - p)Y$ the lottery yielding X with probability p and Y with the complementary probability $1 - p$. Two assumptions seem obvious:

1. If $X \prec Y$ then $pX + (1 - p)Z \prec pY + (1 - p)Z$ for each probability p and each alternative Z.
2. If $X \prec Z \prec Y$, then there is exactly one probability p such that $Z = pX + (1 - p)Y$.

Under these assumptions, there exists a cardinal utility function u such that whenever $Z = pX + (1 - p)Y$, then $u(Z) = pu(X) + (1 - p)u(Y)$. (In this sense, u is *lottery-compatible.*) Moreover, u is unique, up to scale, just as temperature is uniquely defined, up to scale: it is a menial task to change from Fahrenheit to Celsius and back. And in the same sense, as temperatures not only tell us whether today is colder than yesterday, but also whether the difference is large or small, so it is with our utilities. They are the hoped-for cardinal utilities.

The idea of the proof is simple. Let us assume that our alternatives have been ordered already, for instance as

$$N \prec G \prec A \prec F \prec \ldots \prec M.$$

Let us define the utility of our least-favored alternative, namely N, by 0, and that of our most preferred, namely M, by 100. Next, let us consider all lotteries of the form $pM + (1 - p)N$, for any p between 0 and 1. The larger p, the more we like this lottery. For $p = 1$, it is nothing else but our most favored alternative M, and for $p = 0$, our least-liked N. By continuously increasing p, we continuously increase our utility from N to M. For some value of p, we will be indifferent between the lottery $pM + (1 - p)N$ and alternative G. We then set $u(G)$ to be equal to $100p$. And so on, all the way through our list of preferences. "Can it be that no one ever thought of it before?" marveled John von Neumann. Someone had, but this was not noticed until later.

Utility functions are the rock on which rational decision theory is built. This theory offers precise predictions about human behavior, for instance in the context of *expected utility*. Alas, we will see that these predictions are so precise that they can be refuted by experiment. This looks like a letdown. On the upbeat side, however, let us recall that falsifiability is, according to Karl Popper, the hallmark of a serious scientific theory. We learn most from failures. It is not the smallest merit of mathematics that it allows one to be so precise that one can be refuted.

And if these last lines remind you of sour grapes, why, just so. Except that sour grapes were certainly not served at the banquet where the first of these falsifications sprung to light.

Indeed, in 1952, the leading brains of expected utility theory had met for a congress in Paris. Foremost among them was Leonard "Jimmy" Savage, a mathematician and statistician of high merit. During the conference dinner, a French economist named Maurice Allais pleasantly challenged Savage, and with him the whole "American school," with a harmless-sounding quiz. To the dismay of many (but not, maybe, of the French), it turned out that Savage himself did not obey the most elementary rules of decision theory.

Savage conceded it with good grace. He had made a mistake, so what. Maybe the French wine had been too good. But by and by, it became clear

Figure 10.3. Maurice Allais (1911–2010) scores a point.

that the same "mistake" could be found more or less everywhere. A great many of us would fall into the same trap. In due time, the after-dinner test of Allais gave rise to an important field of experimental psychology, maybe even experimental philosophy. As for Allais, he was awarded a Nobel Prize for Economics, many years later.

Puzzled by Lotteries

Allais's quiz presented two pairs of alternatives: *A* and *B* on the one hand, *a* and *b* on the other.

First, choose between *A* and *B*:

A: $1 million with 100 percent probability.
B: $1 million with 89 percent, nothing with 1 percent, and $5 million with 10 percent.

Savage chose *A*. What would you choose? Please take your time to ponder.

Now, choose between *a* and *b*:

a: 0 with 89 percent and $1 million with 11 percent.
b: 0 with 90 percent and $5 million with 10 percent.

Savage chose *b*. Most (but not all!) would decide as he did. But choosing *A* in one case and *b* in the other is irrational, or at least inconsistent.

Indeed, the chance that *A* yields an outcome differing from *B* is only 11 percent (with the remaining 89 percent, *A* and *B* both give you $1 million). In case *A*, these 11 percent give you $1 million, whereas in case *B*, one of the 11 percent gives you nothing, and the other 10 percent earn you $5 million.

In the same vein, with a probability of 89 percent, both *a* and *b* give you nothing. Thus, the chance that *a* yields an outcome differing from *b* is only 11 percent. In case *a*, the remaining 11 percent give you $1 million, whereas in case *b*, one of the 11 percent gives you nothing, and the other 10 percent yield a gain of $5 million.

Hence, the difference between *A* and *B*, on the one hand, and *a* and *b*, on the other, is exactly the same. If you prefer *A* to *B*, you should prefer *a* to *b*. But most people don't.

Note that with a probability of 89 percent, your decision will have had no effect whatsoever on the outcome. In the *A-B* case, you would have received $1 million, and in the *a-b* case, you would have received zilch, but in both cases, there is nothing you could do about it.

Here is an even simpler example. Suppose that you were asked to choose between the following two alternatives:

X: $3000 for sure.
Y: a ticket in a lottery where you can gain $4000 with a probability of 80 percent, or nothing with a probability of 20 percent.

You are sufficiently well-versed to compute the expected value of the lottery *Y*, namely $3200. This is more than what *X* has to offer. And yet... maybe you have a queasy feeling with choosing *Y*. If so, you are risk-averse. Many people have been asked whether they prefer *X* or *Y*. A strong majority, namely 82 percent, opt for *X*. Nothing paradoxical so far.

With the next quiz, things are quite different. Suppose that you have to choose between the following:

x: a lottery where you can gain $3000 with a chance of 25 percent (and nothing with a chance of 75 percent).
y: a lottery yielding $4000 with a chance of 20 percent (and nothing with a chance of 80 percent).

Here, a majority of 70 percent opt for *y*. And why shouldn't they? Well, because the second test offers the same alternatives as the first, except that the probabilities for a win are cut down by the factor four.

In other words, alternative x is nothing but the lottery $pX + (1 - p)Z$, where Z is the outcome "you get nothing at all" and $p = \frac{1}{4}$. Similarly, y is nothing but $pY + (1 - p)Z$. Whoever prefers X to Y (a majority of 82 percent) should also prefer x to y. But only 30 percent do! This means that more than half of the people violate the first assumption, that no-brainer of rational decision theory.

Similar findings have been reproduced all over the world. The ingenious experiments by Amos Tversky and Daniel Kahneman are particularly revealing.

One of their most elegant examples is as follows: You are given $300 and asked whether you prefer *A* or *B*:

> *A* means an additional $100.
> *B* is a lottery, where with equal chance, you gain an additional $200, or not.

What would you prefer? Most people (namely 72 percent) chose alternative *A*.

But suppose that you are given $500 and asked whether you prefer *a* or *b*:

> *a* means to pay back $100.
> *b* is a lottery, where with equal chance, you lose $200, or not.

What would you choose? A majority of 64 percent prefer *b*. Hence, the majority are risk-averse in case *A-B* and risk-seeking in case *a-b*.

Strangely, though, in both cases the alternatives are exactly the same: on one hand, $400 for certain, and on the other hand, a lottery leading with the same probability to a gain of $300 or of $500.

It only depends on how the question is formulated. In the first instance, it is a question about a possible gain, presented as an increment over a baseline of $300; in the second instance, it is a question about a possible loss, a drop from a baseline of $500. Such so-called framing effects are familiar in bazaars. They also used to be encountered in shops, where the price depended on whether you pay cash or credit. The same alternatives can be couched in terms of a cash bonus, or a credit card fee.

It isn't all about gambling and shopping. Some people have jobs requiring them to make snap decisions in matters of life and death: for instance, in the military or medical professions, or among leaders of firefighter units. These people are trained and tested and highly motivated—and yet, as real-time exercises and full-dress maneuvers show, they often fall prey to framing effects. Cognitive illusions and faulty heuristics cannot easily be shaken off.

Most of us are risk-averse about gains and risk-seeking about losses. But gains and losses depend on from where one starts. This status quo can be easily manipulated. Kahneman and Tversky used it as anchor point in their prospect theory, which is currently the best alternative to the doctrine of expected utility maximization.

The Known Unknowns

In lotteries, the probabilities of the various events are usually well known. In usual real life, however, this is rarely the case: we are not told the probabilities. To quote the immortal words of a US politician; "There are known unknowns, and there are unknown unknowns." Frequently, we have not the foggiest idea about the chances of the various alternatives. Nevertheless, decisions must be reached. These are said to be decisions under uncertainty, in contrast to decisions under risk, the latter term being used when the chances are known.

The first philosopher to highlight the difference between risk and uncertainty was John Stuart Mill, the foremost British philosopher of the nineteenth century. He was the son of James Mill, one of Jeremy Bentham's greatest friends. The two philosophers paid the utmost attention to young John Stuart's education, and he dutifully lived up to their expectations, by becoming a hard-nosed empiricist, a legal positivist, a social reformer, and a liberal MP to boot—the first one, incidentally, to plead for the female vote in the House of Commons.

Mill's ideas about the distinction between known and unknown probabilities were taken up by economists such as John Maynard Keynes, and tested by the psychologist Daniel Ellsberg in a series of ingenious experiments.

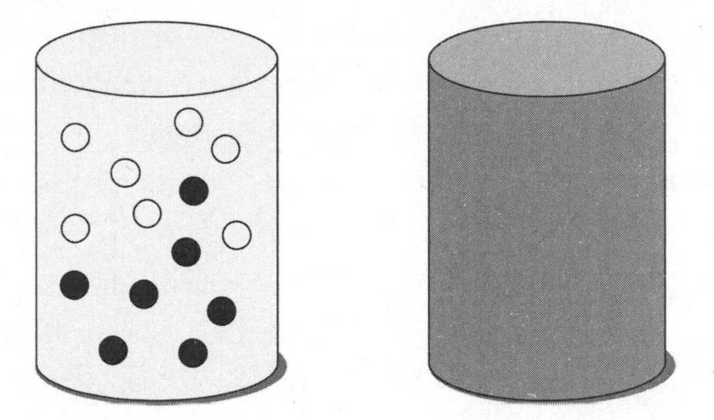

Figure 10.4. The composition of the left urn is known, that of the right urn is not.

For example, imagine that one urn contains ninety balls, more precisely forty-five white and forty-five black balls. You are told about this, and asked to draw one ball. You will receive $1000 if the ball is white, and nothing otherwise.

In a second urn, there are also ninety balls, some white and some black, but you are not told how many of them are white. It could be any number between zero and ninety. Again, you can draw a ball and will receive $1000 if the ball is white, and nothing otherwise.

Now comes the point. You are asked to choose between the left and the right urn before entering the game (Figure 10.4). How would you choose? The vast majority opt for the first option—the urn with the known number of balls. This predilection is not easy to explain. The chances for a win should be the same with both urns—and yet most of us feel a lot better with the known rather than the unknown probabilities.

This preference leads many of us into inconsistent behavior, as can be shown with a similar experiment. You are presented with a new urn, and told that it contains ninety balls. Exactly thirty of them are white, and the other sixty can be gray or black. You are not told how many of them are gray.

First experiment: you may bet on white or on gray, and will win $1000 if you draw a ball of that color. Most people bet on white. There is no reason to think that white is more likely than gray, as the number of gray balls can be anywhere between zero and sixty. Yet, as we know already, most of us feel better with a known probability, namely the $\frac{1}{3}$ chance to draw white.

Second experiment: you may choose between two composite bets, namely "white or black" or "gray or black." Now most choose "gray and black"—and small wonder, since it is an event with a known probability, namely $\frac{2}{3}$, while the other one is not.

If you have decided with the majority in both experiments, you are inconsistent. Indeed, in the first experiment you have deemed white to be more probable than gray. Thus, "white or black" should be more probable than "gray or black." These experiments tell us a lot about our strange ways of handling chance.

A short time after publishing his research, Daniel Ellsberg became world-famous, but for another reason. He leaked the so-called Pentagon Papers, which turned into a cause célèbre in the early 1970s. The scandal is still well remembered by the elder generation, and known to younger people through films such as Spielberg's *The Post*.

The Pentagon Papers shed a sad light on the US administration. Ellsberg became the prototype of the whistleblower. He was indicted under the Espionage Act, and had to face the prospect of 115 years of imprisonment. Fortunately for him, the trial was dismissed for gross governmental misconduct when it became known that Nixon's "plumbers" (so named because they were fixing leaks) had ransacked the office of Ellsberg's psychiatrist. Eventually, the "plumbers" were entrusted with other urgent tasks, and were sent to the Watergate building this time, which they bungled again—whereas Ellsberg became a professor at MIT, highly praised both for his civic courage and his scientific work.

Before leaking the papers, Ellsberg had worked as an analyst for the RAND Corporation and the Department of Defense, dealing on a daily basis with atomic warfare scenarios, doomsday machines, and nuclear strikes. His work must have provided as good an opportunity as any to closely study what it means to make "decisions under uncertainty."

Pascal Places a Bet

Blaise Pascal was haunted by uncertainty. "I look on all sides, and I see only darkness everywhere. Nature presents to me nothing that is not a matter of doubt and concern."

In two furiously scribbled pages entitled "infinity-nothing," which were discovered only after his death, Pascal set out his famous wager. It was published posthumously in his *Pensées*, an anguished dialogue with himself, revolving relentlessly around the frailty of men in an uncertain world.

Pascal had been a mathematical wunderkind. Now in his late thirties, he admitted: "It may be that there are true demonstrations." But this only makes uncertainty worse, said Pascal, because it proves that "it is not certain that all is uncertain."

The question that exercised him most was whether God exists:

> If I saw nothing in the world which revealed a Divinity, I would come to a negative conclusion; if I saw everywhere the signs of a Creator, I would remain peacefully in faith. But seeing too much to deny and too little to be sure, I am in a state to be pitied.

In the end, he decided to believe, or at least to try to believe, and this because he understood that he was playing a lottery, a very particular lottery. Each of us is playing it.

> You must wager. It is not optional. You are embarked.

At stake is your happiness.

> Let us weigh the gain and the loss in wagering that God *is*. Let us estimate these two chances. If you gain, you gain it all. If you lose, you lose nothing. Wager, then, without hesitation that He *is*.

What you can gain, by following the way of the believers, is an eternity of bliss, "an infinity of an infinitely happy life." What you may lose is at most the finite amount of pleasure a libertine can glean in this world. The expected gain from believing is incomparably larger than the gain from not believing in God.

The wager is often misunderstood as a proof that God exists. As such, it clearly fails. Voltaire scathingly remarked that "the interest I have that a thing exists is no proof that such a thing exists." But a proof that God exists was not Pascal's aim. The wager is an argument for embracing religion, by making a decision under uncertainty.

Pascal accepts that some find it very hard to believe in God. Those unfortunates should try, at least. They should start out as if they believed, "taking the holy water, having masses said, etc." Even this little will make you believe, he argues. Today, we would speak of brainwash. We know that it works. In Pascal's term, "it will deaden your acuteness," at which point the skeptic in Pascal cries out: "But this is what I am afraid of!" To which the mystic in Pascal replies: "And why? What have you to lose?"

The whole argument concerns only those who are in two minds about God. Those who assign Him (or Her) probability zero will not be swayed by an appeal to the expected value of the wager.

"Infinity-nothing" presupposes uncertainty. Its convoluted reasoning tells us more about the troubled soul of Blaise Pascal than about God. Which God are we talking about anyway? "An Imam could reason in the same fashion," railed Denis Diderot. And would God really be pleased to acquire followers who are motivated by self-interest?

As a piece of theology, the wager does not lead far. But it is the first explicit application of decision theory, based on a lottery: a step into dark uncertainty with life's happiness at stake. As such, Pascal's existential wager is truly prescient, coming at a time when probability theory still was in swaddling clothes, the notion of expected value hardly conceived, the infinite and the infinitesimal shaky invocations, utility functions not yet dreamed of.

Voltaire was on the wrong track, for once, when he chided Pascal with the words: "The idea of a game, and of loss and gain, does not befit the gravity of the subject."

11

Cooperation

An Eye for Oneself and an I for the Others

Reaching out for the Invisible Hand

Know yourself, said Socrates. The experience is not always pleasant. The self, it seems, is filled with selfishness to the brim. Self-love guides us, said the Duke of La Rochefoucauld, in virtue and vice. And virtues, so he added on another occasion, dissolve in self-interest as rivers dissolve in the sea.

One of the best-known psychologists of today, Jonathan Haidt, writes that "if you want to explain human behavior in two seconds or less, you may say: self-interest." Haidt's very next sentence states that if you had two minutes for the task, you would have to qualify that statement. But to a first approximation, self-interest guides us.

Humans are selfish, yet on the other hand, they are social like few other species. This is something already remarked upon by Aristotle. He viewed humans as *zoon politikon*. There exist some other "political animals," mostly insects such as ants, termites, and bees. Some even build states, but these states look very different from ours, and consist essentially of one extremely tight-knit family, genetically almost homogeneous, a kind of super-organism. The behavior of social insects is hardwired in their genes. Evolutionary biologists feel that with us humans, there must be another reason why our selves fit into communities.

The contest between oneself and the other is an evergreen in philosophy. Self-love often meets with reprobation. In the seventeenth century it had particularly poor press. The mathematician, philosopher, and mystic Blaise Pascal wrote: "We are born unfair; for everyone is inclined towards himself. . . . Selfishness is the cause of every disorder in war, philosophy, economy etc."

The mathematician and philosopher Thomas Hobbes claimed in his book *Leviathan* (1651) that unbridled selfishness must lead to the war of all against all. Clearly, the English Civil War and the Thirty Years' War were conducive to a bleak worldview.

Figure 11.1. Adam Smith (1723–1790) saw an invisible hand.

Not a century later, the philosophers of the Enlightenment had a much more optimistic view, and self-love acquired a better reputation. In his *Wealth of Nations*, the Scottish philosopher and economist Adam Smith opined that the selfish motivations of individuals were channeled into society's welfare *as if* by an invisible hand. "By pursuing his own interest, man frequently promotes that of the society more effectually than when he really intends to promote it." And famously:

> It is not from the benevolence of the butcher, the brewer, or the baker, that we expect our dinner, but from their regard to their own interest. We address ourselves, not to their humanity but to their self-love, and never talk to them of our own necessities but of their advantage.

The idea must have been in the air. Some years before Adam Smith, Voltaire had expressed a similar view in his *Lettres philosophiques*:

> Doubtlessly God could have created beings uniquely interested in the welfare of others. In this case, traders would sail to India out of sheer altruism, and stonemasons would cut stones to please their neighbors. But God arranged things differently.... Our mutual needs are the reason why we are useful to mankind, they are the foundation of every trade, the eternal ties between people.

Adam Smith was just as aware as Voltaire (whom he knew well) that the invisible hand is not *always* at work. All he said was that it *can* promote the common weal.

The shortcomings of the invisible hand are most clearly displayed by simple experiments performed in the game labs of economics institutes all over the world. When you participate in such an experiment, you are normally given some modest sum of money beforehand, as a show-up fee. You sit down in a room, shielded off from the other participants, in front of a monitor explaining the rules of the game. You are paired with a co-player—someone you will never see, and who may be in the next room or in another part of the planet, for all you know.

Here is the game: You have the choice between two buttons C and D. If you click C, you consent to return $5 of the show-up fee to the organizers of the experiment. In that case, your co-player receives an additional $15. If you click D, this will not happen. You are given sixty seconds to decide what to do (if you make no decision, this counts as D). You are told that your unknown co-player faces exactly the same situation, and that this will be your only interaction ever with that person. There is no second round, and you will never learn who the other is. In well-run experiments, which are double blind, the scientists in charge don't know either.

The clock starts ticking.

You don't know how your co-player will decide. Nor does the co-player know what you will do. If you click C, you are effectively donating $15 to your co-player, at a cost of $5 to yourself. Self-interest dictates, presumably, that you should not provide that other person, a complete stranger, with an utterly undeserved gift. Hence, you click D. If your co-player reasons alike, you will both return home with nothing more in your pocket than the fee for attending the game.

You may feel that you have missed an opportunity. For if you had both clicked C, you would both be $10 richer: you would have had to pay $5 for the other's benefit, but would have received $15 in return. Thus, by obeying your selfish interest, you have forfeited $10. Hence, self-interested individuals do not always promote the community's welfare. (As the Nobel Prize–winning economist Joseph Stiglitz famously quipped, the "invisible hand" may be invisible because it does not exist.) More surprisingly, self-interested individuals don't even promote *their own* welfare in this fiendish little experiment.

By the way, some 50 percent of the players click C, which flies in the face of the previous argument. The exact percentage depends on many details, such as the sums of money involved and the player's sex, age, and cultural background.

It may sound strange, but this simplistic experiment has a mathematical origin. It belongs to game theory, which is the mathematics of conflicts of interest. The field emerged in the 1940s, in the United States, as the brainchild of two immigrants from Central Europe who had witnessed how their communities were torn to pieces.

Unpredictable Moves

Oskar Morgenstern was born in Germany in 1902. After World War I, he enrolled at the University of Vienna. His mathematical talent was modest (which is why he had to repeat one class at school), but he sped brilliantly through his studies of economics, obtaining his PhD within three years and winning, as icing on the cake, a three-year traveling grant from the Rockefeller Foundation. This allowed him to spend his postdoc time in England, the United States, France, and Italy.

When Morgenstern returned to Vienna, economist Friedrich Hayek hired him for his newly founded Institute for Business Cycle Research. Soon after, Hayek was appointed as a professor at the London School of Economics. He would become, in due time, the firmest voice against planned economy (which he viewed as "the road to serfdom") and earn a Nobel Prize for Economics. In Vienna, he was succeeded by thirty-year-old Oskar Morgenstern as head of his former Institute.

This was the time of the Great Depression. Morgenstern's task was to provide the Austrian government with economic predictions. His position was fairly incongruous: indeed, Morgenstern's major scientific thesis was that reliable economic predictions are impossible, in principle. In Vienna, nobody minded the paradox. What you proclaimed in a coffeehouse had only little to do with what you did for a job (if, indeed, you were lucky enough to have a job).

Figure 11.2. Oskar Morgenstern (1902–1977).

Figure 11.3. John von Neumann (1903–1957).

Morgenstern's argument was that each prediction would cause the economy to react. This reaction would have to be taken into account by the prediction, a fact that would be anticipated in turn, and so on, in an infinite regress. The bottom line is that predictions about the economy are different from predictions about the weather. Predictions do not influence the weather. The atmosphere does not react to forecasts. The economy does, however, and this leads to a vicious circle.

In his papers and talks, Morgenstern illustrated this doctrine with a literary example: the deadly cat-and-mouse game played between Sherlock Holmes and that fiendishly clever arch-criminal Professor Moriarty. Each one tries to outguess the other, and knows that the other tries to do the same.

We may use instead a less dramatic example. On a cue, two players, Ann and Burt, hold up a number of fingers. If the sum is even, Ann wins, whereas Burt wins if the sum is odd.

Suppose that the experts from the Institute for Business Cycle Research are able to predict the result of that game (via crystal ball or whatever). They forecast, say, that both Ann and Burt will raise an even number of fingers. This means that Ann wins. On learning this, however, Burt will rethink his decision and raise an odd number of fingers instead. This ploy would guarantee his win. But Ann is clever enough to expect it, and therefore she, too, raises an odd number of fingers. But what if Burt anticipates this and counteracts by keeping to his original choice? And so on, and so forth. The vicious circle is well described in Edgar Allan Poe's "The Purloined Letter." It happens whenever two people with opposite interests try to outwit each other. It happens in every game of hide-and-seek. It happens when predators hunt their prey, and it happens when Moriarty hounds Holmes.

Oskar Morgenstern returned time and again to his favorite example. In the end, however, some well-meaning colleague gently apprised him that the brilliant John von Neumann (a shooting star on the mathematical scene) had proved a few years ago that forecasts are possible after all: namely forecasts about the *probability* of this or that turn of events.

John von Neumann, born in 1903 as the son of a Budapest banker, was a mathematical prodigy. As soon as he turned to a new field, he left his mark on it. Mathematical logic, set theory, functional analysis, quantum theory, you name it.... In 1928, while in Berlin, he wrote a paper on "The Theory of Parlor Games" that, as it now turned out, resolved Morgenstern's conundrum. Oskar Morgenstern was deeply impressed. His faith in mathematical methods, which had always been strong, grew immense.

In 1940, the two men met at a party in Princeton. They both had fled Hitler. Von Neumann, as usual, had been quick: having watched the rise of the Nazis, he was among the first scientists to escape from the anti-Jewish laws of the Third Reich. He belonged to the original handful of stellar scientists hired by the newly founded Institute for Advanced Study. As to Oskar Morgenstern, he happened to be in the United States when Nazi Germany invaded Austria in 1938. It was fortuitous: he had not predicted the Anschluss.

In the dingy worldview of the Third Reich, Morgenstern counted as an "Aryan," but his name was bound to arise antisemitic suspicions. On learning that the Gestapo had him on their black list, Morgenstern decided not to return to Vienna to explain his family tree. This was a wise move, but his first expat years were difficult.

Right after they met, Morgenstern and von Neumann started a joint project. It outgrew all expectations, progressing in leaps and bounds. John, during the 1940s, was almost continuously on the road, as a scientific advisor to the US military effort, working on artillery tables, atomic explosions, computers—never mind the topic, he was urgently needed everywhere. And whenever he passed through Princeton, his first visit was to Oskar, to pour out new ideas. What was intended as a small booklet grew into a heavyweight tome. Initially entitled "Theory of Rational Decision," it was renamed *Theory of Games and Economic Behavior*, and became one of the landmark books of the century.

The term *game theory* did much to make the new field popular. Games are associated with pleasure. By the same token, however, the name was bound to cause misunderstandings. Nobody ever became better at poker or chess by studying game theory, just as nobody ever became a better tennis player by reading about physics.

The reason behind the catchy name for the new theory was that its basic terms were lifted from the vocabulary of parlor games: move, player, strategy, payoff. And just as games of chance were key to the vast field of probability, so did parlor games help explore the mathematics of conflicts of interest. The outcomes of such games depend not only on yourself and your luck (if any), but on the decisions of others.

Deviating from the Line

Despite the hype raised by the newborn game theory, it took a while for economists to realize how widely the new tool could be used. Morgenstern and von Neumann had primarily been interested in zero-sum games, where the gain of one player is the other player's loss. Such is indeed the rule in most parlor games. In real-life interactions, however, the interests of the participants need not be completely at odds. The interests are different, usually, but not diametrically opposed. Even in the fiercest of wars, there are situations that both parties want to avoid.

The extension of game theory beyond zero-sum games was mostly based on the work of the young American John Nash. Around 1950, he turned the budding field right side up (somewhat in the fashion of Karl Marx who claimed to have turned Hegel on his feet).

Nash would go on to become one of the great romantic heroes of mathematics, in the "genius and madness" vein. A certified genius he was from early on: already his undergraduate advisor said so, in a one-line recommendation letter to Princeton University. There, freshman Nash stood apart, very much on his own, "apparently trying to re-discover three hundred years of mathematics by himself," in the words of his fellow student John Milnor (another undisputed mathematics genius).

The most important concept introduced by John Nash into game theory was that of an equilibrium (later named after him). The basic idea is simple. If I know the strategy of the other player, I can look for a best reply to it: namely, a strategy from which it would not pay for me to deviate. In most cases, I do not know the other player's strategy, of course. All I can assume is that the co-player, too, will try to find the best reply to *my* strategy. A Nash equilibrium pair is a pair of strategies (mine and that of my co-player) that are best replies to each other. Such an equilibrium condition is certainly a minimal demand on any "solution" to a conflict of interest. Anything else would be inconsistent, since at least one of the players would have an incentive to deviate.

A famous theorem of John Nash says that there always exists such an equilibrium pair. This may seem surprising at a first glance—or rather, plain wrong. Our simple example of the childish odd-even game provides a counterexample, as we have seen. Whatever the outcome, one of the players will have a reason to regret his or her move. But John Nash (following John von Neumann on this point) allows for so-called mixed strategies. Players can use a random device to decide between the alternatives, and opt with such and such probability for one move, and with such and such for another.

It is easy to see that if both Ann and Burt play "odd" or "even" with equal chances, neither of them has any incentive to deviate. Every other pair of mixed strategies provides at least one of the players with an incentive to change to a different strategy.

This simple example is surely a letdown. Nobody can be greatly impressed by the advice "play the Nash equilibrium." You need not consult a mathematics genius to opt for "odd" or "even" with equal chances. But Nash equilibria exist for *every* game, no matter whether zero-sum or not, no matter how many alternative moves are available to the players, no matter how many players are interacting. All that is needed is to give chance a say (by admitting mixed strategies).

In this full generality, the theorem of Nash has proved of huge importance in all social sciences. Every conflict of interest has a "solution" in this sense—an equilibrium that leaves no party with an incentive to change their mind. It seems almost magical, especially as the proof has nothing to do with social interactions at all. It is entirely geometrical.

Indeed, suppose that you have two maps of a country, one smaller than the other. Then, if you place one map on top of the other (Figure 11.4), there will be a point where the two maps coincide: drive a needle through it, and you have pinpointed exactly the same spot on both maps. All you need is that the smaller map lies entirely in the interior of the larger one. Within these bounds, you can shift or turn the smaller map any way you like; you can fold it, or place it upside down, or stretch it (but you are not allowed to tear it). This fixed-point property is a fact of geometry, true in any dimension. A purely geometric insight provides a key result on social interactions.

Having discovered his result, John Nash walked into the office of John von Neumann. As usual, von Neumann was quick to grasp the point. "Oh,

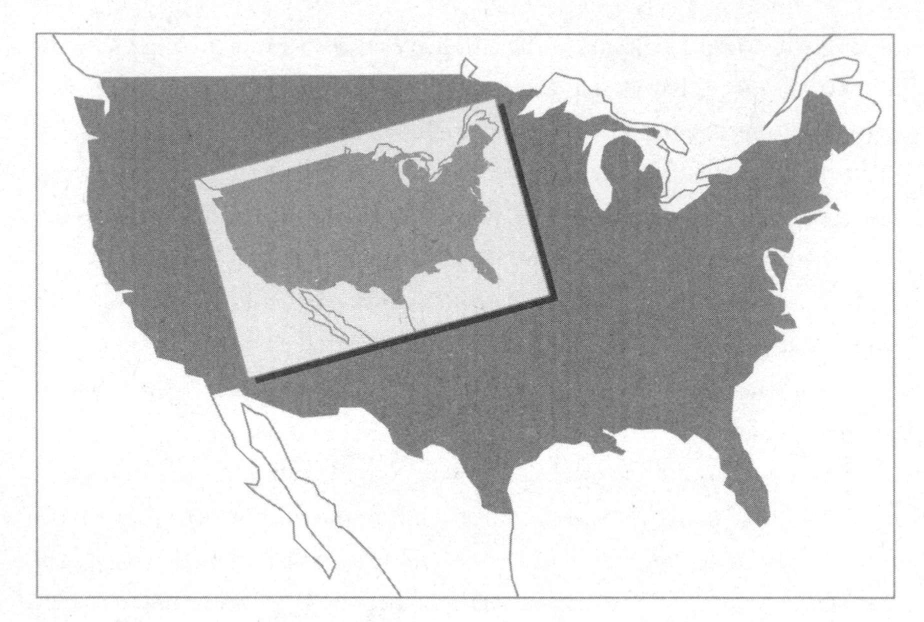

Figure 11.4. Two superimposed maps: they coincide in one point.

this is trivial," he said, and dismissed it. Nash left the office, understandably miffed. His theorem, after all, included John von Neumann's zero-sum result as a tiny special case, and its proof was far more elegant. Later, Nash would somberly comment on the brush-off: "My ideas deviated somewhat from the line (as if of political party line) of von Neumann's and Morgenstern's book."

Being accused of triviality must have rankled. Nash went on with doing game theory (none of it trivial) during the summer months, at the RAND Corporation, that archetypal think tank in California. For the rest of his time, however, back on the East Coast, he grappled and solved some of the hardest problems in geometry and analysis. By the age of thirty, he had made his name: a striking figure on the mathematics circuit, professor at MIT, tall, good-looking, with a ravishing wife—this was when schizophrenia submerged him. He heard voices, deemed himself emperor of Antarctica, resigned from his job, asked for political asylum in Switzerland, and was institutionalized and treated with electroshocks. After years of torment, he seemed to come back, but then sunk again. He kept drifting through the lecture halls of Princeton, a tragic shadow of his former self, leaving cryptic messages behind. However, the institute of mathematics never gave up on him, nor did his wife, Alicia. Some few students gathered around "the phantom of Fine Hall." He was, after all, a mathematical legend already.

Then, slowly, rumors spread that Nash had returned to something close to normal. When the Swedish Academy felt convinced that he was stable enough, he was awarded the Nobel Prize for Economics in 1994. The film *A Beautiful Mind* used the ceremony as its closing scene: a happy ending fit for a Hollywood movie.

Nash lived on, by now one of the most celebrated scientists of his time. He gave lectures on his illness. When asked why he could ever have given credence to his strange delusions—"emperor of Antarctica" and the like—he replied that these ideas occurred to him just like his mathematical ideas. "This is why I took them seriously."

Some twenty years after his Nobel Prize, John Nash received the Abel Prize, its mathematical pendant. On returning home with his wife, the aged couple—both in their eighties—took a taxi from Newark Airport. It crashed, killing both instantly.

Neither would have to mourn the other: a happy ending fit for a Greek myth.

Figure 11.5. John Nash (1928–2015).

The Prisoner's Dilemma

To get more familiar with Nash equilibria, it will be useful to resort to payoff matrices. Let us start with the odd-even game. The payoff matrix is as follows:

	O	*E*
O	(1, −1)	(−1, 1)
E	(−1, 1)	(1, −1)

Both Ann and Burt can choose between two strategies: *O* and *E* for "odd" and "even." Ann can choose between the upper and the lower rows, Burt between the left and the right columns. In the corresponding entry of the 2 × 2 matrix, we find the payoff pairs for the two players: first Ann's (the row player), then Burt's (the column player). If both decide to play *O*, for instance, Ann receives payoff 1 (a gain of $1) and Burt −1 (a loss of $1).

There can be several equilibria. Consider the following payoff matrix:

	C	*D*
C	(40, 40)	(0, 0)
D	(0, 0)	(40, 40)

This describes a so-called coordination game. Ann and Burt should either both play *C* or both play *D*. There are two Nash equilibrium pairs, (*C*, *C*) and (*D*, *D*). But which one to choose? If Ann and Burt have no opportunity to discuss the question beforehand, miscoordination is quite likely.

With the next coordination game, this is different:

	C	*D*
C	(60, 60)	(0, 0)
D	(0, 0)	(40, 40)

Both (*C*, *C*) and (*D*, *D*) are Nash equilibrium pairs, but (*C*, *C*) is obviously better.

Now consider this coordination game:

	C	D
C	(60, 40)	(0, 0)
D	(0, 0)	(40, 60)

Again, (C, C) and (D, D) are equilibrium pairs, but Ann prefers (C, C) and Burt (D, D). Each prefers 60 to 40. Should Ann insist on her preferred solution, or yield? What if Burt yields, too?

Even more puzzling is the following coordination game:

	C	D
C	(60, 60)	(–1000, 40)
D	(40, –1000)	(50, 50)

Again, (C, C) and (D, D) are Nash equilibrium pairs; moreover, (C, C) yields for both players the higher payoff. But should Ann really play C? If she has any reason to fear that Burt plays D (because he is a sadist, or an imbecile), she will lose $1000. Surely it is safer to play D. Incidentally, it is possible that Burt is not an imbecile, but does not know for sure that Ann is no imbecile either. Hence, the equilibrium (D, D) seems safer.

Thus, the problem to coordinate on the "right" equilibrium looms large. (Such a problem does not arise for zero-sum games, by the way.) But there is a still nastier side to the Nash equilibrium. It arises even for games that have only one equilibrium.

In the early 1950s, while John Nash was visiting the RAND Corporation, some postdocs working there on Cold War politics and nuclear deterrence came up with a game that they called the Prisoner's Dilemma. The name stuck, and today, it seems hard to conceive that there was a time when people did moral philosophy without using the Prisoner's Dilemma.

The scenario is well known: two prisoners are accused of a joint crime. The state attorney has no proof, but relies on a cunning ploy. The two prisoners are kept in separate cells, pending investigation, and are offered to turn state's evidence. If both refuse the deal, and remain obdurately silent, they will eventually have to be released, but only after a long pre-trial confinement, say one year. If one of the two prisoners flips, or "sings," he or she will be released right away, while the other inmate gets the full brunt of the law: ten years in the slammer. Thus, it seems like a good idea to confess. But if both prisoners confess, they are not needed for state's evidence any longer and will be sentenced to jail for seven years.

Such a scenario cannot be acted out in a game lab. Moreover, it conveys sinister associations that distract from the point of the game. The dilemma has nothing to do with loyalty, treason, revenge, conscience, nor guilt.

The previously described donation game, on the other hand, has exactly the same strategic structure in all its essentials (minus the *film noir* footage) and serves much better to convey the gist of the problem. It must be admitted, however, that it could never have encountered the blockbuster success of the prisoners' tale.

Consider the payoff matrix of the donation game:

	C	*D*
C	(10, 10)	(–5, 15)
D	(15, –5)	(0, 0)

It is enough to look at it from Ann's point of view only (Burt is in exactly the same conundrum). Hence, we focus on her payoff values:

	C	*D*
C	10	-5
D	15	0

Ann can reason as follows: If Burt chooses *C*, my payoff will be $10 if I play *C* and $15 if I play *D*. Hence, *D* is my best reply to the other's *C*. If Burt chooses *D*, I will lose $5 if I play *C*, but nothing if I play *D*. Hence, *D* is my best reply to the other's *D*. Whatever Burt is doing, my best reply is *D*. Hence, I will choose *D*: no donation to Burt. Burt, however, is in exactly the same situation. No donation to Ann. The two players have found the best reply to each other. The Nash equilibrium pair demands that both play *D*. Both then receive nothing.

It surely would have been better if both players had chosen *C*. Both would receive $10. This is a so-called Pareto optimum: players cannot improve their payoff except at a cost for the other player. To achieve (*C*, *C*), however, they would have to deviate jointly from their equilibrium (*D*, *D*), hand in hand, so to speak. One player is not enough.

By contrast, the Pareto optimum—both players play *C*—is no Nash equilibrium, since each player could improve his or her payoff by switching to *D*, thus harming the other. A figure helps to visualize this (Figure 11.6): the payoff pairs span a parallelogram. If Ann deviates from the Pareto optimum, and Burt does not, then Ann improves her payoff at Burt's expense, and vice versa. A selfish player is one who is not affected by the other's payoff. Two selfish players drag each other into quicksand.

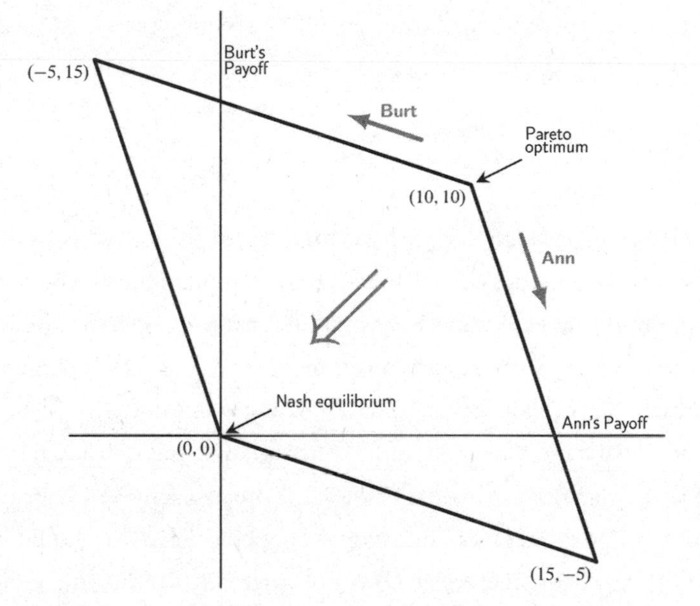

Figure 11.6. From Pareto to Nash in two selfish pulls.

Partners and Rivals

The Prisoner's Dilemma is the most important, but by no means the only, social dilemma. Such a dilemma occurs whenever self-interest conflicts with the common good. It happens frequently that cooperation is useful, and possibly even essential, but that defection is even more promising. (By the way, C stands for "to cooperate" and D for "to defect," but subjects in a game lab should never be told this. The words raise associations that may subconsciously affect their decisions.)

Whenever players are asked, after a donation game, which outcome they would have preferred, most reply: mutual cooperation. A large majority thus seem to consist of conditional cooperators, who would like to play C, if they only were sure that their co-player will also play C. This, at least, is what they say—even those who played D.

Now is the moment to remember that a player's *utility* (the value assigned by that player to an outcome) need not correspond to the amount of money involved. Preferences may depend on many other factors. If Ann were really a conditional cooperator, her "true" payoff matrix might look as follows:

	C	D
C	$10 + X$	-5
D	15	0

Here, X is a positive value that we leave unspecified for the moment, some obscure feel-good bonus possibly. If this X is worth more than the measly $5 that Ann can obtain by exploiting Burt's willingness to cooperate, then D is no longer the best reply to Burt's unknown move, and mutual cooperation seems within reach. The million-dollar question then is: where does that X come from? (It is actually a five-dollar question.) At first glance, the unknown X seems a mere deus ex machina—a makeshift, a stage prop to achieve a happy end.

Crucially, the strategy of conditional cooperation (namely, play C if the other plays C; if not, then play D) is not a feasible option in the donation game, because players do not know what the other will do. But there is a simple modification of the donation game that shows that the wish for conditional cooperation is more than hypocritical lip service or naïve self-deception. Let Burt move first. Ann, then, is aware of Burt's move when she makes her decision. This is the *sequential* donation game (in contrast to the *simultaneous* game, where players decide without knowing the other's move).

Experiments show that if Burt plays *C*, then Ann will most likely reply with *C*. (Indeed, she plays *C* with a probability of 75 percent, considerably more than the 50 percent probability in the simultaneous setup.) Ann seems to feel a certain obligation, or at least an inclination, to return the donation, although she knows that she will never interact with Burt again. Is it gratitude? A feeling of duty? Or mere decency? If Burt plays *D*, by the way, then Ann will almost always also play *D*. In both cases, reciprocation is at work, a strong urge to play tit for tat.

In the sequential game, Burt is less privileged than Ann. He has to commit first, uncertain about Ann's ulterior decision. And yet, the experiment shows that Burt, too, is more likely to play *C* than in the simultaneous version. He also opts for *C* with a probability of (roughly) 75 percent. Apparently, Burt expects that Ann is a conditional cooperator and will reply to his *C* with a *C*. Thus, the reason for Burt to play *C* may simply be what the Romans knew as *do ut des*: "I give so that you give."

In the simultaneous donation game, on the other hand, Burt's uncertainty is larger, because he has to take into account Ann's uncertainty about what he will do. The striking contrast between the simultaneous and the sequential donation game—the frequency of mutual cooperation is almost doubled, from 25 to close to 50 percent—indicates that those who play *D* are motivated less by selfishness and greed, than by distrust and the fear of being cheated.

A conditional cooperator views the co-player as a *partner*. If both you and your co-player approach the donation game in the spirit of partnership, you will both fairly share the best outcome, and receive $10 each. But if your co-player prevents you from reaching that happy result, then he or she will not reach it either. It should never pay for the other to exploit you. You are a willing partner, not a serf. A co-player who opts for *D* should not get away with it.

Some of those who are *not* conditional cooperators may approach the game in a different spirit, and see in the other player a *rival* rather than a potential partner. If your topmost priority is to prevent the other from earning more than you do, the right move is obviously *D*. In the usual parlor games, such as poker or chess, rivalry is the main motivation: you want to beat the other. In the context of an economic interaction, it makes less sense. Why should you care about the other's payoff, since your own income is what matters? And indeed, more than two thirds of the players seem to view the donation game in the partnership spirit. Some of the remaining third may have been swayed toward defection by being told that they play an experimental "game" and that they are in a "game" lab. The word *game* suggests an antagonistic setup. Some of the *D* players, then, may be mere victims of a framing effect.

The Golden Rule and Other Alchemy

The partnership spirit behind conditional cooperation is a hard-nosed, economic approach, quite distinct from the hallowed principle to "do to others as you would have them do to you," which has been known as the Golden Rule since baroque times. The Golden Rule is a highlight of the Sermon on the Mount (Luke 6.31, Matthew 7.12), but that was by no means its first release. The principle seems to be a fixture of every world religion. Here are some examples from *Wikipedia*:

> Talmud: "What is hateful to you, do not do to your neighbor. That is the entire Torah."
>
> Mahabharata: "Do not to others what you do not wish done to yourself.... This is the whole Dharma."
>
> Zoroaster: "Human nature is good only when it does not do unto another whatever is not good for its own self."
>
> Buddha: "Hurt not others in ways that you yourself would find hurtful."
>
> Confucius: "Do not do to others what you do not want done to yourself."
>
> Mohammad: "None of you is a believer until you desire for another that which you desire for yourself."

These are all blessed words. If they guide your actions in the game lab, you will always play *C*, which makes you an unconditional cooperator. Yet if that's what you are, you belong to a rare breed. In the sequential donation game, very few reply to a *D* with a *C*. Economists discard the Golden Rule; theologians extol it; philosophers keep their distance.

Immanuel Kant never failed to stress that the Golden Rule must not be confused with his own Categorical Imperative: "Act only according to that maxim whereby you can at the same time will that it should become a universal law." This imperative has been hailed as "the Golden Rule with umlauts," but the two principles have a very distinct flavor. They are, admittedly, similar in being abstract. They never prescribe any specific action, but invoke a meta-rule transcending the perspective of the I. As Kant explained, a criminal sentenced by a judge might hopefully try to plead for a milder verdict with an appeal to the Golden Rule; by contrast, invoking the Categorical Imperative would merely confirm the judge in his verdict.

While the Golden Rule prescribes unconditional cooperation, the Categorical Imperative could act as a guide for conditional and unconditional cooperators alike. Indeed, if Earth was entirely peopled with cooperators, it would make little difference whether they were conditional or unconditional. Only the presence of defectors allows us to tell them apart.

Figure 11.7. David Hume (1711–1776) expected no return.

Jean-Jacques Rousseau contrasted the Golden Rule with "that other maxim of natural goodness a great deal less perfect, but perhaps more useful: do good to yourself with as little harm as you can to others." In the context of game theory, this agrees with conditional cooperation. You do no good to yourself if you reply to a co-player's *D* with a *C*.

"That other maxim" may be useful, according to Rousseau—indeed, it is utilitarian in spirit, and supports the Pareto optimum—but where is its justification? That conditional cooperation is good for society is beyond doubt. But why should a selfish individual opt for it? This conundrum was known to David Hume already, a friend of Adam Smith, Rousseau, and Voltaire. The Scottish philosopher distrusted sentimentalities, and was well aware that human nature pays little heed to the teachings of religion founders. Hume used a homely parable:

> Your corn is ripe today; mine will be so tomorrow. 'Tis profitable for us both, that I should labor with you today, and that you should aid me tomorrow.

This down-to-earth scenario describes a sequential donation game—all that is missing is the payoff matrix. Hume goes on:

> I have no kindness for you, and know you have as little for me. I will not, therefore, take any pains upon your account; and should I labor with you upon my own account, in expectation of a return, I know I should be disappointed, and that I should in vain depend upon your gratitude. Here then I leave you to labor alone. You treat me in the same manner. The seasons change; and both of us lose our harvests for want of mutual confidence and security.

Without benevolent feelings such as the "kindness" or "gratitude" mentioned above, there should be no cooperation. In the sequential donation game, as acted out in the lab, players do not know each other and have no reason for kindness or gratitude. This, then, is the riddle: why does it happen so frequently that Ann feels motivated, indeed almost obliged, to reply to Burt's *C* with a *C*? Why does Burt count on it, more often than not?

Hume shows the way. If partnership breaks down "for want of mutual confidence and security," then confidence and security are what is needed for conditional cooperation.

How can security be achieved? Best by a binding contract. This is a topic for later, because it requires a social contract. And what about mutual confidence? Everyone knows that confidence is the most important factor for a flourishing economy. Trust is what oils the wheels of trade and industry—where does this trust spring from?

Play It Again, Sam

At the end of Hume's philosophical tale, "the seasons change." The opportunity for mutual help is over. What Hume fails to mention is that the seasons repeat. This alters everything. If the game is played not only for one round, but again and again, it is not "more of the same": its structure is dramatically different.

Actually, players need not even be sure of a further round: it needs only to be likely enough. In that case, whoever cheats the other must reckon with a payback; and whoever helps may hope for one. The repeated Prisoner's Dilemma game offers opportunities for retaliation. This raises the specter of the talion law, which is even older than the Golden Rule: an eye for an eye, a tit for a tat.

To study this in the game lab, we may ask players to play the donation game for several rounds. If we wish to study how the shadow of the future affects the players, they should not be told which round is the last.

The number of possible strategies, in a game of many rounds, is huge: such a strategy is a program telling the player how to act in each round—press the *C* or the *D* button—depending on all that happened in the previous rounds.

Let us first explore a simplified scenario, where players *do* know the number of rounds—six rounds, say. We will ensure, however, that they commit themselves right at the start, by offering them only two strategies to choose from. One consists in relentlessly playing *D*, which is the Nash equilibrium strategy for the one-round donation game. This strategy of unconditional defection is denoted by AllD. The only other option offered to the players is to choose TFT, the tit-for-tat strategy. It consists in playing *C* in the first round and from then on repeating the co-player's previous move: after a *C*, play *C*, and after a *D*, play *D*. Only TFT and AllD are on the menu: players must decide to follow one or the other.

Ann's payoff, in this case, looks as follows:

	TFT	AllD
TFT	60	−5
AllD	15	0

Indeed, if both players opt for TFT, they will mutually cooperate in each round, each time winning $10. If both players opt for AllD, there will be no donation ever: economic standstill. If a TFT player meets with an AllD player, only the first round yields some action: the TFT player will be cheated of $5, and the AllD player collects $15, as the wages of sin. From then on, we are back to economic standstill.

The best reply to TFT is TFT. The best reply to AllD is AllD. We are faced with a coordination game. It is best to do whatever the co-player is doing. Thus, there are several Nash equilibrium pairs to choose from. This is quite different from the one-shot donation game, where *D* was the best reply to *every* move of the co-player.

TFT is a partnership strategy: the co-player can exploit it, but can only do so at a cost. TFT is not the pinnacle of conditional cooperation, as it can be exploited in the first round. But with more and more rounds, this initial setback weighs less and less.

In a certain sense, moreover, TFT promotes its own success. To better understand this, we have to switch our lens to a wider angle, and see individuals as members of a community.

Let us imagine a population of players and assume that from time to time, one pair is chosen at random and asked to play the repeated donation game, with TFT and AllD as the only options. The overall success of your strategy will depend on the composition of the population. If most play TFT, you should play TFT. If most play AllD, you should play AllD. It does not pay to deviate from what the others are doing. You should swim along with the stream.

On principle, such coordination has nothing to do with ethics. Indeed, it is not a moral command to drive on the right-hand side of the road. But in a country where all other drivers use the right-hand side, you should do so too, simply from self-interest. This convention is nothing but an agreement. It is a social norm. Should you prefer to dissent, you do so at your peril. That you imperil others too can, at best, be a meagre consolation.

In a community where almost everyone plays TFT, both the many TFT players and the few AllD dissidents encounter mostly co-players using TFT. Hence, TFT players earn almost always $60 (and on rare occasions, lose $5), whereas AllD players earn almost always $15 (and on rare occasions, nothing). If the community is dominated by TFT, a TFT player should not switch strategy, but an AllD player should.

If the community is dominated by AllD, the opposite holds. Both social norms, namely the good one (everyone play TFT) and the bad one (everyone play AllD), are cul-de-sacs.

In the one-shot donation game (only one round), this is different: a *D* player always fares better than a *C* player, no matter how the population is composed.

The mathematics of such social learning—imitating strategies that fare better—is the topic of evolutionary game theory. In its simplest version, it considers how fictitious populations evolve. The individual players are programmed to play this or that strategy. They meet randomly, play the game, accumulate payoff (depending on their strategy and that of their co-player), and then, from time to time, switch to another strategy that fares better (if there is any). Thus, the frequencies of the strategies in the population evolve as a function of their payoff. However, that payoff is a function of the frequencies: it depends on whom the players are likely to meet. This makes for a feedback loop between the strategies' frequencies and their payoff.

If the one-shot donation game is played, *D* players will invariably have the upper hand, and in the end, the population will consist only of defectors (Figure 11.8). But if the game is the repeated donation game, with only TFT and AllD as possible alternatives, then the outcome depends on the initial conditions. There is a threshold: if the initial number of TFT players is above the threshold, they will take over; if not, they will yield to AllD (Figure 11.9).

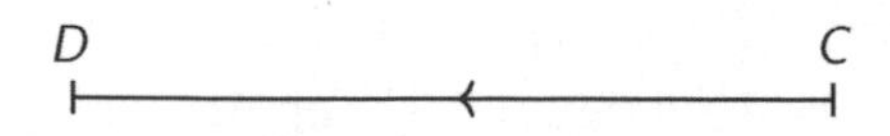

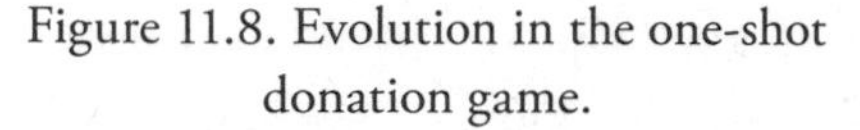
Figure 11.8. Evolution in the one-shot donation game.

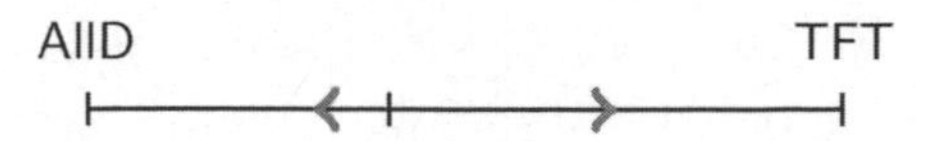

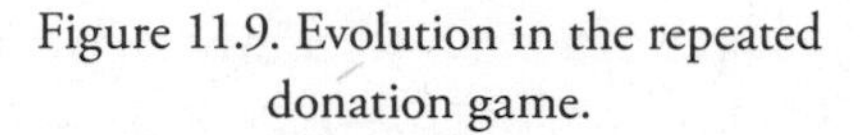
Figure 11.9. Evolution in the repeated donation game.

We don't have to restrict ourselves to two alternatives only. There is an amazing wealth of strategies for the repeated donation game. It became a spectator sport, among philosophers and game theorists alike, to set up computer tournaments and watch the population evolve via social learning. Sometimes, one strategy takes over; sometimes a stable mixture of several strategies emerges; sometimes they fluctuate endlessly. The complexities seem boundless.

TFT often does well, although not invariably so. The frequent success of TFT is all the more remarkable as TFT players can never have a higher payoff than their co-players. They are never the first to defect on the other. They can be cheated in one round, but then they switch to playing *D*, and resume cooperation only when they have exploited the other in turn and restored the balance.

The main weakness of the tit-for-tat strategy is its vulnerability to errors. In the silicone world of a computer tournament, errors are not likely to occur. In real life, however, it may well happen that players occasionally make a wrong move. They may be distracted, or ill, or simply broke. If two TFT players are engaged in a repeated donation game and one plays *D* by mistake, then the other will retaliate in the next round. The strict TFT program then demands that the first player plays *D* again, this time not by mistake, but because the strategy prescribes it. The other player will reply with *D*. Thus, the two players are locked in a series of backbiting defections that will cost them dearly.

It turns out that a variant of TFT, one that occasionally forgives a coplayer's defection, overcomes mistakes much better. Thus, social learning—merely imitating what fares best—can lead to the emergence of *forgiveness*, a central idea in morality.

If the population becomes all too forgiving, however, then all-out defectors can invade and take over. Another strategy is safer. It consists in repeating the move from the previous round (which was *C* or *D*) if the payoff was positive ($10 or $15) and switching to the other move if it was not ($0 or –$5). The strategy is named Win-Stay, Lose-Shift and embodies the simplest learning rule. It is well known to all animal trainers, but little regarded by moral philosophers.

By now, shelf-loads of books and many hundreds of research papers deal with the game theoretic intricacies of the repeated Prisoner's Dilemma game. The main message is: repetition promotes cooperation, because it allows for *reciprocation*. (We have seen that even the simple step from a one-shot simultaneous to a one-shot sequential donation game allows the second player to reciprocate, and effectively doubles the amount of mutual cooperation.)

This does not explain the curious fact that in the simultaneous one-shot donation game, half of the players opt for *C*, despite having been explicitly told that they will interact for this one round only. Is it because many of the players have a sneaking suspicion, conscious or not, that they might, after all, meet the other person again? For thousands of generations, our forebears spent their lives in small tribal and village societies, where it was almost unavoidable to bump into each other time and again.

Anthropologists report that in hunter-gatherer societies, meetings with strangers are extremely rare, and not unlikely to be fatal (one-shot interactions in the most literal sense, possibly via poisoned arrow). Such instinctive distrust of all aliens is mentioned jokingly by Adeimantus, a brother of Plato, in the latter's *Republic*:

> Why, a dog, whenever he sees a stranger, is angry; when an acquaintance, he welcomes him, although the one has never done him any harm, nor the other any good.... He distinguishes the face of a friend and of an enemy only by the criterion of knowing and not knowing.

This led Socrates to the pleasant aside that the dog must have the soul of a philosopher, "a lover of knowledge," because he likes whom he knows and dislikes whom he doesn't.

Temptations for the Invisible Man

To know or not to know: the philosophical dog points the way to another experiment in the game lab. Again, we need a community of players: several dozens, if possible even hundreds. Each plays many rounds of the simultaneous one-shot donation game. They are told that they will never encounter the same co-player twice. As usual, they interact via computer screen only, and know that their "real persons" will remain completely anonymous.

The main point in the new experiment is that it will be run in two versions, or treatments, to use the jargon. In one treatment, players know what their co-player has played in the previous rounds. In the other, they don't. It will come as no surprise that most players are more inclined to cooperate if they know that they are matched with a co-player who, so far, has mostly or always played *C*. Such a co-player appears trustworthy. For this same reason, many players will think twice about defecting on their current co-player, for this makes it less likely that their next co-player opts for *C*.

In the treatment that allows players to build up a reputation, cooperation emerges almost invariably. The co-player, while anonymous, is not a complete stranger. Something is known about him or her. The philosophical dog wags his tail.

This treatment promotes cooperation, just as with the repeated donation game. No pairing lasts for longer than one round, however. No interaction is repeated. As before, reciprocity is the key, but it is not a direct reciprocity: it is indirect. In the words of game theorist Ken Binmore, direct reciprocity (Figure 11.10) works according to the maxim "I'll scratch your back if you scratch mine," whereas indirect reciprocity (Figure 11.11) uses the much subtler "I'll scratch your back if you scratch the back of others." In the former case, you are guided by your own experience with the co-player; in the other case, by the experience of third parties. This requires, of course, that such experiences can be communicated. Harvard biologist David Haig hit the nail on the head when he said: "For direct reciprocity, you need a face; for indirect reciprocity, you need a name."

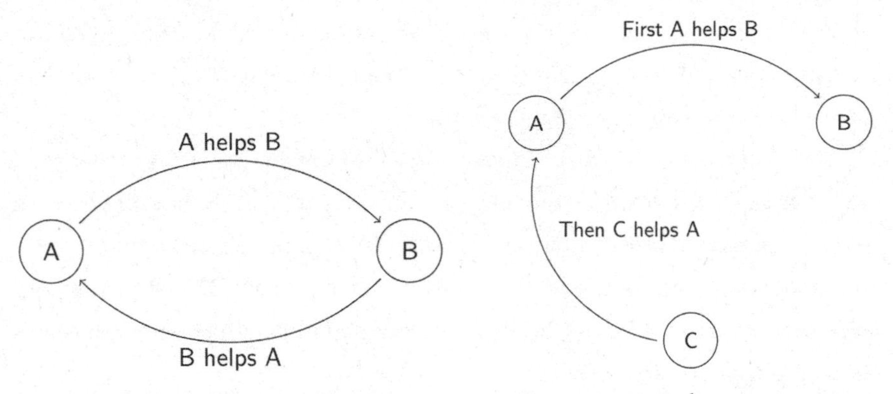

Figure 11.10. Direct reciprocity.

Figure 11.11. Indirect reciprocity.

Information promotes confidence. That point is nicely confirmed by the story of eBay, a story proudly told on the company's homepage. This Internet platform is a tool for trading on a global scale, and under conditions of strict anonymity. The setup is the very opposite of our ancestral environment: in an eBay exchange, the partners usually do not know each other, are unlikely to ever interact again, and may live thousands of miles apart. Yet, they rarely cheat on each other.

The reason is obvious. After the business transaction, each participant can rate the other, and this praise or blame becomes common knowledge. In such a fashion, each member of the eBay community acquires a reputation, which can be good or bad. The corresponding points or stars—colored stars, indeed, as in kindergarten or grade school—are directly displayed on the feedback forum.

In the first version of eBay, which started out as AuctionWeb, the feedback forum did not yet exist. AuctionWeb was a flop. Mutual suspicion prevailed. The public at large started to trust the new way of doing business only after the platform's founder, a young Paris-born programmer named Pierre Omidyar, introduced his rating system. He did so with crisp words: "Give praise when it is due; make complaints when appropriate." This maxim made his fortune: a golden rule indeed.

Needless to say, economic transactions have always been based on reputation, well before the feedback forum. Reputation grows on information.

Whenever people meet, they gossip. They probably did so from their first campfire onward. And gossip is mostly about third parties. This is why we are so concerned about our reputation.

It is well stated in Plato's *Republic*: "Parents always encourage their sons to act rightly. But why? Not for the sake of the good cause, but for the sake of their character and reputation." It is interesting, incidentally, that among the many meanings of the word *character*, one designates the enduring core of a person's traits, and another a recommendation by a former employer. A good reputation brings credit.

As usual, a game theoretic analysis shows that what we tend to take for granted deserves a closer inspection. The new treatment of the donation game via indirect reciprocity turns out to be surprisingly complex. Each player now has a reputation, a score in the eyes of the others. It sums up his or her former decisions and informs the decisions of the future co-players. For indirect reciprocity to work, Ann must be prepared to play *D* against Burt, if Burt is known to defect. By playing *D*, however, Ann risks to blemish her own score in the eyes of her next co-player, Conny. Hence, Conny should know that Ann's defection was justified. She should know Burt's score too, and hence whether or not Burt's actions were justified. Even in a small population, this seems to require considerable algorithmic complexity, and almost Machiavellian skills. But once again, to forget (or to forgive) makes cooperations easier.

The view that to do good must be good for something may sound cynical, but today's economists have no problem with it. Reputation management has turned into a hot topic. Most moral philosophers, however, tend to brush off the concern with one's reputation.

The great exception, as so often, is Plato. In his *Republic*, he lets his brother Glaucon tell the tale of one Gyges, a shepherd by trade, who discovers a ring in an old tomb split open by an earthquake. Gyges makes off with the ring, and notices after a while that by turning the ring on his finger, he can become invisible at will. Having understood that the ring enables him to pursue his self-interest without harm to his reputation, Gyges enters the sleeping chamber of King Candaules, wows the queen, kills the husband, and becomes king in turn, by a turn of his ring.

Glaucon challenges Socrates to refute the cynical lesson:

> All men believe in their heart that injustice is far more profitable than justice. . . . Anyone obtaining this power of becoming invisible, and never doing any wrong or touching what was another's, would be thought by the lookers-on to be a most wretched idiot, although they would praise him to one another's faces . . . for the highest reach of injustice is: to be deemed just when you are not.

At some point brother Adeimantus takes over:

> If, though unjust, I acquire the reputation of justice, a heavenly life is promised to me. Since then, as philosophers prove, appearance tyrannizes over truth and is lord of happiness, to appearance I must devote myself. . . . No one has ever blamed injustice or praised justice except with a view to the honors, glories and benefits which flow from them.

Socrates rose to the bait and attempted to refute the misanthropic views advanced by the two brothers; and many philosophers down the ages have followed suit. Did they succeed? A person who does the right thing merely in the hope of looking good in the eyes of the others will enjoy less prestige than one who does the right thing from inner conviction.

But where does that inner compass, conscience, come from? Conscience has been said to be "the inner voice that warns us that someone may be watching."

Amusingly, it is well documented that players of the donation game are more cooperative if there is a picture—a mere picture—of an eye on the screen. It need not even be a realistic picture. In fact, it has been shown that three dots on the screen increase cooperation if they can be interpreted as a face, but not otherwise (Figure 11.12). We see faces in the most unlikely patterns. Someone may be watching.

Does this mean that the concern with our reputation is what, consciously or unconsciously, makes us conditional cooperators?

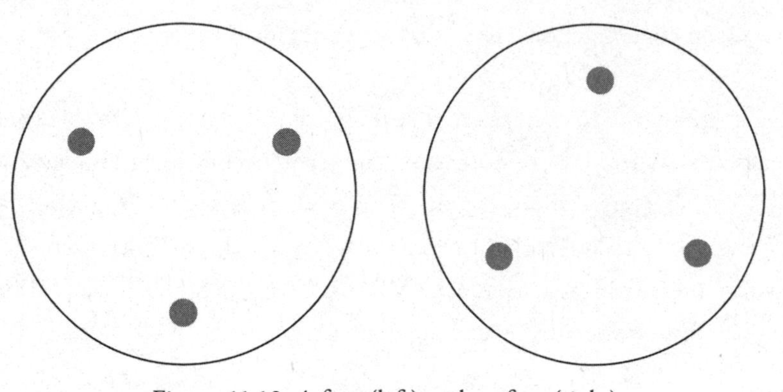

Figure 11.12. A face (left) and no face (right).

Charles Darwin knew that this issue is hugely important for our social life. Just like Aristotle, he saw the parallels between human societies and the colonies of insects such as bees, wasps, and ants. In each case, there is a strong tendency for mutual help. But, so Darwin said, "man's motive to give aid no longer consists solely of a blind instinctive impulse [as with the insects], but is largely influenced by the praise and blame of fellow-men."

Our concern with reputation sits deep. Darwin was fascinated by the "most peculiar and most human of all expressions," namely blushing. "It is the thinking of what others think of us, which excites us to blush."

"Self-attention" is the term used by Darwin. A large part of our I is in the eyes of others. Despite pointers by Plato and Darwin, it took game theory a long time to discover that reputation is more powerful than rationality in guiding our behavior.

12

Social Contract

Punish or Perish

A Late-Bloomer's Fling with Geometry

The saddest encounter of a philosopher with mathematics began as a love story, in the England of 1628. The scene: a gentleman's library; on a lectern, Euclid's *Elements*, open. Enter Thomas Hobbes, forty years old. He ambles by, gives the book an idle glance, sees proposition 47 (the theorem of Pythagoras), and exclaims: "By God, this is impossible."

The scene is related by one John Aubrey, antiquarian, who explains in a confidential aside that Hobbes would now and then resort to an oath, "by way of emphasis." Aubrey reports:

> So he reads the demonstration of it, which referred him back to such a proposition; which he read. That referred him back to another, which he also read. *Et sic deinceps*, that at last he was demonstratively convinced of that truth. This made him in love with geometry.

Hobbes pursued his love affair most assiduously. Geometry became in his view "the only science that it hath pleased God hitherto to bestow on mankind." He had labored in many fields of natural philosophy and law, and now felt convinced that the geometrical method was the only one worth following. "Men begin at settling the significations of their words; which

Figure 12.1. Thomas Hobbes (1588–1679) took a bleak view.

settling of signification, they call definitions, and place them in the beginning of their reckoning." Hobbes decided to build his whole philosophical system on this method.

As the tutor of British noblemen, he was able to journey extensively through Europe, meeting luminaries such as Galilei, Descartes, and Father Mersenne while crisscrossing the continent. Eventually, some ten years after his first encounter with geometry, Hobbes became the mathematics instructor of the Prince of Wales, later Charles II. The job augured well for his future career. Hobbes entertained high hopes of becoming one of the foremost mathematicians of his time.

His enthusiasm, however, far outpaced his power. When Hobbes published a method for squaring the circle, which by then already was a byword for the impossible, the mistakes of the overambitious neophyte were mercilessly pointed out by John Wallis, an eminent professor of mathematics at Oxford University and the co-founder of the Royal Society. Wallis denounced Hobbes's proof as "a shitty piece" (by way of emphasis). Hobbes, not one to mince his words either, promptly replied in kind. Moreover, he firmly refused to discern any fault in his proof. His steadfastness in clinging to his errors clashed oddly with his professed admiration for the scientific method. "For who," he once wrote, "is so stupid as both to [make a] mistake in geometry, and to persist in it, when another detects his error to him?" Yet he persisted.

Although he never acknowledged his mistake, Hobbes kept producing other proofs for squaring the circle with ruler and compass, which often contradicted each other and were all equally wrong. (Two hundred years later Ferdinand von Lindemann proved that the task is indeed impossible.)

Wallis pointed out the errors in about a dozen "quadratures of that pseudo-geometer" Hobbes. He published a pamphlet with the title "Due Corrections for Mr. Hobbes: or Schoole Discipline for Not Saying His Lessons Right." Do you judge that Hobbes deserves such a lengthy refutation? asked Christian Huygens. Aye, it must be done, replied Wallis: for Hobbes's whole hateful philosophy is based on it. To support his outrageous views on church, university, and state, Hobbes "takes the courage from his mathematics." He cannot be allowed to continue. Some mathematicians must stand up and show how little Hobbes understands of it. "We should not be deterred from this by his arrogance, which we know will vomit poison and filth against us."

The more he was proved in the wrong, the more stridently Hobbes defended his claims, convinced that his entire philosophical impact was based on his credentials as a geometer. He disparaged the newfangled methods of algebra and analysis of which Wallis was a master, and crusaded to preserve the purity of geometry.

In time, Hobbes became the laughingstock of mathematicians all over Europe. One of his circle-squaring proofs led to the conclusion that $\pi = \sqrt{10}$; another, that $\pi = 3.2$. Such blunders were a far cry indeed from the formula found by Wallis,

$$\frac{\pi}{2} = \left(\frac{2}{3}\right)^2 \times \left(\frac{4}{5}\right)^2 \times \left(\frac{6}{7}\right)^2 \times \ldots,$$

which is both miraculous and true (*and* lacking no squares). Age did nothing to mellow Thomas Hobbes. He took to arguing that from Euclid on, geometers had misunderstood the meaning of "point" and "line." There are no immaterial things. A point is a body, like everything else: we simply disregard its magnitude.

The last act of the drama had Hobbes on his deathbed, proffering with his last breath his conviction that the theorem of Pythagoras is wrong—the very theorem that had caused his infatuation with geometry, half a century earlier.

However, the bitter tale ended ironically: for though Hobbes had persuaded himself that his whole reputation was based on his work on geometry, philosophers discovered that they could very easily separate his political philosophy from his cantankerous geometric ramblings.

Hobbes, who witnessed the beheading of Charles I, had found a way to legitimize absolute monarchy. In his *Leviathan* from 1651, Hobbes claimed that it was based, not on divine right, but on a covenant between citizens, to end "the war of all against all" by agreeing on a sovereign.

The power of this sovereign was huge. He could, for instance, "burn all books of geometry," as Hobbes contended in a grim aside, "and thereby suppress, if not dispute, that the three angles of a triangle should be equal to two angles of a square, if he deemed this contrary to his right of dominion."

A Chain of Thoughts

A century later, the sovereign was not necessarily a king anymore. Geneva-born Jean-Jacques Rousseau, with the Swiss example right before his eyes, daringly suggested that the people themselves could fill the role of sovereign. It did not matter, for the success of his views, that Rousseau was never very clear about the "general will" of the people. The first lines of his book *The Social Contract* became a meme in an age of revolutions: "Man is born free, and everywhere he is in chains."

The slogan is often misquoted: "Man is born free, *but* everywhere he is in chains." This is missing the point. The two half-sentences are not opposed to each other. The social contract posits that the members of a community submit *freely* to an authority forcing them to behave.

Kant picked up the contractual perspective in 1784 with his *Idea for a Universal History in Cosmopolitan Aim*; and in the late twentieth century, contractarians such as John Rawls and David Gauthier dominated political philosophy.

Figure 12.2. Jean-Jacques Rousseau (1712–1778) had a vision.

Whiffs of a social contract theory can be traced back to antiquity. They are encountered in the writings of Epicurus, Cicero, Lucretia, and, as will hardly surprise, Plato. In *The Republic*, Plato has his brother Glaucon say that men, having tasted the comforts and discomforts of lawlessness, "think that they had better agree among themselves to have neither; hence there arise laws and mutual covenants; and that which is ordained by law is termed by them lawful and just."

The social contract is a great favorite with philosophers, but a bugbear with historians. When was such a contract ever signed? A handful of episodes offer some distant resemblance, such as the Rütli Oath of the Swiss (which is a mere myth), or the US Declaration of Independence. These are at best sporadic events in human history, isolated singularities within thousands of years of mindless muddling on, same as yesterday or the day before, interrupted only by explosions of cruelty and rage: an endless litany of coercion, submission, serfdom, and slavery. Even today, we are never asked whether we agree with the laws of the state in which we happen to be born. (In *Crito*, Socrates, when just about to drink his cup of poison, argues that by having remained as an adult in his native Athens, he had tacitly accepted the laws of the community. This is a generous take on what is in fact an almost inescapable bondage.)

The social contract, thus, is nothing but an abstract fiction; a brainchild, like the geometric figures that obsessed and irked Thomas Hobbes. Abstract models or thought experiments are no part of the material world. They belong to mathematics. The social contract, in particular, belongs to game theory, the mathematics of conflicts of interest. It is the epitome of methodological individualism: a doctrine that attempts to reduce society to the aggregate behavior of its members, somewhat in the same way as Boltzmann explained the physics of gases by the wriggling of their molecules.

There are many game theoretic models of the social contract. We will pick a path that leads, step by step, from the social trap via the war of everyone against everyone to the social contract.

The Stag Hunt

All traditional social contract scenarios start with what is called the "state of nature" or, in current jargon, the original position. It is doubtful whether such a background story is necessary, but it is entertaining to compare the various scripts: for example, the paranoid cutthroat version of Thomas Hobbes with Jean-Jacques Rousseau's account of a paradise for noble savages.

What we know today is that the earliest human communities were groups of hunter-gatherers. Only very few subsist today. They strike us with a near absence of hierarchy and personal property. Moreover, they know hardly any division of labor, except between men and women. Warfare in the modern sense was unknown: instead of battles, hunter-gatherer communities engaged in occasional raids. Yet these could, in due time, lead to the extinction of entire tribes.

Cooperation was reduced to common efforts in such raids, the raising of young, defense against predators, and hunting. This last topic is nicely picked up in Rousseau's *Discourse on Inequality.* His celebrated Stag Hunt parable describes a joint undertaking to hunt a deer, by encircling its hideout and preventing its flight.

Rousseau said: "Everyone well realized that he must remain faithful to his post; but if a hare happened to pass within reach of one of them, we cannot

Figure 12.3. A stag hunt without hares.

doubt that he would have gone off in pursuit of it without scruple." So you go for the hare. Your selfishness jeopardizes the success of the remaining deer hunters. It is a social dilemma: by pursuing your own good, you reduce the common good.

The modern way to handle such parables is by using game theory. In its simplest version, the Stag Hunt is described by the interaction of two players only (although part of its relevance shows only for larger groups of participants). So, let us return to Ann and Burt, and assume that their options are to hunt stag (C) or to hunt hare (D). Ann's payoff matrix could be of the form

$$\begin{array}{ccc} & C & D \\ C & 15 & 0 \\ D & 5 & 5 \end{array}$$

If Ann (who chooses the row) plays D, she obtains a modest payoff, say 5 (the utility of "the hare"). This payoff is independent of what Burt (who chooses the column) is doing. You need no help if you hunt a hare. If, on the other hand, Ann plays *C*, and hunts stag, the outcome depends on what Burt is doing. If Burt also plays *C*, and joins the stag hunt, Ann's payoff (namely, her share of the stag) is larger than what the hare can offer. If, however, Burt plays *D*, and goes after hares himself, then Ann's chance of bringing the stag down is nil, and she has just wasted her time.

This differs from the donation game, aka Prisoner's Dilemma, where Ann's payoff matrix is

	C	*D*
C	10	-5
D	15	0

In that case, Ann should always play *D*, if she wants to maximize her payoff. The only Nash equilibrium is for both Ann and Burt to defect.

In the stag hunt game, by contrast, there are several equilibria. Ann should play *C* if Burt plays *C*, and *D* if Burt plays *D*. This is a coordination game, with the same structure as the *repeated* donation game.

The same two equilibria (all hunt hare, or all hunt stag) also prevail if there are more than two participants in the game. A hare hunter's payoff is independent of what the other players are doing. A stag hunter's payoff depends on how many co-players join the stag hunt. If no one else joins, the payoff is lower than the payoff that a hare hunter can expect. However, if many join the stag hunt, the payoff will be higher. If all others hunt stag, it is best to hunt stag. If all others hunt hare, it is best to hunt hare. Playing *C* (hunting stag) is a kind of speculation, a venture: it is advantageous if enough of my co-players contribute, and a waste of time if they don't. Hunting stag requires trust.

Similar situations abound, whether among hunter-gatherers or in big-city life. It is good to cooperate, but only if the others cooperate, too. Will they? Under conditions of social learning, with players imitating successful strategies, one of two social norms evolves: all play *C*, or all play *D*. The outcome depends on the initial position, on that nebulous "state of nature."

Mutual Aid and the Common Good

The donation game can also be extended to more than two players. The British economist Robert Sugden described it as the mutual aid game. It can easily be played in a game theory lab.

Suppose that each of six players can decide whether or not to invest $5 into a common pool, knowing that his or her contribution will be multiplied by a factor of three, and then divided among the *other* players. My contribution will provide $3 to each of my five co-players. If we all contribute, each of us will receive a total of $15 at a cost of $5, and hence will net $10. If I am the only one to contribute, however, then I will receive nothing from my co-players and lose $5. The outcome thus depends on how many co-players contribute. My payoff is maximal if I am the only one not to contribute: a net gain of $15 at no cost at all. The others will gain $12 each, which is not bad, but less than my gain. No matter how many contribute, they will always fare worse than the free riders.

To sum up, it appears that I should not contribute. Yet if all think so, no one will get anything at all. Self-interest is self-defeating.

This social trap is very similar to the donation game. In fact, it *is* the donation game, if the number of players is reduced to two. With larger groups, however, the game offers new twists. Indeed, the simple procedure of reciprocation is of no help: if some of my players contribute, and some do not, then whom should I reciprocate with? If I do not contribute, then I harm contributors and non-contributors alike.

The mutual aid game, played round after round, is very similar to a mutual insurance scheme. The difference is that with insurance, I have to contribute regularly, but will only rarely be in need of help. In the mutual aid game, both giving and taking occur in each round.

Anthropologists have documented that mutual help groups can emerge spontaneously, without any incentive from authority. To give an example, lightning strikes a shed in Montenegro and kills all the sheep. The shepherd is faced with ruin, but the community saves him. Each of his neighbors gives him a sheep, and the shepherd is back in business. Of course, he will be expected to help in his turn, should an occasion arise.

Here is another example, dating back 200 years, to the time of Manchester Capitalism: Long before the first trade unions, workers spontaneously formed "sick clubs." They paid into a common account to help each other out in case of illness. (An aside: there seems to be no kind soul, whether in Montenegro or in Manchester, who multiplies the contributions by three before sharing them out. This multiplication factor, however, mirrors the fact that someone in need, like the thunderstruck shepherd or the ailing mill hand, profits from the gift far more than what the donor is losing.)

A minor variant of the mutual aid game is the common good game: I can decide, as before, whether to contribute my $5 or not. The sum of all contributions is multiplied by some factor, say three again, and constitutes the common (or public) good. It is shared among *all* players, irrespective of whether they contributed or not. (The difference with the mutual aid game is that in the common good game, I receive a share from my own contribution in return.)

The common good game is a stylized version of many such "games" in real life. Here are some examples: Young couples may decide to take turns escorting their toddlers to the playground. Some of the parents may be free riders who regularly manage to skip their turn. Our sturdy ancestors may have joined forces for a mammoth hunt. The free riders were those who followed the maxim of never being closest to the mammoth. If all were to act like this, the hunt would never succeed. The defense of a fortification requires cooperation. A free rider hiding behind the others benefits from a success just as much as the others, but puts all at risk. The cleaning of a communal kitchen is a less adventurous pastime; but it, too, offers plenty of scope for a social dilemma.

The tragedy of the commons is proverbial. The commons are grazing land belonging to the whole village. These grounds were frequently overexploited and thereby ruined: for if a farmer sends an extra piece of cattle to the commons, then its milk and meat benefit him alone, whereas the cost to the grassland is borne by all. Nowadays, there are not many commons left. The "common goods" include clean air, rich fishing grounds, and public transportation. They always offer scope for exploitation by free riders.

What happens in the game lab? Hundreds of experiments have studied mutual aid or common good games, in many variations. Frequently, players are not restricted to contributing either a full share or nothing, but can chose any amount between, say, $0 and $20. In the first round, the game is usually such as can be expected from the donation game. Some contribute more, some less, and the average amount will be about half of the full contribution. From then on, in round after round, the contributions decline almost invariably.

Are the players learning to be selfish? Are they imitating those who gain more, namely the free riders? Or are they simply fed up with being exploited?

This experiment has been repeated in many places (Copenhagen, Minsk, Samara, Chengdu, Riyadh, etc.). There is a considerable geographic variability, interesting for students of ethnic prejudice. Yet, the overall trend is clear. Contributions decline in round after round. The game offers less and less prospects for a gain. The social trap closes with a snap.

Retaliation and the War of All Against All

The remedy seems obvious. The free riders need to be punished. In economic experiments of the public good or mutual aid type, this can be achieved by a simple variation of the game.

Each round, now, consists of two phases. Phase one is the former game: players decide to contribute or not. Phase two offers an opportunity for the players to punish the exploiters in their group. The free riders are sanctioned: they have to pay a fine. This fine does not land on the accounts of the players who penalized the free riders. On the contrary, these players have to pay themselves a fee for imposing the punishment. Fees and fines are collected by the experimenters.

In the jargon of game labs, this type of sanctioning is named *peer punishment*: players impose penalties on the free riders, at a cost for themselves.

Indeed, punishing someone is usually expensive, in real life: it costs time and energy, and comes at a risk, since punished players are apt to retaliate, rather than meekly conform. Sanctioning is a costly business, as we learn frequently from the political news.

Despite these drawbacks, the effect of peer punishment is quite remarkable, as was shown in a much-touted game lab experiment run by Ernst Fehr and Simon Gächter. They had their subjects play, for the first six rounds, the usual public good game, without punishment. As a result, we see what we expect: in the first round, players invested on average some 50 percent of their game money in the common good. From then on, contributions declined, round after round.

Then, after six rounds, the players were offered the possibility to punish free riders. Immediately, contributions jumped up. They were larger than in the initial round—and this, even before the first punishment had been meted out. Better still, contributions increased in the following rounds (see Figure 12.4). In the end, almost everyone cooperated to the full, and hardly anyone needed to be punished.

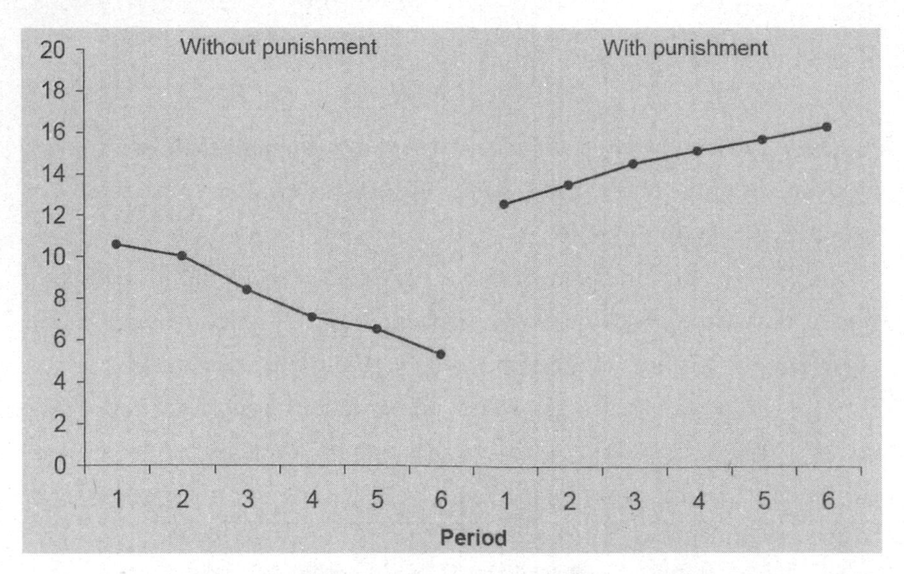

Figure 12.4. Contributions to the public good game without and with punishment.

This result is surprising. Profit maximization should lead to a second-order social dilemma. Indeed, the effect of peer punishment benefits all players, by increasing the average contributions. But the cost of punishing free riders is borne by the individual punisher. Why not simply contribute to the common good, and leave the task of punishing free riders to the other players? (This strategy can obviously arise only if more than two players are involved in the game.) Whoever acts in such a way is a second-order free rider. If all players adopt this option, there will be no punishment, and consequently the first-order free riders—those who do not contribute to the public good—will take over. Indeed, they have nothing to fear.

One possible solution of the conundrum would be to also allow for the punishment of the second-order free riders. This, however, makes third-order free riding possible, and raises the specter of an infinite regress.

Let us leave these theoretical objections aside, for the moment. It is well documented that in most experiments, many players are willing, and even eager, to engage in the costly punishment of free riders. Some do it with pleasure. One might suspect that they reckon with a long-term effect of their sanctions: they expect that free riders, once punished, will reform, and thereafter will sagely contribute in the following rounds. Yet, such an expectation cannot explain all. Some experiments are arranged in such a way that for each round the groups are newly formed. This ploy guarantees that players never meet their previous co-players. Players are informed of this. They may possibly reform free riders by punishing them, but know that they will never meet them again, and thus will never benefit from their own (costly) decision to punish. Yet, they punish, and do so with passion. In most of the usual game lab experiments, boredom prevails: but introduce punishment, and the interest quickens perceptibly.

Q and A sessions after games of public good with punishment show that the motivation to reform free riders by penalties comes only second, at best. The foremost impulse is simply revenge. Players are irked at being exploited by free riders and want to retaliate. The cost of punishment plays a minor role. Revenge seems to be a very natural drive. It is irrational to a high degree. Small children kick the door they banged against.

Vindictiveness is usually viewed as base and destructive. However, the experiments by Fehr and Gächter suggest that it plays a positive role in an economy. It is remarkable, by the way, that vengeance is mostly described in economic terms, and indeed even as bookkeeping. "*Wir rechnen noch ab!*" "It's payback time!" "*Il va me payer cher!*" "*Un règlement de compte.*" Similar idioms occur in many languages.

The need to retaliate is obviously a deep-seated drive. "Revenge is sweet," as they say. We enjoy revenge even vicariously, from second hand: countless films and novels deal with vengeance, and entertain the millions.

As any good Darwinist knows, pleasurable drives usually have some survival value. This makes one ask how we profit from vengefulness.

The most probable reason is that if it becomes known that we are prone to retaliate, others will think twice about slighting us. Just as with indirect reciprocity, reputation plays a key role in punishment. Anger and indignation are loud: they broadcast something. The lowliest gangster demands respect. You cannot treat me that way. I will not take this from you.

We are entering a problem zone here. In the early lab games on public good with punishment, one player sanctions another, and this is it. *Basta.* The punished player takes it literally sitting down. Such a situation is completely unnatural. The punished players are unlikely to meekly accept the sanction. They want to hit back. As soon as the rules allow for penalized players to retaliate, costly vendettas spring up, even in the anonymous, almost clinical environment of the game lab, where "to punish" means merely to reduce the modest sums of money on the players' accounts. In real life, the spiral of destruction can be murderous.

As the philosopher John Locke noted in his *Two Treatises of Government* from 1689: "Such resistance [to punishment] many times makes the punishment dangerous, and frequently destructive, for those who attempt it."

Modern experimental economists were highly surprised, and some even scandalized, when they observed *asocial punishment*: exploiters punish cooperators, in retaliation or even as a preemptive measure, to intimidate them. Such reactions can lead indeed to a war of everyone against everyone. Let us listen to what Thomas Hobbes has to say:

Figure 12.5. John Locke (1632–1704) feared passion.

> [So that] in the nature of man, we find three principal causes of quarrel. First, competition; secondly, diffidence; thirdly, glory. The first makes man invade for gain; the second, for safety; and the third, for reputation.

What Hobbes names "glory" is the wish to be respected. The "diffidence" is fear, which leads to aggression and preemptive strikes to anticipate the enemy. Competition, finally, is based on selfishness. We find all these causes of quarrel in the game theory model. Selfishness undermines cooperation; fear leads to attack; the wish for respect prevents any yielding. Hobbes stresses that "war" does not consist only of actual battle, but of being ready for it. The mere preparedness is ruinous.

Hobbes continued:

> In such condition there is no place for industry, because the fruit thereof is uncertain, and consequently no culture on earth; no navigation... no arts; no letters; no society; and, which is worst of all, continual fear, and danger of violent death; and the life of man, solitary, poor, nasty, brutish and short.

And short! What a blessing that the torture of such a life lasts only for a limited time.

To take the law into our own hands, as in peer punishment, means anarchy. On Earth, we find it only in those corners that authority cannot reach: in the legend-crusted Wild West, among nomads, in ill-run jails—or in the fiction that baroque philosophers agreed to call the "state of nature."

To quote Locke and *Two Treatises of Government* again:

> For everyone in that state [of nature] being both judge and executioner of the law of nature, man being partial to themselves, passion and revenge is very apt to carry them too far, and with too much heat, in their own cases.

Onward with Locke:

> It is this makes them so willingly give up everyone his single power of punishing, to be exercised by such alone, as shall be appointed among them; and by such rules as the community, or those authorized by them to that purpose, shall agree on.

Here is the social contract. Players submit to an authority (a sheriff, a lord, a police force). This step too can be mimicked by a stylized game that is a variation on the mutual aid game.

Each round consists of three stages. In the first stage, players may contribute to a punishment pool, or not; in the second stage, players may contribute to the mutual aid funds (the common good), or not; in the third stage, finally, the free riders—namely those who failed to contribute to the punishment pool or the mutual aid funds—are punished. The punishment is all the more severe the better the punishment pool is endowed.

This game was first introduced by the Japanese psychologist Toshio Yamagishi. It works well, unsurprisingly enough: most players cooperate. The punishment pool is the equivalent of a police force. The better that force is equipped, the more likely free riders will be spotted. Players are required to pay for the police, and thus cover the cost of punishment up front, before the mutual aid game is even played, and hence *before* it is known whether there will be any free riders to be punished.

This so-called pool punishment has considerable advantages, compared with peer punishment. It is more objective and less personal, thus making retaliation less likely. Moreover, it allows the spotting and punishment of the second-order free riders (those who contribute to the common good, but not to the punishing). However, pool punishment has a serious drawback. If all players cooperate, round after round, the police have nothing to do. Nevertheless, it must be paid; and such a tax reduces the economic advantage of the mutual aid. By contrast, the costs of peer punishment arise only when needed. Moreover, it requires communication and coordination to establish a punishment pool, whereas peer punishment needs no more than a lone, vengeful soul.

The Importance of Hunting Hares

Evolutionary game theory allows us to study cooperation and the social contract in mathematical models, which are simple thought experiments. Let us consider fictitious populations of players who can opt between various strategies. From time to time, a randomly chosen sample of players engages in a game. Players accumulate more or less payoff; the amount depends on their strategy, and on what the other members in their sample are doing. Occasionally, the players can adapt, by switching to another strategy, preferentially one that is doing better. Players interact only within their current sample, but they can imitate anyone in the community. Thus, the toy population evolves by social learning—a myopic, payoff-driven adaptation.

If the game is a mutual aid game, pure and simple, without punishment of any kind, then cooperation is doomed. It is just as much doomed as in the two-player version, which is the donation game. Free riders always do better. They are imitated preferentially, and eventually make up the whole population.

The plot line changes in a surprising way if the players from the sample are offered to play the mutual aid game, but left free to pull out. They are not obliged to take part. They can decline, stand aside, and do something else instead, some activity whose payoff does not depend on others. Philosophy buffs will recognize that extra option as nothing else than hunting hares, in Rousseau's parable.

With this third alternative, we have defined the so-called voluntary mutual aid game. The players selected in the random sample have three strategies at their disposal:

1. Don't participate.
2. Participate and contribute.
3. Participate, but don't contribute.

If you have chosen the third option, you are exploiting those who chose the second—the contributors. This is something that the nonparticipants—those who go for the first option—don't do: they rely only on themselves. We assume that the payoff, for the nonparticipants, lies somewhere between the payoff obtained in the mutual aid game if all participants contribute, and the payoff if no one contributes, which is zero. A technical point must be added: a single would-be participant volunteering for the mutual aid game cannot play all by himself or herself, but must hunt hare. Mutual aid needs several participants, who each decide, independently, whether to contribute or not.

The three strategies in the voluntary mutual aid game cyclically supersede each other, in a way reminiscent of the Rock-Paper-Scissors game. As known by children across the world, Rock beats Scissors, Scissors beats Paper, and Paper beats Rock.

In the same cyclic vein, a population of nonparticipants (1) will be invaded by contributors, who (2) will be overcome by defectors, who (3) will yield to nonparticipants (1) in their turn (Figure 12.6).

Indeed, if enough players are willing to participate and to contribute, they do well. More and more hare hunters will imitate them. They participate and contribute whenever they are offered an opportunity. Once enough

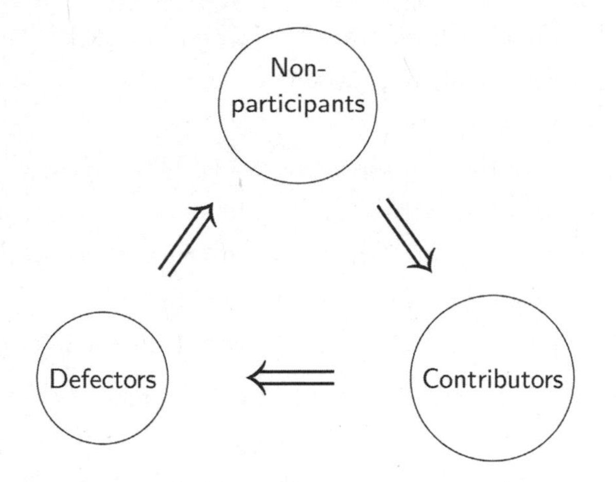

Figure 12.6. A Rock-Paper-Scissors cycle.

of them are around, the free riders cash in on the suckers. With each additional free rider, however, participation becomes less alluring, for good and bad alike. In the end, those who don't participate will do better. Nobody will want to play the mutual aid game any longer. This standstill persists until, by chance, a handful of players are sampled who want to participate and to contribute. Cooperation booms right away; but it takes only a short while until free riders spread. And so it goes on: long periods when nobody wants to play, interspersed with bursts of contribution, which quickly turn out to be bubbles because free riders undermine the game. Whenever one strategy dominates in the population, another is set to take over.

The voluntary mutual aid game can be viewed as a blend of the two best-known social dilemmas: the Stag Hunt and the Prisoner's Dilemma. Put together, they weaken the springs of the social trap. Admittedly, long-term cooperation is not achieved. Yet, short-term bursts of cooperation recur.

Things improve even more when a punishment option is introduced into the game. Which form of punishment? As may be expected, peer punishment proves less stable than pool punishment, since it can be subverted by second-order free riders. But if the game is voluntary, free riders can always be overcome in the end. (See Figure 12.7.) This is in striking contrast to the compulsory game, where the free rider trap cannot be escaped. The voluntary aspect of the commitment (players participate because they hope to do better) is not a polite bow to democratic feelings: it is an essential ingredient of the strategic ploy.

To sum up, social learning leads spontaneously to a social contract. Players are not required to be rational, to communicate, or to coordinate. Myopic imitation of the currently successful behavior is enough to lead players to voluntarily commit themselves to cooperation. They freely bind themselves: "Mutual coercion, mutually agreed upon," to quote the famous recipe proposed by ecologist Garrett Hardin for overcoming the tragedy of the commons.

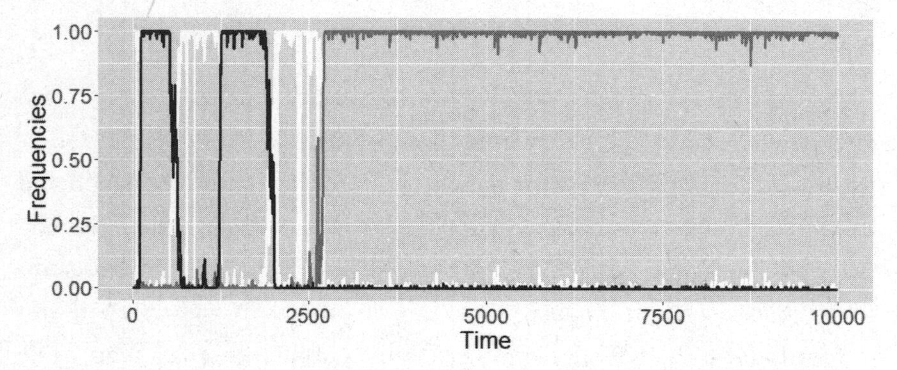

Figure 12.7. A computer simulation of social learning (the imitation of successful strategies) in an artificial population. The black line indicates the frequency of peer punishers; the gray line, the frequency of pool punishers, who eventually take over. A cooperative regime is only possible if the players have the option to abstain from the joint enterprise. (These "hare hunters" are shown in white.) If this option is not available, then the social trap closes: social learning leads quickly to the unbroken domination of free riders.

So much for the toy societies of evolutionary game theory. They are mere models, "myth mixed with math." What about real life? In almost every part of the civilized world, some authority reigns. Anarchy is a pipe dream. A few philosophers, like Pierre-Joseph Proudhon and Peter Kropotkin, tried their best, but all in vain. Sometimes, however, the law-enforcing authority is far away. In such cases, it can happen quite spontaneously that a sheriff is hired—or a referee, a janitor, a security team, someone to keep order.

History tells us that even the outcasts need a social contract. Apparently, the mafia emerged, one and a half centuries ago, in a prison in Naples, as a task force to suppress quarrels between inmates.

Anthropologist Elinor Ostrom made it her life's work to document how, all over the world, so-called small-scale societies of herders, fishers, nomads, and hunter-gatherers spontaneously create simple institutions to enforce rules for cooperation, fair sharing of fishing grounds, regular irrigation of fields, and sustainable forest management. Ostrom received a Nobel Prize for Economics for her work.

In Ostrom's view, "institutions are tools to set up incentives for overcoming social dilemmas." Such incentives can consist in rewards, but far more often, they are penalties. Taken in this sense, the social contract is not a founding myth propagated by philosophers, but can arise bottom-up, on a small scale, even today.

Figure 12.8. Elinor Ostrom (1933–2012).

13

Fairness

To Hold and to Share

Fair Sharing

A hoary joke dating back to the bygone days of the German Democratic Republic has Chairman Erich Honecker and the almighty head of the Soviet Union, Leonid Brezhnev, taking a relaxed stroll through the woods. They suddenly stop and stare: a heap of gold coins lies at their feet. They pick it up and count. A hundred gold coins! It seems incredible. And far and wide no one to watch them. Their eyes meet. "Let us share brotherly," says Brezhnev. "Oh no!" exclaims Honecker. "Oh no, no, no! Fifty-fifty!"

Fair sharing is a thorny problem, as brothers must have known since Cain and Abel. Sharing offers the first opportunity for kids to school their sense of fairness. Five-year-old children learn already how to fairly share a piece of cake with their friend. Ann splits the cake, then Burt chooses a slice. Since Ann can be sure that Burt will leave her the smaller part, she must make sure that this part is as large as possible—namely one half. Ann is maximizing her minimal payoff.

Ann grows up and graduates with a major in economics. Before moving to another state, she wants to sell her car, but not for less than $3000. Burt is interested: he has been looking for a used car and is willing to pay up to $5000. A bargain can be struck. But at what price? Ann would prefer $5000,

Burt $3000. The task is reduced to splitting the $2000 that make up the difference.

Is there a fair solution? Fifty-fifty, as in kindergarten days? Does it still make sense to maximize the smaller part of the split? In real life, of course, the two players will not state their true price limits. But this is not the point here. The real question is whether it makes sense at all to speak of a fair bargaining price. For a long time, economists didn't think so.

"Fairness" is even more problematic than "truth." Nobody contests that it is an excellent thing. But so much can be meant by it. What is a fair action? A fair auction? A fair punishment? A fair society? A fair draw?

Philosophers are well aware of that multiplicity of contexts, and some of the more analytic brand are cautious enough to look at simple examples first, before turning to the question of what a fair world should look like. Let us start with an example that has been used by the Canadian philosopher David Gauthier in his 1986 book *Morals by Agreement*. It is about Ann and Burt, who want to reach a contract for their mutual benefit. Ann owns $20,000 and Burt $80,000. They want to invest their money in a bank.

Usually banks offer a constant rate of interest for savings, let us say (optimistically) 3 percent: an interest rate that is independent of how much is invested. In contrast to this, Gauthier's bank offers a rate of interest that increases with the investment: namely 3 percent on your first dollar, 4 percent if your investment is $10,000 or more, 5 percent if it is $20,000 or more, and so on. For every $10,000 the interest rate grows by 1 percent.

If Ann invests her $20,000, she receives an interest rate of 5 percent, which nets her $1000 by the end of the year. If Burt invests his $80,000, he profits from a higher interest rate, namely 11 percent: in one year, he gains $8800. If both players put their savings together and jointly invest $100,000, they benefit from an interest rate of 13 percent. This investment scheme is obviously good for both of them. Ann's $20,000, invested at 13 percent, earns her $2600, a nice jump up from the previous gain of $1000. As for Burt, he now invests his $80,000 at 13 percent and thus gains $10,400 (instead of the $8800 from before).

End of the exercise. No need to think any further, except if one is philosophically minded.

Indeed, we have just seen that both players profit equally from jointly investing: each by an extra gain of \$1600. They profit by the same amount, although Burt has invested four times more than Ann into the joint account. Ann increases her interest rate by 8 percent, Burt by 2 percent. Is this fair? Shouldn't Burt deserve more?

Let us denote Ann's payoff (her gain from the investment) with u and Burt's with v (Figure 13.1). If they invest their savings separately, the pair of their payoffs is $(u_0, v_0) = (1000, 8800)$. If they invest jointly, they will jointly gain \$13,000. A solution of the bargaining problem of Ann and Burt is therefore a pair (u, v) satisfying $u + v = 13{,}000$. In addition, we must have $u \geq 1000$ and $v \geq 8800$, since none of them would be willing to gain less by adopting the new scheme. Thus, Burt can ask, at best, for \$12,000, and Ann at best for \$4200. The sum of these claims exceeds the \$13,000 to be divided. Obviously, there must be some concessions from the maximal demands.

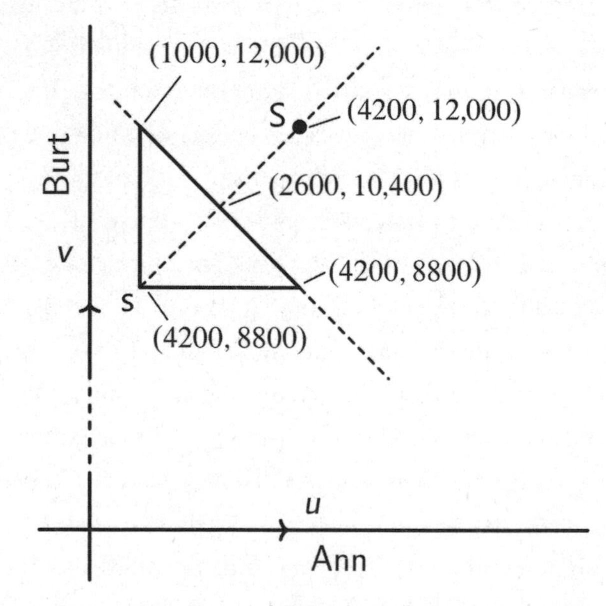

Figure 13.1. Ann and Burt share their winnings in Gauthier's investment example.

If Ann and Burt agree on some solution (u, v), Ann has conceded $4200 - u$ and Burt $12{,}000 - v$. Their concessions are equal if u is halfway between 4200 and 1000, meaning $u = 2600$ (and consequently, $v = 10{,}400$).

These are the same numbers as before: we now see that they correspond indeed to a fairness principle: The two partners concede equally much.

We could also view it in the following way: the least Ann can obtain from the scheme is $1000, for she could get this sum all by herself; the most she can demand is $4200, for otherwise Burt would not be interested in jointly investing. Indeed, the least Burt can obtain from the scheme is $8800. By the same token, he cannot ask for more than $12,000, for otherwise Ann would not be interested in a joint investment. The minimal demands correspond to the point $s = (1000, 8800)$ and the maximal demands to $S = (4200, 12{,}000)$. The latter pair is not feasible, since only $13,000 exists to be shared. The two points s and S define a segment in (u, v)-space, and the bargaining solution proposed by Gauthier, namely (2600, 10,400), is obtained by gliding downward, along the segment from S to s, until one arrives at a feasible sum. It is as if the players would, step by step and hand in hand, reduce their aspirations until their wishes can be realized.

The procedure resembles haggling in a bazaar. Is there a principle lurking behind it? Compared with their maximal demands, Ann and Burt concede equally much. They both have the same maximal amount to concede, namely $3200 each. Gauthier claims that if the maximal concession of Ann had been twice as large as that of Burt, Ann would have had to concede twice as much. Ann and Burt should give up the same fraction of their maximal demand. In other words, compared with their maximal demands, the *relative concessions* of Ann and Burt should be equal.

Is this what justice demands?

Nash Strikes a Bargain

Gauthier's little exercise belongs to bargaining theory. This theory was created by twenty-two-year-old John Nash in 1950, the same year he developed his concept of a game theoretic equilibrium. Before Nash, the prevailing view was that asking for a "just" solution of a bargain had no meaningful answer. The outcome could be anything compatible with the minimal demands of the bargainers; in practice, it was deemed to be the outcome of a psychological tug-of-war beyond calculation.

Nash reformulated the question. Two players, called Ann and Burt as before, have the opportunity to enter a binding agreement. (Such a commitment, incidentally, presupposes a social contract: there must be an authority able to enforce the binding agreement.) By this contract, they can agree on a payoff pair (u, v) (payoff u for Ann and v for Burt). The set of all possible (or feasible) payoff pairs is a subset M of the plane. We assume that this subset M is bounded (no one can ask for the sky), and that it is convex: for any two points P and Q in the feasible set M, the segment joining these endpoints also belongs to M. Indeed, the two players can agree on a lottery (outcome P with probability p and Q with probability $1 - p$) and thus reach any point on the segment PQ. As usual, the payoffs are expectation values, and they are measured by their utilities.

Each point in the feasible set M can be realized, if Ann and Burt both want it. But it is conceivable, of course, that players do not agree on a solution and break off negotiations. In this case, each of them obtains a payoff that is independent of what the other player is doing. The corresponding point is denoted as the *status quo point* $s = (u_0, v_0)$. Players will obviously not agree to a bargaining solution that yields them less than what they can obtain by the status quo. This is similar to the ploy used in the voluntary mutual aid game: players can withdraw from the negotiation and see what they can do by themselves: we recognize the hare hunting option.

Whatever payoff values u and v will emerge from the bargaining, we can be sure that $u \geq u_0$ and $v \geq v_0$. Therefore, the set M of feasible solutions is in the positive quadrant whose bottom-left corner is at s.

The stage is prepared. What should one request from a fair solution? First and foremost, no doubt, that it is symmetric. If the two players exchange their roles, which means that the two components of the payoff pair are permuted, this should not affect the bargaining solution, except that the two payoff values of the solution are likewise permuted. In other words, if we mirror the bargaining problem by replacing Ann with Burt, and Burt with Ann, that is, by reflecting both the bargaining set M and the status quo point s on the diagonal $u = v$, we should obtain the mirror image of the solution.

We may imagine that a referee works out the fair solution, and only then learns who is Ann and who is Burt. In the words of philosopher John Rawls, whose book *A Theory of Justice* from 1971 achieved an extraordinary impact on political philosophy, the referee reaches the decision behind a "veil of ignorance."

Veil of ignorance: this expression, which is not unpoetic, has become a capsule into which the whole philosophy of John Rawls is squeezed. Such compression is a professional risk for successful thinkers. To mention Schopenhauer is to evoke "the world as will and representation." Leibniz stands for the best of all possible worlds, Darwin for the struggle for survival, Einstein for the relativity of all and everything, and Heidegger for the nothing that nothings. Such knee-jerk associations often serve as a quick and ready veil to hide ignorance. This, however, is not what Rawls meant by his "veil of ignorance."

Back to the bargaining solution of Nash, in addition to symmetry, Nash required that the solution is independent of the scale used to measure the utilities of the two players. Thus, we can transform the scale of the utility function u in an affine linear way, and do the same with that of v, without changing the outcome. It should have no more repercussion than the use of Celsius instead of Fahrenheit would have on a heat wave.

The third condition required by Nash is Pareto optimality. Indeed, an outcome that leaves one player the option to improve his or her lot without thereby harming the co-player is hardly a reasonable solution. No utility should be left unused. The solution, therefore, must be on the *Pareto boundary* of the bargaining set M, which is that part of the boundary where any step in the upper-right direction leaves the set of feasible payoff pairs.

The fourth and last condition is independence of irrelevant alternatives. If we replace the feasible set M by a proper subset N, we are clearly faced with a new bargaining problem. But if N contains the old bargaining solution, that should be the new bargaining solution, too. It is as with a horse race: the horse that finishes first against a large field of contenders should also finish first if some rivals are disqualified.

Symmetry, invariance of scale, optimality, and independence of irrelevant alternatives—these are very general conditions required from a bargaining solution. Surprisingly, they suffice to determine it exactly: the solution is the unique pair (u, v) in M that maximizes the product $(u - u_0)(v - v_0)$.

We can imagine the Nash solution at the upper-right corner of a rectangle whose sides are parallel to the coordinate axes and whose lower-left corner is the status quo point s. Then, this rectangle (with sides parallel to the coordinate axes) has the largest possible area. All parallel-sided rectangles whose diagonal leads from s to another point in the bargaining set M have smaller areas.

Nash discovered his solution for the bargaining problem almost at the same time as Ken Arrow discovered his dictator theorem, shortly after 1950. In both cases, it had been possible to derive a far-reaching result of philosophical impact from very simple mathematical assumptions. In both cases, the independence from irrelevant alternatives plays a key role.

Independence of irrelevant alternatives is the condition that is most strongly contested. Let us consider (Figure 13.2) the two bargaining sets M and N, both with the status quo point $s = (0, 0)$. The triangle M is spanned by (0, 0), (100, 0), and (0, 100). The trapezoid N is spanned by the four corner points (0, 0), (100, 0), (50, 50), and (0, 50). For both sets, the Nash solution is (50, 50). For M this looks certainly like a fair solution. But what about N? Ann seems to have been served a raw deal by the Nash solution. Indeed, Burt receives his maximal feasible payoff, and has to make no concession at all, whereas Ann concedes half of her maximal feasible payoff.

An even more glaring injustice seems to be going on in Figure 13.3. Here, M is spanned by (0, 0), (200, 0), (200, 140), and (0, 200), and N is spanned by (0, 0), (200, 0), (150, 150), and (0, 200). The Nash solution is (150, 150) for case N and (200, 140) for case M. That the solution is fair in case N can hardly be disputed; but in case M, Burt must feel cheated. His payoff is reduced, despite the fact that the feasible set M is larger. For each given payoff

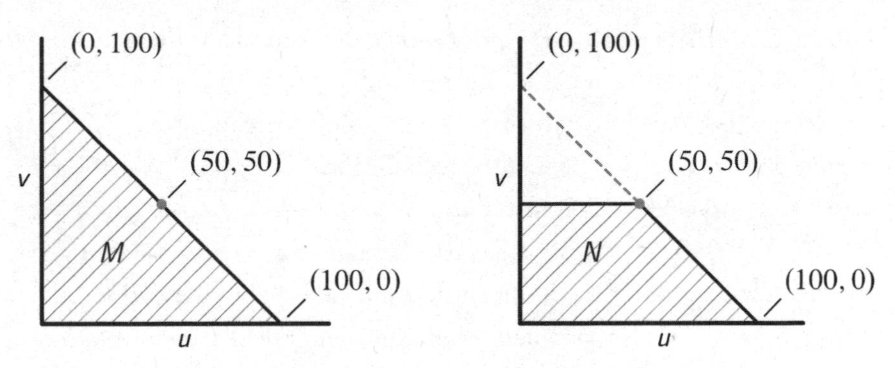

Figure 13.2. The Nash solution (50, 50) looks fair for the bargaining set *M*, but not for *N*.

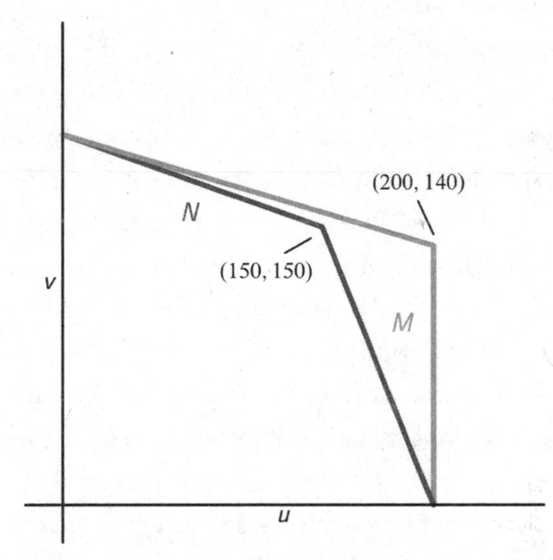

Figure 13.3. Two Nash solutions: Burt loses from an enlarged bargaining set.

value for Ann, Burt could obtain a higher payoff for *M* than for *N*. Yet, his Nash share becomes smaller.

This is why the two Israeli mathematicians Ehud Kalai and Meir Smorodinsky proposed another solution, which is based on the idea of the maximal feasible payoff. The independence of irrelevant alternatives is replaced by another, no less obvious principle: whenever the status quo point and the maximal feasible payoffs remain unchanged, but the bargaining set *N* is contained in M, then the larger set M should not lead to a worse bargaining outcome, neither for Ann nor for Burt.

If this is postulated, then there is again a unique bargaining solution. It is obtained by considering the smallest rectangle with sides parallel to the axis, with the bottom-left corner at the status quo s and the upper-right corner $S = (U, V)$ given by the maximal payoff values for each of the two players. Usually, that point S will lie beyond the Pareto boundary. The Kalai–Smorodinsky solution is obtained by gliding along the diagonal from S to s until the feasible set is reached. In this case, both players have conceded an equally large fraction of their maximal feasible payoffs (namely $U - u_0$ for Ann and $V - v_0$ for Burt).

If we return to Gauthier's example, we see that the philosopher follows the guiding principle of Kalai and Smorodinsky: equal concessions. In this example, however, the Nash solution leads to the same outcome. This also holds for the triangular bargaining set M from Figure 13.2. For the bargaining set N, however, the solutions differ: Nash yields (50, 50) and Kalai and Smorodinsky (200/3, 100/3). In Figure 13.3, finally, the approach by Kalai and Smorodinsky yields for M the approximate solution (154, 154), which leaves Burt better off than with the Nash outcome.

With that, we have not yet properly settled what the status quo actually should be. In our examples, the answer was obvious. If there is no agreement, no player receives any piece of the cake, no car will be sold, and Ann and Burt are reduced to the payoffs that they can guarantee for themselves.

But what about the game described by the following matrix, where C and D are merely two alternatives open for the row player Ann and the column player Burt (the letters, in this example, have nothing to do with "cooperate" and "defect"):

	C	D
C	(1, 2)	(9, 3)
D	(6, 6)	(2, 1)

If Ann opts for C and Burt for D, then Ann receives a payoff of \$9 and Burt receives \$3, etc. It is easy to see that (C, D) and (D, C) are equilibrium pairs. In both cases each player has found the best reply to the other player's move. But Ann will certainly prefer equilibrium (*C*, *D*) and Burt equilibrium (*D*, *C*). Which solution should the two players agree upon? This is a bargaining situation, if we assume that the two players can conclude a binding agreement. The Pareto optimal set is the segment from (6, 6) to (9, 3). This is where the solution should be found.

But where is the status quo? One might say that if the two players cannot reach an agreement, both will earn nothing. This would mean that (0, 0) is the bargaining point. But no strategy pair in our game can lead to this outcome. The players have only the choice between *C* and *D*. What should they consider as status quo?

It is sometimes suggested to take the maximin solution as status quo. This is how both players can guarantee a minimal payoff for themselves. In other words, each player opts for whatever strategy maximizes his or her payoff under the condition that the worst happens. A bit of calculation shows that Ann can be sure of payoff $\frac{58}{13}$ by playing *C* with probability $\frac{4}{13}$ and *D* with probability $\frac{9}{13}$. This is the best expectation value Ann can secure for herself. Similarly, the best Burt can guarantee for himself is $\frac{8}{3}$ (by playing *C* with probability $\frac{1}{3}$). The status quo point would accordingly be the payoff pair $\left(\frac{58}{13}, \frac{8}{3}\right)$, and from there, one can obtain the corresponding bargaining solutions, be they Nash or Kalai–Smorodinsky. But why should the maximin solution be the right status quo? It is not even an equilibrium. If Ann knows that Burt will opt for *C* with probability $\frac{1}{3}$, then she would play *C* with probability 1!

Moreover, to count on the worst possible outcome is something for pessimists. Many of us are more upbeat. And is all that computing worth the while? How much simpler if both players could agree to let a coin toss decide between (6, 6) and (9, 3); or even simpler, to settle on (6, 6), which is a fair split of the cake presented on a silver plate? And finally, since when can players be sure that they know the payoff of their co-player? It is in utilities, not money. Who would be dumb enough to let their own payoff be known, let alone to trust what the other player says?

An Ultimatum

Some goods are easier to divide than others. A pint of beer can be shared with a friend; sharing a cloak is more difficult. Yet a Roman officer named Martinus became famous for cutting his cloak in half and sharing it with a pauper. It seems a far-fetched measure, but resulted in his promotion to sainthood. Today's economists would have advised using a lottery. Saint Martin could have tossed a coin to decide whether to give away the whole cloak, or to keep it.

Money can be easily divided, whether it is a sum of 100 gold coins or 100 US cents, as in the notorious Divide the Dollar game: two players Ann and Burt are offered a dollar if they agree on how to share it. They must submit their bids separately and independently: they are not allowed to communicate beforehand.

The players are asked to write down on a slip of paper how many cents they demand for themselves. If the sum does not exceed 100 cents, then they get what they wanted. Otherwise, they don't receive anything. The obvious solution is, of course, for both of them to ask for 50 cents—and this is indeed the usual outcome in experiments. But it is far from being the only Nash equilibrium pair: indeed, such a pair is obtained by every two demands summing up to 100 cents. If Ann asks for 40 cents and Burt for 60, none can improve their payoff by deviating unilaterally. If they ask for less, they receive less; if they ask for more, they receive even less, namely nothing.

If the two players know each other, they may engage in a guessing game. But if they don't—if they are separated by a veil of ignorance—then the players have a good reason to opt for fifty-fifty, since it is the only symmetric Nash equilibrium pair.

However, even a small modification of the symmetry assumption makes the outcome much less obvious. This is the so-called ultimatum game, which has been studied in hundreds of experiments all over the world.

The scene is a game lab. The cast consists, as so often, of two players who do not know each other and know that they will never meet again. First, the experimenter decides, by throwing a coin, who of the two players will make an offer, and who will reply; the former player is said to be the Proposer, the latter the Responder. In this pointedly fair way, the roles are distributed.

The experimenter explains the rules, maybe via computer screen (anything to make the interaction seem less personal). The two players will have the opportunity to share $10 between them. The Proposer will have to suggest how to split the sum: an offer of so much for you, and the rest for me. If the Responder accepts, this is how the sum will be divided. In that case, the game is over, and the players go their separate ways. Otherwise, the Responder rejects the offer. This, too, marks the end of the game: the experimenter pockets the $10 and neither of the two players receives anything. No second offer. No haggling. No negotiations. The main difference from Divide the Dollar is that the second player decides in full knowledge of the first player's decision. The symmetry between the two players is broken.

Obviously, the status quo point is (0, 0), and each split of the $10 among the two players is a Pareto optimal point of the bargaining set, as well as a Nash equilibrium. Not that there is much bargaining going on. One offer, and one decision—a yes or a no. That's why the game got its name: an ultimatum.

Responders who want to maximize their income are bound to accept every positive offer, small as it may be. One dollar is better than nothing. Therefore, Proposers who want to maximize their income should make a minimal offer, knowing that it will be accepted, and keep the rest for themselves. The prediction is clear. Yet in reality, such a minimal offer is hardly ever proposed, and almost never accepted. An overwhelming majority of Proposers make "fair" offers, namely $5 or at least $4. Offers of $2 or $1 are rarely made and almost always rejected out of hand.

But what is "fair" in this context? This is where John Rawls's "veil of ignorance" reappears. An offer is fair if the two players would agree on it before even knowing who will propose and who will respond. If they are risk-averse, they will maximize the smaller share of the $10, since they may well end up with it. Fair players divide fifty-fifty. Should we assume that players take the trouble to imagine how things would have looked to them before the veil of ignorance was lifted—before they were apprised that they would act as the Proposer or the Responder? This seems unlikely.

The ultimatum experiment has been repeated in many places (in Tokyo, Ljubljana, Chicago, Zürich, . . .), with results that agreed more or less: the fair solution is far more frequent than the rational solution, which is the minimal offer.

In due time, game theorists became tired of always testing the same kind of people: mostly jean-clad undergrads provided by departments of economics. They got some anthropologists interested, who took to playing the ultimatum game in small-scale societies such as the Machiguenga (hunter-gatherers from the Amazon Basin), the Hadza (nomads in Tanzania), or the Lamalera (fishers in Indonesia). At this stage some substantial cultural differences showed up. The hunters and gatherers are the least fair of all: they offer on average about 25 percent of the sum (little, by our standards, but still far from the minimal offer). The Lamalera, by contrast, offer more than 50 percent (and oddly, often refuse such offers—apparently due to their complex gift-exchanging traditions). The "fairest" (in our sense), the closest to the fifty-fifty norm, are groups living in modern Western cities such as Los Angeles and Chicago. Some economists think that societies become fairer as they become more familiar with the concept of a free market and traditional habits of haggling.

Many variations of the ultimatum game have been tested. For instance, if the Proposer is determined not by a flip of the coin, but by better performance in a game of skill, then low offers are more frequent and get accepted more easily. The inequality in the two roles is felt to be justified. If Responders are told that the offer is made by a computer, they are willing to accept much smaller offers. They are not felt as a personal slight. If a Proposer is told that five Responders will compete for the offer, that offer will be much smaller, and will be taken up nevertheless. By contrast, in pairwise encounters, self-esteem insists on a fair share.

One conceivable explanation of the fifty-fifty norm is maladaptation. The ultimatum game is usually played under conditions of strict anonymity, but many of us are prone to suspect, more or less subconsciously, that we are being watched. This inkling may simply be the remnant of a formerly useful trait: in small tribal and rural societies, such as those our ancestors hung around, for thousands of generations, everyone knew almost everything about everyone else. Thus, we may find it difficult to imagine that our decisions remain secret for long. We act as if chances are good that the other members of our community will learn what we are up to.

Once it becomes known that I have accepted a low offer, I have given myself away, and must expect that I will very likely be faced with low offers in the future. It is better to reject a small offer, even if it comes at a cost to myself, and thereby acquire the reputation of someone who insists on a fair share.

Once more, reputation is the key. There are obvious parallels to the mutual aid game with peer punishment, and to indirect reciprocity. If I reject an unfair offer, I am punishing the Proposer. I do so at a cost, but this cost is small—it amounts to the low share that has been offered to me. The rejection costs a lot more to the Proposer. In addition, I will have proved that I am not to be trifled with. As a matter of fact, I have proved it only to myself: the rules of the game lab, as explained by the experimenter, make sure that no one is watching. But this does not preclude that I feel scrutinized.

This explanation can be tested with a variant of the ultimatum game, by means of a mathematical model. Let us imagine a fictitious population of players engaging in an ultimatum game for round after round, never against the same co-player twice. Players are sometimes in the role of the Proposer, sometimes in the role of the Responder.

All players are characterized by their strategy, which consists of two numbers: namely the percentage p that the player is willing to offer, as Proposer, and the aspiration level q on which the player insists as Responder. Let us furthermore assume that the players can learn from each other, by adopting the strategy of a co-player who does better. Once again, this is social learning in action. We start from any random distribution of p and q values, and allow, from time to time, a new strategy—a new pair (p, q)—to enter, as a minority, into the population.

What the computer simulation in Figure 13.4 shows is that the value of the aspiration level q drops down—low offers will be accepted more easily. In response, the size of the offer p drops to 0. And then, the evolution by

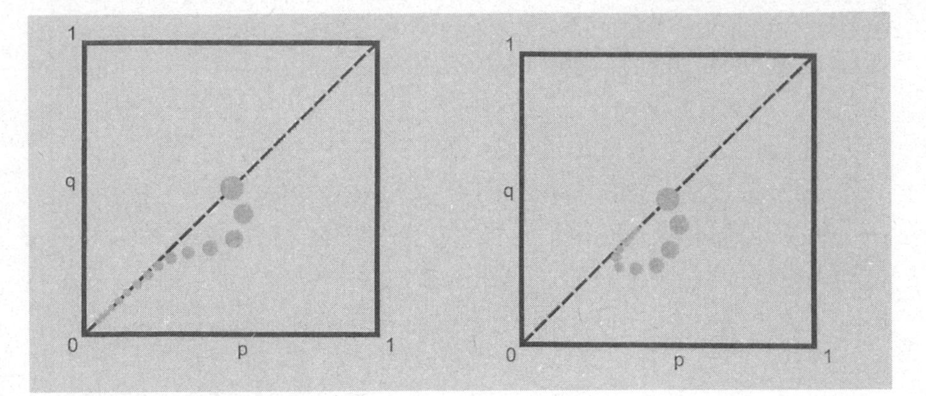

Figure 13.4. The ultimatum game without (left) and with (right) reputation. The population starts from the large circle and evolves along the decreasing circles, in one case to unfairness and in the other to fairness.

social learning freezes. The values p and q are very close to zero. The fictious population learns to behave like *homo economicus*, that cardboard figure exclusively guided by maximizing personal monetary interest.

In a second treatment, we repeat the whole show, but this time under the assumption that players know what their co-players have played in the previous round, and are opportunistic enough to make a lower offer if they know that the co-player has accepted such an offer before.

At first, in this second version, social learning proceeds just as before: the aspiration level q nosedives, and the mean offer p follows suit, until both values are rather small. But in contrast to the previous treatment, social learning has not reached its end yet. Slowly, very slowly, the average values of p and q start to grow and creep along the diagonal into the region between 40 and 50 percent. The population has learned how to behave fairly.

It is obviously inconceivable that our fictitious players (or more precisely, their behavioral programs) can form any ideas about justice as fairness, or the like. They are simply imitating behavior that proves successful, and are aware of what their co-players have been doing. Just as with indirect reciprocity, reputation leads to behavior that appears to be good for the society: in one case by being helpful, in the other case by being fair.

Fight and Convention

If the players don't agree on a solution to their conflict of interest, they have to fall back to their status quo payoff—whatever that is. At worst, it means to fight. In civilized societies, this is frowned upon. But a few centuries ago, dueling and feuding was all in. In a "state of nature," or in a schoolyard or in a prison cell, fists may fly. Egos get broken. Noses, too.

In a less crude way, the same issue arises frequently in daily life: whether to escalate a conflict or not. This can be described by another little game. In this scenario, the two players have to choose between sharing a sum, which means conceding some of it, or fighting it out. For the benefit of the squeamish, we will only consider a thought experiment.

Let us denote the first alternative by *C* and the second by *D* and assume that the object at issue has a value of \$10. Whoever loses an all-out fight loses \$100, say (the cost of a broken nose). We assume that both players are equally strong, and thus equally likely to win the fight. If both players opt for *C*, they share the \$10 fifty-fifty. If one player opts for *D* and the other for *C*, the latter quits with payoff zero and the other ends up with the whole sum of \$10. If both players choose *D*, the conflict escalates. One player ends up with the \$10 and the other with the doctor's bill, and thus the expected value for their payoff is $\frac{10 - 100}{2} = -45$. This yields the following payoff matrix:

	C	*D*
C	(5, 5)	(0, 10)
D	(10, 0)	(–45, –45)

Both strategy pairs (*C*, *D*) and (*D*, *C*) are Nash equilibrium pairs. This is an anti-coordination game. Each player should avoid making the same choice as the other player. But, who is the one and who is the other? The initial situation is completely symmetric, yet the two outcomes are not.

On closer inspection it turns out that there is one more equilibrium pair, and this one is symmetric: both players can play *D* with the same probability, namely 10 percent. This comes at some risk. With a probability of 1 percent (10 percent of 10 percent), both players play *D*. The conflict, in that case, will escalate and lead to a costly fight. In 18 percent of the conflicts, however, one player turns tail and leaves the field to the other player, the one who played bully: brazenness triumphs in that case. Yet, the good news is that with a likelihood of 81 percent, the conflict will be bloodlessly resolved by a coin toss.

Interestingly, this "solution" that reduces escalation comes up, almost by itself, in evolutionary game theory. Let us once more imagine a fictious population of players, each one equipped with a strategy—this time, a mixed strategy. It is defined by the propensity x of the player to escalate. Players with various x values encounter each other at random, play the game, and obtain their payoffs. From time to time, a player can adopt the strategy of a player who is doing better; and on very rare occasions, a player tries out a new value of x. It is easy to guess what the computer simulation will show. If most players play C, it is better to play D: the opponent will almost certainly run off. The D strategy is better, and will therefore spread. However, when the likelihood to encounter a D player becomes too large, it is worth playing C and avoiding a bloody outcome. Social learning leads to an equilibrium; the probability of meeting a co-player prepared to escalate will converge to 10 percent, the value given by the Nash equilibrium. The population learns to avoid conflict—but not completely.

This little game has a history. Students of animal behavior had long been fascinated by conflicts within species (about food, or females, for instance), and puzzled by the fact that many of these species had evolved ways to resolve such conflicts by so-called ritual fighting. They roar frightfully, their crests swell, their fangs are bared—but very often, one of the animals gives up before blood is shed. Wolves, for instance, instinctively refrain from biting other wolves; similarly, crows do not hack at each other's eyes. It is very rare that a stag rams his antlers into the unprotected flank of a rival and wounds him seriously.

For many years, this widespread tendency for ritual fighting was explained by the fact that it obviously serves the good of the species. However, this is bad thinking. Inherited traits spread when they provide their bearers with more offspring. Natural selection is based, first and foremost, on individual reproductive success, not on the good of the species.

Stags who do not obey the rules of ritual fighting can kill their rivals, or drive them away. They multiply faster. But their offspring inherit their dad's propensity to escalate fights. By spreading within the stag population, they will encounter their like more and more often. Yet, their strategy is not the right response against itself: it is self-defeating. This is how their frequency is held in check: the good of the species is a collateral bonus, but not the reason for the prevalence of ritual fighting.

The British theoretical biologist John Maynard Smith was the first to use this type of argument, and to lay the foundation of evolutionary game theory. Individuals inherit some strategy (in this case, a propensity x to escalate). The payoff measures the increase in reproductive success. Higher payoff means more offspring, who inherit the type of their parent. Thus, frequencies regulate themselves.

Figure 13.5. John Maynard Smith (1920–2004).

It was soon found out that evolutionary games are a very flexible instrument. They worked for stags, beetles, insects, plants, and even bacteria! For the first time, game theory was rid of the assumption that players are rational and know that the others are rational, too (a "common rationality" assumption that even for humans is rarely justified). The strategies of evolutionary game theory are simply behavioral programs. Successful strategies spread, either because they are imitated more often or because they are inherited more often.

The first theoretical prediction of evolutionary game theory was that conflicts escalate less easily in species that are armed more heavily. This was brilliantly confirmed. The more deadly the weaponry of a species (whether teeth, claws, or horns), the more frequent the inhibition against using it within the species. On the other hand, weakly armed species (for instance, the proverbially peaceful dove) have no inhibition at all to put each other to death; they merely are unable to do it, under normal conditions, because the weaker dove can escape. If doves are kept in a cage, they can torture each other to death. Doves are not adapted to being at each other's mercy. Wolves are.

The basic message for moral philosophy is that social norms can develop even in the absence of rationality. Konrad Lorenz, one of the founders of the field of animal behavior (and an adherent of the flawed "good of the species" thinking) described such social norms as "moral-like behavior."

Interestingly, as soon as we allow for asymmetries between the contestants, very different norms can evolve. Indeed, let us assume that the players can be in different roles: strong or weak, young or old, male or female, resident or intruder. In such cases, conditional strategies can evolve, for instance: "If you are the stronger, escalate; if you are the weaker, run away." This is clearly a Nash equilibrium strategy: it does not pay for the stronger to deviate from it, and it pays even less for the weaker.

There just remains a small point: namely how to decide who is the stronger and who the weaker. A large part of the behavior of animals in inner-species conflicts can be viewed as a series of tests to find out just this. Rival stags, for instance, ogle each other, do a lot of roaring, and run for a while close to each other on parallel course, almost shoulder to shoulder—all this helps them learn about their rival. Such rituals can last for a long time. They usually end with one of the contestants deciding to concede and give up. On the rare occasions that this is not the case, the last resort is a direct trial of strength, a pushing match with interlocked antlers: again, this does not lead to fatal injuries, in general.

Game theory has been used to describe contests between owners of a territory and intruders. The conditional strategy "if owner, escalate to the hilt; if intruder, concede" leads, once again, to a Nash equilibrium. It does not pay for the owner to deviate from it. It does not pay for the intruder either: any deviation can lead to death or serious injury—and the price isn't worth it.

This conditional strategy, which John Maynard Smith facetiously named Bourgeois, has been observed in many species, be they birds, mammals, insects, or fishes. Even butterflies display territorial behavior, when "territory" is nothing but a small sunny spot in a forest. Whichever butterfly is the first to discover such a spot will act as if it owns the place, and chase away all would-be intruders. But if the biologist craftily covers the resident butterfly with a dry leaf, for example, then the next butterfly alighting nearby will, within a few minutes, feel at home. If the leaf, then, is removed, both butterflies seem to think that the place is theirs, and try to repel the other. This leads usually to escalated contests: protracted spiral flights that cost both butterflies a lot of energy, somewhat like a war of attrition. Similar experiments on humans can be conducted on the sun terraces of a holiday resort. Guests going for a swim, or for a drink at the bar, tend to leave a towel on their deck chair to indicate that it is theirs. Remove the towel behind their back, and wait for the next guest to occupy the deck chair. When the previous holder returns from the pool, this may lead to dark looks, and occasionally even to an escalated conflict.

The Bourgeois strategy describes a very familiar human trait. Most of us are ready to defend our property and to respect that of others. Where does this social norm come from? Many thinkers have dealt with this question.

Jean-Jacques Rousseau ascribed to the notion of property a truly crucial role in history:

> The first man who, having enclosed a piece of ground, bethought himself of saying "This is mine," and found people simple enough to believe him, was the real founder of civil society. From how many crimes, wars and murders, from how many horrors and misfortunes might not any one have saved mankind, by pulling up the stakes, or filling up the ditch, and crying to his fellows: "Beware of listening to this impostor."

In the eyes of Karl Marx, private property was the main tool for dominating others. It had to be destroyed. "Expropriate the expropriators!" was the conclusion of his reasoning. His anarchistic predecessor Pierre-Joseph Proudhon blew into the same horn: "Property is theft!" Both philosophers rebelled against the seemingly unbeatable phalanx of all those who held ownership rights as natural and obvious, or at least as useful, if not downright sacred.

A well-known slogan states that "property is nine-tenths of the law." It is not merely that property and ownership are secured by law and custom, but by a web of social regulations. Economic experiments have documented a deep-seated endowment effect on the individual level. An object becomes more valuable in our eyes when we own it. Normally, I would not be ready to pay $200 for an opera ticket. But if I got it as a gift, I would not be willing to sell it for $200, either. This is a common tendency. Richard Thaler received the Nobel Prize for Economics for discovering it. Before him, the endowment effect was more or less taken for granted. We prize our property, in part because it is valuable, no doubt, but in part also because it is *ours*. The sparrow in my hand is better than the dove on the roof. The trait must have to do with loss aversion: the widespread propensity to count losses as higher than gains.

Small children grasp at anything. The instinct to hold tight may be a leftover from the tree-dwelling past of our ancestors. The writer Elias Canetti sees in that instinct the germ cell of grab-and-hold capitalism. Be that as it may, kids have no problem grasping the notion of "mine." Moreover, many animal species seem to have a sense of ownership. This holds particularly for those species that own a territory, a nest, or a cache. In this respect, Rousseau is wrong when he thinks that property is the invention of some scoundrel. It is quite conceivable that a tendency to claim something as "one's own," and to defend it, is not a mere cultural trait but has a biological basis.

This leads us back to evolutionary games. We have seen that in an owner-intruder game, the Bourgeois strategy entails a Nash equilibrium. It fits with our social norms. Curiously enough, there is another equilibrium, its mirror image, which is also stable: escalate if you are a challenger, concede if you are the owner. It serves just as well to avoid escalated conflicts. Moreover, it can even be supported by an appeal to fairness. You have had time to enjoy your property, now it is another's turn.

John Maynard Smith named this the Proudhon strategy. In real human societies, the Proudhon strategy seems never to have emerged, except fleetingly. There are many attempts to explain its glaring lack of success—none generally accepted. Obviously, evolutionary game theory still has some way to go.

To apply mathematical reasoning to ethics is an old dream. Baruch Spinoza, for instance, wrote a treatise on ethics entitled *Ethica, Ordine Geometrico Demonstrata* (or *Ethics, Demonstrated in Geometrical Order*). Up to a few decades ago, it was taken for granted that such a mathematical underpinning of ethics must be based on rationality—for isn't reason the domain of mathematics? Evolutionary game theory, however, is mostly based on a naturalistic view of ethics as an anthropological or even biological phenomenon. This perspective threatens to loosen the grip of philosophy on ethics, and to hand it over to the sciences. Such a science-based perspective—ethics as a branch of psychology—is hardly new, but it is surprising that so much of its recent progress makes use of mathematical tools that do not even mention rationality.

In the early years of game theory, it was frequently discussed whether it was a normative or a descriptive theory—whether it applies to what agents should do, or to what they actually do. Game theory, however, is a branch of mathematics. As such it is as little normative or descriptive as, for instance, algebra. It merely helps to explore the consequences of this or that assumption. That players are rational is one such assumption.

The rationality doctrine leads to many consequences that are nowhere to be found in real-life interactions. One telling example is backward induction. If two players know that they will play a donation game for six rounds, the outcome of the last round seems a foregone conclusion: both will defect, since this is their dominant strategy—it is better no matter what the co-player does. Nothing in round five will affect this outcome. To all intents and purposes, it is as if round five were the last round. But then, its outcome is also a foregone conclusion: both players defect. And on it goes, step by step back to round one. Backward induction dictates that rational players will never cooperate, and remain stuck with payoff zero. In experiments, however, this outcome is rare. Players will mostly cooperate, with the possible exception of the last or next-to-last rounds. In such a game, it is simply too silly to be rational.

The prominent role of rationality is due to a force of habit alone. Some 200 years ago, mathematicians lost their creed in a unique set of geometrical axioms. Why should anyone expect a unique set of axioms for game theory?

PART IV

14

Language

Speaking in Ciphers

A Double Bill

In the spring term of 1939, the University of Cambridge offered two distinct courses on "Foundations of Mathematics"—an intellectual extravagance of sorts. However, there was little danger of waste, even less of redundancy: the two lecturers, Ludwig Wittgenstein and Alan Turing, were each known to go their own ways.

Figures 14.1 and 14.2. Ludwig Wittgenstein (1889–1951) and Alan Turing (1912–1954) on parallel courses.

Ludwig Wittgenstein was turning fifty, an expat professor of philosophy in Cambridge. Alan Turing had not yet reached his thirties. He was a fellow at King's College who now held his first lecture course, for the modest fee of 20 British pounds. His topic was the "foundations" in the classical sense, as understood by modern mathematicians: meaning axioms and logic. In the wake of Hilbert and Gödel, Turing had electrified the field with his seminal paper on computation and the decision problem. Here was a worthy successor to the Cambridge trio of Russell, Whitehead, and Ramsey, who had done so much of the groundwork.

Wittgenstein went after a different game. He had no truck with the usual spiel that mathematicians dished out to each other about the foundations of their science, perhaps even fooling themselves to believe it. Wittgenstein was from Vienna, and as little inclined to take words at their face value as Sigmund Freud or Karl Kraus. Wittgenstein wanted to know what mathematicians *really do*.

Everything in Wittgenstein's biography is spectacular. His father was a steel baron and patron of the arts, Habsburg's answer to Andrew Carnegie. Ludwig grew up in a palace, as the youngest of eight. He started out in aeronautics at a time when the conquest of the air took wing. In 1912, however, the uncommonly intense young engineer switched tracks and enrolled in the University of Cambridge for philosophy. His teachers were Bertrand Russell and George W. Moore, the heralds of analytical philosophy. Within a few months, they took down his dictations on logic. Shortly after, he retired to an isolated hut in Norway to pursue his thoughts undisturbed.

When World War I started, Wittgenstein volunteered for the Austrian Army. In between spells at the front, he finished his *Logical-Philosophical Treatise*, coolly stating in the preface that he considered its truth to be "unassailable and definitive." After the catastrophic defeat of the Central Powers and a one-year spell in an Italian POW camp near Monte Cassino, he returned to forlorn, destitute Vienna, and donated his vast inheritance to his surviving siblings (three of his brothers had taken their lives). He earned his living as a teacher, at elementary schools located in the backwoods of Lower Austria. His booklet, renamed *Tractatus Logico-Philosophicus*, appeared after agonizing delays.

Wittgenstein was done with philosophy. Hadn't he solved the problems, in their essentials? Refusing to engage with busybodies, he snubbed the persistent attempts of the Vienna Circle, that avant-garde group of philosophers and mathematicians, to get close to him and soak up his words.

As a teacher, Wittgenstein was highly motivated, but prone to fits of classroom rage: slapping his pupils, or pulling their hair, their ears, whatever came to hand. His career was brought to an abrupt end when he knocked out an eleven-year-old. A chastened Wittgenstein returned to Vienna after six years of school service. Next, he named himself an architect and directed the construction of a modernistic town house for his sister. And at last, having finished with the workers and craftsmen, he condescended to meetings with select members of the Vienna Circle. Some turned out to be worth talking to. It gradually transpired that there was still something left to do in philosophy.

In 1929, now aged forty, Wittgenstein took the train back to Cambridge and submitted his *Tractatus* for a PhD. Some time before, he had claimed that nobody can do philosophy for more than ten years. Nearer to the truth is that nobody can do *without* philosophy for more than ten years. All through the 1930s, Wittgenstein wrote and discussed tirelessly, in Cambridge, in Vienna, or back in his old Norwegian hut, but he published nothing. This did not stop the University of Cambridge from appointing him to a chair in philosophy. They knew a legend when they met one.

Only a handful of select disciples were allowed to see Professor Wittgenstein. The rest were kept at bay. "In certain circles," wrote Ernest Nagel, a young philosopher from the United States, "the existence of Wittgenstein is debated with as much ingenuity as the historicity of Christ has been disputed in others." Those who wished to attend his lectures had to undergo an interview with Wittgenstein. Nagel was turned down: Wittgenstein said that he wanted no tourists. Turing, however, was accepted: Wittgenstein could make use of a mathematician who was unafraid to come out of his corner and take it. This is how Turing came to stand answer to Wittgenstein for all that had gone wrong with the foundations of mathematics, be it set theory, formal systems, or metamathematics.

Here is a sample of their exchanges:

> Wittgenstein asked Turing: "How many numerals have you learned to write down?"
>
> Turing, sensing what was to come, replied cagily: "Well, if I was not here, I would say countably infinite!"
>
> Wittgenstein: "How wonderful—to learn infinitely many numerals, and in so short a time!" And with Turing still so young!
>
> Turing conceded: "I see your point!"
>
> Wittgenstein: "I have no point!"

And so it went on.

Like all the other students, Turing had had to promise beforehand never to skip any of Wittgenstein's classes (of which there were two per week). On March 19, 1939, however, Turing excused himself. Wittgenstein was nettled, acidly remarking:

> Unfortunately Turing will be away from the next lecture, and therefore that lecture will have to be somewhat parenthetical. For it is no good my getting the rest to agree to something that Turing would not agree to.

Turing took it without flinching. He knew how to keep mum. For some time already, he had been earmarked by the British Secret Service. Sometimes he had to leave Cambridge to follow ultra-secret courses on cryptanalysis. Everyone knew that war loomed around the corner, and MI6 was worried about Germany's military cipher. Turing's former PhD advisor Max Newman had brought up his name. It proved a brilliant hunch. The two of them would eventually devise some of the first proto-computers, rattling dinosaurs of machinery, to decode German top-secret messages. (See Chapter 5.)

By 1939, however, Alan Turing had only conceived a purely hypothetical computer to investigate the limits of formal systems. The abstruse automaton would play an important role in his lectures on foundations. In the wake of Gödel's demonstration of undecidable mathematical propositions, these last ten years had seen breathtaking progress.

Forever the contrarian, Wittgenstein saw things in a completely different light. He explicitly stated: "My task is not to speak on Gödel's proofs etc. but to speak *past* them." These last ten years, for him, had been taken up with establishing the philosophy of language.

Wittgenstein's guiding rule was: "The meaning of a word is its use in the language" (though only "for a large class of cases"). To examine that use more closely, he had devised the method of language games, "to bring into prominence the fact that the speaking of language is part of an activity, of a form of life." His task, as a philosopher of mathematics, was to describe these games, not to explain them. Games have rules, and players need not always be aware of them. Wittgenstein wanted to uncover them, patiently, one by one. He did not share the view that there lurked a seamless entity behind what is called "mathematics." Instead, he spoke of the "colorful medley" of mathematics. Just as astronomy deals with a wide variety of phenomena (planets, radio waves, galaxies, dark matter) that have little in common except being out there in the sky, so mathematics cannot be reduced to a single object or a single method. It is a motley.

To follow in Wittgenstein's wake, we may imagine landing, like ethnographers, on some unknown shore of Archipelago Mathematics, and observe how the natives—the mathematicians—communicate with each other. This may help us grasp the rules by which they live, their "form of life" (to use one of Wittgenstein's favorite terms).

Speaking Mathese

Each science has its language. So has mathematics, evidently. In addition, mathematics *is* a language. This is a widely accepted view. Here are quotes from two of the foremost mathematicians of our time, Yuri Manin and Alain Connes. The former says: "The basis of all human civilization is language, and mathematics is a special form of linguistic activity." The latter goes even further: "Mathematics is unquestionably the unique universal language."

Ever since Galilei, physicists have taken this view for granted: "The universe is written in the language of mathematics: the letters of this language are triangles, circles and other mathematical figures."

Goethe agreed (in his fashion): "Mathematicians are a kind of Frenchmen: whenever you talk to them, they translate it into their language and right away it is something quite different."

Languages are systems of signs. They serve to communicate—that is, to transmit information by means of symbols. Languages have their vocabulary (signs and words). They have syntax (the rules for combining these signs, usually in a linear array) and semantics (the meaning of the text). Last but not least, they need the biotope of a community using them.

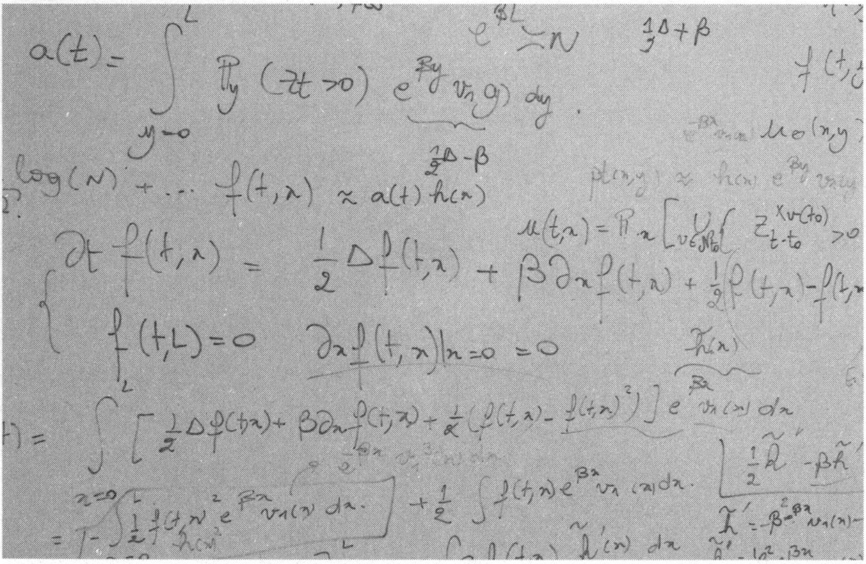

Figure 14.3. Do not erase!

To start with the vocabulary, the first point to note is that most communication, in mathematics, is in writing. Mathematicians use the spoken word, of course, for instance when lecturing to students or conversing with colleagues. It never takes long, however, before written signs are resorted to, be it on a blackboard in some seminar room, or on a paper napkin in a restaurant, or on the marble top of a coffeehouse table.

Mathematical symbols are continuously evolving. The triangles and circles mentioned by Galilei showed up in Euclid's *Elements* already. Complicated geometric figures, graphs, and diagrams play a major role in many mathematical texts.

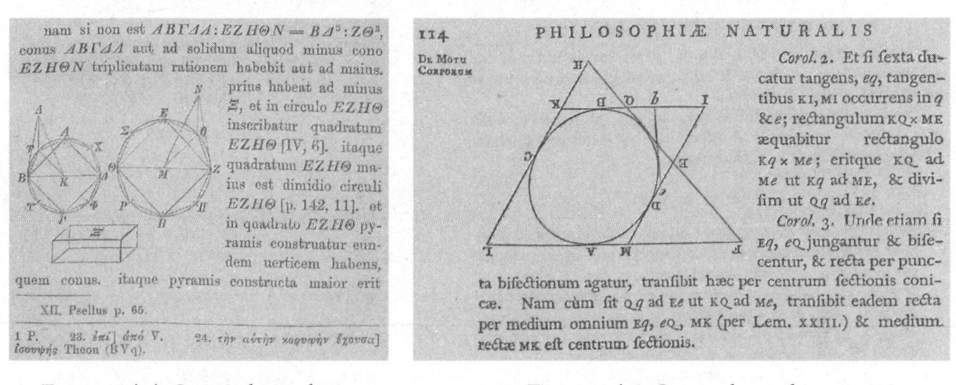

nam si non est $AB\Gamma\Delta\Lambda : EZH\Theta N = B\Delta^3 : Z\Theta^3$, conus $AB\Gamma\Delta\Lambda$ aut ad solidum aliquod minus cono $EZH\Theta N$ triplicatam rationem habebit aut ad maius. prius habeat ad minus Ξ, et in circulo $EZH\Theta$ inscribatur quadratum $EZH\Theta$ [IV, 6]. itaque quadratum $EZH\Theta$ maius est dimidio circuli $EZH\Theta$ [p. 142, 11]. et in quadrato $EZH\Theta$ pyramis construatur eundem uerticem habens, quem conus. itaque pyramis constructa maior erit

XII. Psellus p. 65.

1 P. 23. ἐπί] ἀπό V. 24. τὴν αὐτὴν κορυφὴν ἔχουσα] ἰσοϋψής Theon (BVq).

114 PHILOSOPHIÆ NATURALIS

De Motu Corporum

Corol. 2. Et ſi ſexta ducatur tangens, *eq*, tangentibus KI, MI occurrens in *q* & *e*; rectangulum KQ × ME æquabitur rectangulo K*q* × M*e*; eritque KQ ad M*e* ut K*q* ad ME, & diviſim ut Q*q* ad E*e*.

Corol. 3. Unde etiam ſi E*q*, *e*Q jungantur & biſecentur, & recta per puncta biſectionum agatur, tranſibit hæc per centrum ſectionis conicæ. Nam cùm ſit Q*q* ad E*e* ut KQ ad M*e*, tranſibit eadem recta per medium omnium E*q*, *e*Q, MK (per Lem. XXIII.) & medium rectæ MK eſt centrum ſectionis.

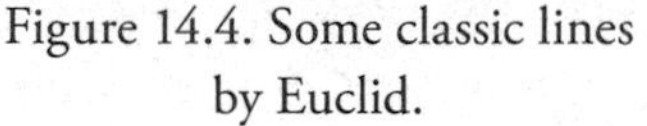

Figure 14.4. Some classic lines by Euclid.

Figure 14.5. Some classic lines by Newton.

From early cuneiform writing onward, numerals display a stunning variety of shapes. Increasingly, however, even the most exotic alphabets have been invaded by Arabic numerals.

Children are trained for years in the proper use of the ten digits. As a former teacher in elementary schools, Wittgenstein was well-versed with the repetitive nature of this "drill," as he used to call it:

> Suppose the pupil now writes the series 0 to 9 to our satisfaction. And this will only be the case when he is often successful, not if he does it right once in a hundred attempts. Now I continue the series and draw his attention to the recurrence of the first series in the units; and then to its recurrence in the tens. . . . And now at some point he continues the series independently—or he does not. But why do you say that? So much is obvious! Now, however, let us suppose that after some efforts on the teacher's part he continues the series correctly, that is, as we do

it. So now we can say he has mastered the system. But how far need he to continue the series for us to have a right to say that?

Wittgenstein's father, the steel tycoon, had his children brought up by private tutors. For this reason, Wittgenstein had never been a first-grader himself—he only taught them. This made him appreciate all the more the stupendous amount of knowledge absorbed by kids. Heinz von Foerster, a pioneer of cybernetics, related a charming anecdote. When he was ten, he was asked by Wittgenstein what he meant to do as an adult. "I will be a researcher," said Heinz. "Oho!" said Wittgenstein. "You have to know a lot to become a researcher." "But I *do* know a lot," insisted Heinz. "Yes," said Wittgenstein earnestly, "but what you do *not* know is how right you are."

Children become habituated to mathematics during their early years. After computing with numbers, they start computing with letters. This technique was unknown to the ancient Greeks (who were particularly handicapped, in this respect, because they used the letters α, β, γ, . . . for the numerals 1, 2, 3, . . .). Nowadays, most mathematics books are filled with formulas. They mean the same everywhere, and are always written from left to right, even where normal writing is from right to left or from top to bottom.

In the beginning, the signs used in the formulas entered mathematics at a snail's pace. The zero was introduced into the Western world in the High Middle Ages only, via India and the Islamic world. For the decimal system, or any kind of place value system, the zero is essential. The plus sign showed up in the time of the Gothic cathedrals. Bishop Oresme may have been one of the first to use it, for his harmonic series that does *not* sum up. The minus sign came into use only a hundred years later. Next came signs for the square root and parentheses. Only in 1557, at about the beginning of the Elizabethan Age, did the equal sign appear.

It was devised by one Robert Recorde, who decided to use (in his own words) "a pair of parallels, or twin lines of the same length, because no two things can be more equal." The argument is odd, but Recorde's design has certainly proved successful. It seems inconceivable today to do mathematics without a sign for equality. Not having anything like it, the Greeks had to write: "The surface area of a sphere is four times as large as the area of the greatest circle contained within that sphere." We learn that $A = 4\pi r^2$.

In some of the least defensible statements of his *Tractatus*, Wittgenstein claimed that "the essential of the mathematical method is working with equations" (6.2341), and that moreover "the propositions of mathematics are equations" (6.2). This may with some goodwill hold for algebra, but certainly not for analysis, which crucially relies on inequalities (which are not negations of equalities!). Moreover, little is gained in translating a proposition such as "to each prime there is a larger one" into an equation, say

$$\text{cardinality(Primes)} = \text{cardinality(Natural Numbers)}.$$

Not long after the equality sign, the letter x was introduced as a symbol for the unknown. It proved to be another inspired choice, looking like the sign pirates use to mark the spot where treasure is hidden.

To quote *How to Solve It* by George Polya:

> *The use of signs appears to be indispensable to the use of reason.* . . . Mathematical notation appears as a sort of language, *une langue bien faite*, a language well adapted to its purpose, concise and precise, with rules which, unlike the rules of ordinary grammar, suffer no exception. If we accept this viewpoint, SETTING UP EQUATIONS appears as a sort of translation, translation from ordinary language into the language of mathematical symbols. (Emphasis and capitals from Polya.)

From 1600 on, mathematical signs proliferated:

$$\times, \pm, \div, >, \infty, \neq, \propto, \partial, !, \approx, \%, \Delta, \nabla, \equiv, \int dx, \{\,\}, \aleph, \mathfrak{R}, \gg, \vee, \in, \ldots.$$

Contemporary mathematicians are not only conditioned to think in terms of formulas (which can be viewed as a code), but they can print them out using TeX, an exquisite program requiring another code. TeX was offered by Donald Knuth as a gift to the mathematical community. He aimed for a

```
\begin{align*}
q(x) &= (\Lambda \ast p)(x) = \int _{\bR } p(x-t) \Lambda (t) \, dt  =\int
_{\bR } \Big(\sum _{j=0}^n \frac{(-t)^j} {j!} p^{(j)}(x) \Big) \Lambda (t) \, dt \\
&= \sum _{j=0 }^n  \frac{(-1)^j} {j!} \mu _j \, p^{(j)}(x) = F(D)p(x)  \, .
\end{align*}
```

$$
\begin{aligned}
q(x) &= (\Lambda * p)(x) = \int_{\mathbb{R}} p(x-t)\Lambda(t)\, dt = \int_{\mathbb{R}} \Big(\sum_{j=0}^{n} \frac{(-t)^j}{j!} p^{(j)}(x)\Big)\Lambda(t)\, dt \\
&= \sum_{j=0}^{n} \frac{(-1)^j}{j!} \mu_j\, p^{(j)}(x) = F(D)p(x)\, .
\end{aligned}
$$

Figure 14.6. A formula written in TeX (above), and how it appears in print (below).

design "that would be usable in a hundred years." No modest aim, for sure. Yet, the first forty years are over, and TeX shows certainly no signs of faltering.

Moreover, mathematicians are using software such as Mathematica or Maple that allows not only numerical but also symbolic computation, with built-in packages for plotting functions and figures, manipulating them, programming algorithms, and doing statistics. Such software is fantastically efficient in speeding up routine tasks. It turns the heuristics of mathematics (which is, and always has been, based on experimenting with lots of examples) into a form of recreation, almost. Handling such programs requires, again, a translation from one language to another, from a coded command to a formula or a figure.

However, not every mathematician works with formulas and figures. Moreover, a mathematical text is not a sequence of formulas. It needs syntax. Some first-year students are slow to accept this. In their examination papers, they list one formula beneath the other, without comment, as if the strings of symbols have fallen from the sky. When told that something is missing, they are capable of pulling out their smartphone and showing a photo of the blackboard, which documents that their professor had written this selfsame sequence of formulas. Indeed, lecturers normally do not write down all they say; however, what they say is what it takes to proceed from one formula to the next. Without this text, the lines are gibberish.

A typical example for the syntactic mayhem likely to be found in college examination papers is

$$x^2 + 2x + 5 = 2x + 2 = -1 = 4 = \text{Minimum!}$$

This student had the right thing in mind. When faced with the task of finding the minimum of the quadratic function $x^2 + 2x + 5$, the first step is to take its derivative, namely $2x + 2$; the second step is to note that it is zero for $x = -1$; the third is to compute the corresponding value of the quadratic function, which is 4. This yields indeed the minimum. Replacing these steps by equal signs garbles the structure of the argument, its syntax.

A mathematical text appears most often as a mixture of formulas and plaintext. A run-of-the-mill example can look like the next example (never mind the content):

Proposition: *If $f: X \to Y$ is continuous and $U \subseteq Y$ open, then $f^{-1}(U)$ is an open subset of X.*

Proof: Let x be an element of $f^{-1}(U)$. Then $f(x) \in U$. Therefore, since U is open, there exists $\eta > 0$ such that $u \in U$ whenever $d(f(x),u) < \eta$. We would like to find $\delta > 0$ such that $y \in f^{-1}(U)$ whenever $d(x,y) < \delta$. But $y \in f^{-1}(U)$ if and only if $f(y) \in U$. We know that $f(y) \in U$ whenever $d(f(x),f(y)) < \eta$. Since f is continuous, there exists $\theta > 0$ such that $d(f(x),f(y)) < \eta$ whenever $d(x,y) < \theta$. Therefore, setting $\vartheta = \theta$, we are done.

The style is minimalistic: *then, therefore, if and only if, but, whenever.* Most other words, for instance *continuous* and *element,* belong to a mathematical lexicon. The words of this lexicon are sometimes newly hatched, for example *ergodic* and *homological.* Others, such as *field* and *ideal,* are borrowed from common language and provided with a different meaning. The word *open* in the previous text is an example. Students quickly learn that a set can be both open and closed, or neither one nor the other. They also have to get used to other particularities of the mathematical jargon; after a time, they accept without demur that "all present kings of France are bald." Indeed, there is no king of France, at present, and therefore no counterexample. They also learn that "*A* or *B*" does not preclude that *A* and *B* both hold; or that "*A* implies *B*" is not quite the same as in the colloquial sense. Indeed, common mortals use it only when *A* has something to do with *B*, whereas mathematicians use it to stand for "*A* is false or *B* is true." Thus, "blue is red implies $1 < 2$" holds true, and so does "$1 > 2$ implies that blue is red." This use of implication, which shuns all causal connection, was familiar to the stoics already, whose school of logic rivaled that of Aristotle for a while. It gave rise to a dispute between Diodorus and Philo, won by Philo. It was his convention concerning "*A* implies *B*" that survived in the long run.

Colloquial "mathese" of this sort can cause problems at first, but students quickly get the hang of it. Thus, "if and only if" is not a particularly emphatic "if" but states a necessary and sufficient condition. "It is obvious that..." means "you have to work this out for yourself"; and "it can easily be seen" announces that it will cost you an hour to verify the details. Most nerve-racking, to outsiders, is the addictive use of the byword *trivial*.

Talking to Machines

Mathese is a naturally evolving language suited to mutual understanding between mathematicians. Almost orthogonal to its purpose is the construction of a language to speak with machines. Mathese is meant to communicate with the like-minded, whereas *machinese* addresses the no-minded. It is a rather exclusive language, as of yet. Most mathematicians have other priorities than to master it. But a dedicated minority of scientists, fascinated in equal parts by the foundations of mathematics and the art of programming, labor hard at letting computers have the last word, so to speak.

In their view, mathematics has undergone three major revolutions. The first was the introduction of proof: due to Thales, probably, and since Euclid's time a sine qua non for mathematics. The second was the introduction of rigor in proof: due to Cauchy and Weierstrass, who set a new standard in the nineteenth century. The third was the introduction of a thorough and complete formalization. This third revolution is currently underway. It aims to display all logical steps so explicitly that they can be checked mechanically, not to say mindlessly.

It takes arduous work to make a traditional proof digestible for a computer proof assistant. All assumptions must be made explicit; no "abuse of notation" is ever tolerated; nothing can be left to be read "between the lines." Whether "2" means a natural or a real number must be spelled out. Each alternative case must be treated in repetitious detail without ever appealing to analogy with previous cases. The factor by which such an expanded version of the proof is longer than the original proof is called the *de Bruijn factor*. It is more or less constant across various mathematical fields, roughly in the order of four to one (which seems surprisingly small). Typically, it takes one expert one week to translate one page of "normal" mathese into a computer script.

```
let GAUSS_LEMMA_SYM = prove
 (`!p q r s. prime p /\ prime q /\ coprime(p,q) /\
           2 * r + 1 = p /\ 2 * s + 1 = q
           ==> (q is_quadratic_residue (mod p) <=>
                EVEN(CARD {x,y | x IN 1..r /\ y IN 1..s /\
                                 q * x < p * y /\ p * y <= q * x + r}))`,
  ONCE_REWRITE_TAC[COPRIME_SYM] THEN REPEAT STRIP_TAC THEN
  MP_TAC(SPECL [`q:num`; `p:num`; `r:num`] GAUSS_LEMMA) THEN
  ASM_SIMP_TAC[] THEN DISCH_THEN(K ALL_TAC) THEN AP_TERM_TAC THEN
  MATCH_MP_TAC EQ_TRANS THEN EXISTS_TAC
   `CARD {x,y | x IN 1..r /\ y IN 1..s /\
              y = (q * x) DIV p + 1 /\ r < (q * x) MOD p}` THEN
  CONJ_TAC THENL
   [CONV_TAC SYM_CONV THEN MATCH_MP_TAC CARD_SUBCROSS_DETERMINATE THEN
    REWRITE_TAC[FINITE_NUMSEG; IN_NUMSEG; ARITH_RULE `1 <= x + 1`] THEN
    X_GEN_TAC `x:num` THEN STRIP_TAC THEN
    SUBGOAL_THEN `p * (q * x) DIV p + r < q * r` MP_TAC THENL
     [MATCH_MP_TAC LTE_TRANS THEN EXISTS_TAC `q * x` THEN
      ASM_REWRITE_TAC[LE_MULT_LCANCEL] THEN
```

Figure 14.7. A sample of machinese (in this case, HOL Light).

Talking mathematics to computers seems a mad pursuit, since they will not understand. But it serves two purposes, one having to do with the daily growth of the "tree of mathematics" and the other with its foundation. The former is, in some sense, sociological; the other philosophical.

Most mathematicians do not use theorem verification in their day-to-day working lives. This is because they want primarily to understand what is going on. Though it is possible, with some training, to understand formalized statements, it is a mind-numbing chore to go through the lines of formalized proof, and it does not provide any insight whatsoever. With time, no doubt, proof assistants will become more user-friendly—which means better translatable into mathese, and thus closer to the "form of life" of mathematics and hence its "foundations" in Wittgenstein's sense.

It does not require the visionary power of Bacon to predict that translation from mathese to computer script will eventually be automatized. This is when proof assistants will take over the reviewer's work: no paper will be published without their okay. Theorem or not? Computers will give the thumbs-up, as with Hales's proof of the Kepler conjecture (Chapter 5). Just as journals nowadays are sending their style sheets to prospective authors, so they will send their proof verifiers.

But why stop at that? Much brainpower is devoted to devising computer programs for finding new proofs, rather than translating old ones, and (even more ambitiously) for generating new theorems and new conjectures. Needless to say, some branches of mathematics are more amenable for such tasks than others.

To be fully convincing, a proof discovered by computer requires that it can be translated back into mathese, into a proof digestible by humans. Here, too, science fiction has landed in reality. Consider the text that was presented in the box earlier as a typical sample of mathese (*For a continuous function, the pre-image of an open set is open.*) It was not written by a mathematician. It is the output of an automated theorem prover that is human-oriented rather than machine-oriented. The program was developed by Tim Gowers and colleagues. Mathematicians, as blind tests have shown, are not aware that the text was produced by artificial intelligence.

Semantic Descent

Just as composers do not think in terms of strings of tadpole-like musical notes, so mathematicians do not think in strings of symbols. Even the most devoted formalists agree that mathematicians associate some meaning to their concepts. They may, with Hilbert, playfully pretend that they could refer to "beer-glass, chair, table" just as well as to "point, line, plane," but this is merely bravado, not meant to fool anybody.

The semantics of mathematical language is based on years of familiarization with sharply defined concepts. To pick up an example used by Tim Gowers, all mathematicians know what a Hilbert space is. Their stereotypical answer: it is a vector space that has an inner product and is complete. So what is a vector space? (The answer takes about half a page.) And what is an inner product? (The answer takes two lines.) And what does "complete" mean? First, to be precise, it is understood that "complete" refers to the metric that is defined through the norm induced by the inner product. Well, then, what is a norm, what is a metric? Having settled this, we are ready for the answer: "complete" means that every Cauchy sequence has a limit. Good, but what does it mean to have a limit? And what is a Cauchy sequence? In this answer, epsilontics rears its head: for every real number epsilon larger than 0, there exists a natural number N such that etc. etc. And the game is still far from having reached an end.

The point here is that mathematical concepts are defined in terms of other concepts, which in turn rely on other concepts, and so on. This is why the language is so hard to learn. If children want to know what *arbre* means, they are told to which word it corresponds in their mother tongue; or else, a tree is pointed out to them. Such an explanation is based on another language, or on some extra-verbal means. However, if people want to know what a Hilbert space is, they can only be told *within* mathematics. This presupposes years-long training that in turn is based on drills undergone during elementary education and even before.

Proofs consist in deriving theorems from other theorems. Definitions are based on other definitions. This has to end at some stage, of course, namely with the axioms—propositions about concepts that are accepted as given. In today's mathematics, axioms are almost always formulated in terms of set theory. The customary picture asserts that set theory and mathematical logic are the roots of the huge tree of mathematics, with its thousands of branches. They are the foundations of mathematics in the usual sense, as understood, for instance, in Turing's lecture course.

To this, adepts of Wittgenstein reply: set theory is only one mathematical discipline among many, a highly sophisticated special field that does not seem more basic than the others. Does it really deserve its status as the "root" of mathematics? Roots serve to provide a tree with a foothold and with nutrients. Is this the case here?

In Wittgenstein's sense, the foundations are rather to be found in the rules that children are trained to follow automatically. Over the centuries, this strange language game has known many vicissitudes and crises, as with every human creation. Time and again, the development of mathematics has led to situations where the commonly accepted rules proved insufficient or contradictory, and had to be modified in some appropriate way. If there is anything mysterious about it, it is that the modification that eventually emerged always proved to be uniquely right—not in some profound sense (which may well exist, by the way) but simply as a matter of historical fact. This stands in striking contrast to what happened in the history of religious or philosophical teachings, or for that matter in the evolution of real languages branching out and splitting off.

To the foundations, as understood in this sense of language philosophy, belongs the act of counting, for example. How did the insight arise that two eyes, two mountains, and two days have something in common? (In certain languages, various "twos" are denoted by different words, by the way.) How many computations did each of us perform before grasping not only the steps of whatever rule was used, but the notion of computing? And what did mathematicians think they were doing in the centuries before "foundational" concepts emerged, such as function, set, or recursion?

Cosmic Esperanto

It is frequently claimed that mathematics is a universal language—the unique universal language, according to Alain Connes. This may be understood in two ways: as a language that is spoken everywhere, or as a language that can be used for everything.

That mathematics does not care about national borders is obvious. In this, it is not different from chess, for example. Short-lived attempts such as the Third Reich's "German mathematics" heralded by some Nazi professors were ridiculous from the outset (Figure 14.8).

Figure 14.8. Tribal mathematics.

Figure 14.9. The international tribe of mathematicians.

The fact that nomenclature is fairly internationalized is obvious, too. Such is the rule with all sciences. In the humanities, it is not uncommon that various schools of thought assail each other with gusto and perseverance for generations. In mathematics, as in physics, such ideological trench warfare is rare and short-lived. The German poet Hans Magnus Enzensberger has hailed mathematics as "the purest of all humanities," but while humanities rarely know consensus, it is as widespread in mathematics as in physics or chemistry. Possibly even more so: although there exist hundreds of mathematical disciplines, the unity of mathematics is jealously guarded. It is agreed that the most striking feature of mathematics is provided by the unexpected connections between different branches, such as those between complex numbers and Euclidean geometry (to mention only a very down-to-earth example).

As an outer sign of this unanimity, a World Congress of Mathematicians (Figure 14.9) is held every four years, an analogue to the Olympic games; there is no world congress of physics, or biology, in contrast. A mathematics talk, at such a world congress, is understood usually by a small minority only, yet the other attendants sit through it with admirable fortitude. Moreover, the internal rank ordering within mathematicians (who tops whom) is astonishingly consensual, quite in contrast to what is encountered, for instance, in economics. In this "sociological" sense too, mathematics is more unified than other sciences.

To laypersons, mathematical texts often appear as encrypted; yet mathematics shares little resemblance with a secret science. In the days of Pythagoras, this must have been different, if legend has to be believed. Today, however, there are only a few disciplines, such as cryptology, where some discoveries have to be kept secret, at least for a while. It must be very irksome to the discoverers.

Mathematics, hence, is a global language. But is it universal? Apparently, yes. If we want to get in touch with aliens, what else should we use? The sci-fi thriller *A for Andromeda*, written in the early 1960s by cosmologist Fred Hoyle, is based on such a scenario.

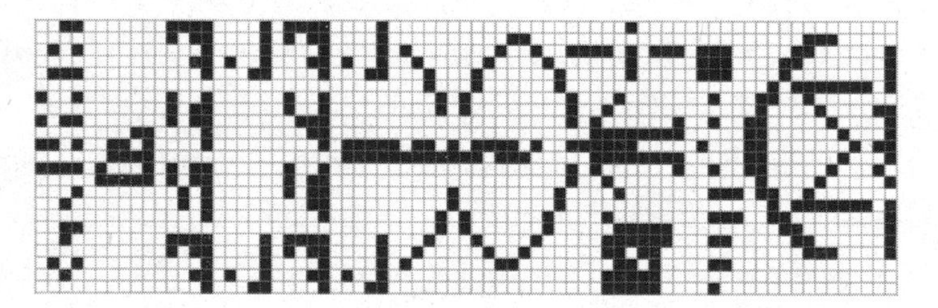

Figure 14.10. A crossword puzzle? No, the message from Arecibo to the aliens.

Things did not stop at fiction, by the way. In the year 1974, a message was sent to the aliens (Figure 14.10), apparently on the surmise that they would be eager to fraternize with us. The message was broadcast from Arecibo, an observatory in Puerto Rico. It consisted of 1679 bits. The idea was that the hypothetical receiver would no doubt quickly understand that the number can be factored into two primes, 23 and 73. The bits can therefore be arranged in a 23 × 73 matrix. The leftmost columns of the decoded message show the numbers 1 to 10 (in binary representation, of course, since we cannot count on the aliens having ten fingers). The following columns show the number of protons in atoms of hydrogen, nitrogen, oxygen, carbon, and phosphor, the elements making up DNA, and some additional information about that molecule. Farther on comes a minimalistic picture of a human silhouette and another number, namely 14. Whoever reads this is supposed to understand that we are about fourteen times the size of the wavelength of the signal. Finally, the message closes with a few things about how to locate our solar system and the Arecibo observatory. (Yours sincerely, wasting away—)

All this is encoded in a few numbers. It is obvious that the authors of the message view mathematics as a universal language. The aliens need have no fingers, no ears, no sense of music—but they must know a bit of arithmetic to get our meaning. And thus, since 1974, the signal pulse spreads unstoppably across interstellar space, saying a lot about us humans—at least for those who read between the lines.

There is a postscript to the story: on December 1, 2020, two cables snapped and the giant Arecibo radio telescope, that miracle of engineering, collapsed.

Apologizing for Nothing

When mathematicians speak with others than aliens, computers, and themselves, what do they speak about? Applications, presumably. This leads to the second sense in which a language can be universal: universally applicable. Here things become downright paradoxical. Indeed, whoever takes the formal aspect seriously must be inclined to say: it speaks of nothing. Set theory is based on the empty set, and the propositions of logic are mere rules for using signs.

Yet, we all know that mathematics plays center stage in a tremendously wide range of applications. For some mathematicians, this aspect is of secondary relevance, if not downright disagreeable. In his *Apology of a*

Mathematician, the British number theorist G. H. Hardy, the purest of the pure, snubbed all applications, seeing them as mere collateral damage. Fortunately, Hardy added, there exist fields that escape all possible applications, and probably always will. His examples include general relativity and number theory. *Tempora mutantur!* Nowadays, Einstein's field equations are used in GPS, and number theory in all email platforms. Credit cards are coded via prime number factorization.

It is often held that World War I was the war of chemists, World War II was the war of physicists, and a coming Armageddon will be a war of mathematicians. Already in the last World War, both sides deciphered the coded messages of their enemies by means of cryptanalytic methods devised by mathematicians, such as Poland's Marian Rejewski, England's Alan Turing, and Sweden's Arne Beurling.

The efficiency of mathematics has a weird touch. It is based on two reasons that are probably linked. One is the existence of many surprising, sometimes almost spooky, cross-connections between seemingly quite different theories. The other is the use of abstraction. This property is often held as a blemish against mathematics, leading straight to distraction and ruin. On the contrary, abstraction is the secret of its success: the readiness to delete hundreds of details, but to envision other possibilities—to compare what is with what can be, and even with what cannot be. Mathematicians are virtuosi of the thought experiment and the "what if."

Mathematics has been described as "the ultimate of technology transfer." To give just one example, branching processes were introduced first to describe the extinction of family names; then, they were applied to populations of microorganisms; then, to nuclear chain reactions. Here is another example: the same methods can deal with the stability of electrical circuits, chemical reactions, or mechanical control. The question always reduces to whether or not an eigenvalue is on the right-hand side of the complex plane. (What is an eigenvalue? Now here we are, again. First, you must know about vector spaces and linear operators. . . .)

Not only is nature written in the language of mathematics, as Galilei said, culture is too. All apps are based on sophisticated algorithms. So are our smartphones and credit cards. Stock exchanges are driven by computer programs; photos and films and concerts are digitally stored and transmitted. Visualization and imaging are huge industries. Our daily use of email and the Internet is based on signal processing, which means coding, which means mathematics. Digitalization is so ubiquitous that we are not even aware of it—except when we have mislaid our smartphone.

Leibniz already had envisioned a *characteristica universalis* as the basis of a *scientia universalis*: things should correspond to signs, and relations between things to relations between signs. Leibniz wrote to a friend:

> The only way to rectify our reasonings is to make them as tangible as those of the mathematicians, so that we can find our error at a glance, and when there are disputes among persons, we can simply say, *Calculemus*, Let us calculate, without further ado, in order to see who is right.

Utopian? Dystopian? Whatever. The program of Leibniz acts as a guide through the growing range of mathematical applications.

Is there any domain forever closed to mathematics? Or is it rather that mathematics can apply to everything whereof one can speak? "Whereof one can speak" is meant in the sense of Wittgenstein's *Tractatus*, of course—not, however, in the broader sense of his *Philosophical Investigations*. Our daily language serves not only for representing, but also for expressing things. This is where the language of mathematics reaches its limit. There are things whereof one cannot speak. One can then be silent thereof—that's what the *Tractatus* enjoined. One can also swear, or call for help, or engage in other language games. But wherever one speaks "about" something, mathematics can help, as a language that makes it "impossible to be imprecise."

15

Philosophy

Plato's Shadows in Jurassic Park

A Circle Meets a Turning Point

Once upon a time, or more precisely a hundred years ago, philosophers and mathematicians gathered every second Thursday at a quarter past six in a derelict little seminar room, and quickly proceeded to fill it with tobacco smoke. Their meetings would last for two hours or more and were then (this being Vienna) relocated to a plush, brightly lit coffeehouse just around the corner, with billiard tables and huge wall mirrors. More smoke, and more debates, mostly on the foundations of science. The discussions lasted late into the night, and went on even as the members dispersed, walking home in small groups through the silent streets. The quietest and youngest member of the group, a student named Kurt Gödel, did not even have to walk home. For a while, he lived just a few floors above the café.

The founder of the group was fiftyish Hans Hahn, a tall man with a loud voice, a mathematics professor of considerable standing. He had always been fascinated by the philosophy of mathematics. Hahn was a hard-nosed empiricist, and therefore saddled with a fundamental problem: How is the empiricist position compatible with the application of logic and mathematics to reality? Hahn founded a discussion group, together with his brother-in-law Otto Neurath, the left-wing director of a museum propagandizing for Red Vienna. When Einstein's favorite philosopher, Berlin-born Moritz Schlick, was appointed to a chair at the University of Vienna, he was entrusted with presiding over the discussion group. Some very talented postdocs and students joined, such as Karl Menger, on his way to found dimension theory, and Rudolf Carnap, a former student of Frege who had just completed *Der Logische Aufbau der Welt* (*The Logical Structure of the World*), no less. Now he was looking for a job.

After a few years of philosophical discussions, the Schlick Circle came out into the open: the fearlessly radical avant-garde of the scientific worldview, with their own journal, book series, and international meetings. They knew what "branding" meant, and did it with gusto. They called themselves the Vienna Circle, proselytized for logical positivism, and loved to vex all those who had no head for mathematics. The times were modern: beards were out, and the *Bauhaus* in. Metaphysics was derided as the pathetic refuge of benumbed reactionaries. From behind his spectacles, young Gödel listened politely. He hardly ever spoke during these meetings. Even his closest friends misunderstood his silence as approval.

For the philosophy of mathematics, this was the best of times. Three great schools, headed by formidable thinkers, were contending with each other—and most remarkably, many mathematicians actually cared! Philosophical issues were the order of the day. In the wake of Frege and Russell, the logicists claimed that mathematics was entirely grounded in logic. The followers of Hilbert, the formalists, attempted to describe mathematics as the manipulation of signs according to well-determined rules. In L. E. J. Brouwer's wake, the intuitionists insisted that mathematics deals with mental constructs and nothing beyond, least of all the actual infinite. Each of the "big three" philosophies of mathematics vied for supremacy.

In 1930, Hans Hahn decided that the time was ripe for a showdown. His Vienna Circle organized a summit meeting where the claimants would thrash it out. The tournament was held in Königsberg, now Kaliningrad, on the shore of the Baltic Sea—portentously, in the town of Kant. The three movements were championed, not by their figureheads, but by very capable seconds. Carnap (by now a flag-bearer of the Vienna Circle) spoke for the logicists, Arend Heyting for the intuitionists, and John von Neumann for the formalists.

In Hans Hahn's somewhat biased perspective, the contest ended with the logicists up front, by a comfortable margin. But the high point of the meeting, its fifteen minutes of immortality, occurred with Kurt Gödel's surprise announcement of his first incompleteness theorem.

Hahn was deeply impressed by this exploit of his former student. He held that "it will have its place in the history of mathematics." Most gratifyingly, it had sprung forth from his Vienna Circle, that fountainhead of the scientific worldview.

This scientific worldview had little patience with metaphysical superstitions and other theological leftovers. It made short shrift of the abstruse Platonic myth that the world of true being is a world of ideas inaccessible to our senses. Hahn did not mince his words: "This Platonic position is entirely metaphysical and seems wholly unsuitable as the foundation of mathematics"—*entirely* metaphysical, *wholly* unsuitable!

For Hahn, as for Frege and Russell, mathematics was logic. Moreover, Hahn was totally won over to the view so forcefully advanced by Wittgenstein in his *Tractatus*:

> 6.1. The propositions of logic are tautologies.
> 6.11. Therefore the propositions of logic say nothing.

Hahn concurred: "Logic therefore does not say anything about the world; it only has to do with the way I *talk* about the world." Moritz Schlick shared the same conviction: "Logical conclusions express nothing about facts. They are merely rules for using our signs." The *linguistic turn* in philosophy was gathering momentum.

What is true for logic must also hold for mathematics. It consists, says Hahn, in tautologies "that show how something that we could have said could be said in various other equivalent ways." This is indeed a radical stand, as Hahn admitted readily: "The whole of mathematics, with all its hard-earned struggles and frequently surprising results... dissolved into nothing but tautologies."

The notion sounds absurd enough to squelch the whole idea once and for all, so it seems. But hold it, said Hahn:

> This argument overlooks just one minor detail—namely, that we humans are not omniscient. An omniscient being would, of course, know instantly everything that is implied when a set of propositions is asserted. . . . Such a being would know immediately that what is meant by 24 × 31 and 744 is the same thing.

And thus: "An omniscient subject needs no logic, and contrary to Plato we can say: God never does mathematics."

Moritz Schlick seconded this opinion and celebrated what he hailed as "the turning point in philosophy": "Today, no one can be of the opinion that 'propositions of arithmetic' can say anything about the real world." They are not "about" anything at all; and even the term "propositions of arithmetic" deserves to be put in quotes, well removed from ordinary meaning. "Propositions of arithmetic" are no decent propositions at all, but mere pseudo-sentences.

Carnap agreed: mathematics is the analysis of a "linguistic framework" that we are free to choose. Which framework proves most useful is a pragmatic, not a mathematical, question, the offshoot of conventions about the use of symbols. Discussions about the reality of mathematics are meaningless metaphysical babbling.

The linguistic turn seemed set to carry the day. But in the 1940s, Kurt Gödel, who in the meantime had acquired a towering reputation, started to voice his opposite conviction: namely "the view that mathematics describes a non-sensual reality which exists independently both of the acts and the dispositions of the human mind and is perceived, and probably perceived very incompletely, by the human mind."

Bertrand Russell, who met Gödel in 1944 at a dinner party in Princeton, was aghast. He had encountered "an unadulterated Platonist." Hans Hahn, by then dead for ten years, must have turned in his grave. The onslaught of the Nazis had scattered the survivors of the Vienna Circle to all winds, but they had kept on spreading the good word, advocating logical empiricism with remarkable success. And now it came to light that with young Gödel, a cuckoo's egg had been hatched in the Circle!

During the 1950s, Gödel went to great pains to refute Carnap's views on mathematics. He wrote and rewrote an essay entitled "Is Mathematics a Syntax of Language?" He answered with a resounding NO!, but to his editor's despair, he never published it. After Gödel's death, however, his huge *Nachlass* (with six versions of the essay) made it abundantly clear that he had always been a true Platonist.

Even Kurt Gödel owned up that "the Platonists' view is rather unpopular among mathematicians." On this point, there was common consent. In 1940, the eminent historian of science E. T. Bell wrote that "according to the prophets, the last adherent of the Platonic ideal in mathematics will have joined the dinosaurs by the year 2000." Today we know that, as a matter of fact, the dawn of the new millennium saw Platonism back in strength among mathematicians. Talk about joining dinosaurs!

Indeed, the current view is that most of today's working mathematicians are not camp followers of any of the big three, but rather closet Platonists, guided by a more or less subconscious feeling that there is an objective mathematical reality "out there," waiting to be discovered step by step. These mathematicians see themselves as explorers. Mathematical entities—groups, polygons, primes—are just as real for them as toads and crocodiles are for zoologists.

Gödel can hardly be held responsible for that shift in opinion, although he certainly lent prestige to Platonic views. Neither is Plato accountable for the trend. He had no philosophy of mathematics in anything like the modern sense; but then, neither is such a philosophy likely to be found behind the daily toil of mathematical research.

Few believe that our immortal souls are remembering mathematics by anamnesis. However, the silent majority of mathematicians are convinced by their own day-to-day experience, whether with discoveries or with dead ends, that mathematics exists independently of us. "This is a notion that is hard to credit, but hard for a professional mathematician to do without," wrote Robert Langlands, one of their foremost representatives. These mathematicians feel that they have no choice. This is because mathematical entities are so obstinate. They cannot be talked around. Therefore, they must exist.

"Why do you have to climb Mount Everest?" "Because it is there," George Mallory said. (And he is still there, on Mount Everest.)

Mathematics: Is It for Real?

A philosopher can hardly pick a better field than the philosophy of mathematics. It is filled to the brim with big questions. To start by quoting Kant: "How is pure mathematics possible?" To proceed: What is mathematics? What is it about? What do mathematicians mean by "truth" or "existence"? What is a proof? What makes us so sure? What are numbers? What are sets? What is logic? What is discovered and what is invented? Why is mathematics useful? Why is it so unique?

But first of all: Why should we care?

It is not merely that most mathematicians blissfully ignore the philosophy of mathematics, and that some even scorn it. Among the philosophers themselves, there is a growing concern to justify their doings.

"Why is there a philosophy of mathematics at all?" asks Ian Hacking, one of its most prestigious and perceptive practitioners. It is a touchy question. The answers of philosophers often have an apologetic touch; the answers of mathematicians tend to be condescending.

There was a time when philosophers dared to set mathematicians right about their business. Plato complained about the "most ludicrous language" used by the adepts of geometry. Hobbes scolded them for being muddleheaded about the notion of a point. Schopenhauer derided their benighted concerns about parallel lines. Wittgenstein despaired of the "hocus-pocus" of set theory. It was all in vain. The mighty caravan of mathematics moved on, and the barking receded in the darkness. By now, philosophical hecklings from the sidelines are rarely heard. Mathematicians need not to be told their business. There is no need for any extra-mathematical endorsement. No backseat driving by philosophers, please. This is a view that many wise philosophers, in the wake of Willard Van Orman Quine and Penelope Maddy, have quietly adopted. It seems that the reputation of mathematics has become all too intimidating.

Assuredly, there have been profound philosophical crises in the history of mathematics: the scandal about irrational numbers, the slippery nature of infinitesimals, the paradox of Russell, to name but a few. In each case, however, the issue was resolved by mathematicians, with no help from outside.

(This autarky of mathematics does not imply that mathematicians do not care about philosophical questions, by the way. In fact, many of them are ready to spend hours—usually late hours—discussing whether mathematics is discovered or invented. Their readiness declines with age, as a rule, but more from resignation than because they have figured out a solution to the conundrum.)

The consensus is that the philosophy of mathematics embraced by mathematicians is their private affair and does not interfere with their work. One never hears: "Don't trust the theorems of this man! He is nominalist!" or "When she turned naturalist, she promptly lost her edge."

The high-water mark of mathematicians' interest in the philosophy of mathematics was reached with the showdown of the "big three." Since that time, the excitement has receded. It's back to work—and in the terse formula of Reuben Hersh, "The working mathematician is a Platonist on weekdays, a formalist on weekends." That sums it up.

Philosophers, needless to say, are much more careful in labeling themselves. Roughly speaking, mathematical realists are those who hold that the propositions of mathematics are true or false, and that what makes them true or false is something external. In other words, objective mathematical truth has an existence of its own, in an eternal, objective way: discovered, not invented. Platonists (in the modern-day sense) go one step further and believe in the existence of mathematical objects ("out there," as some are apt to add).

The word *existence* is, of course, a most treacherous term. A stone exists; so does a cloud, a dream, a tale, a program on TV, a program on a computer, a code, a rhythm, a whirl, a fancy. So why not a number? An "eternal existence" is harder to describe. And what about "out there"? Out where? In reply, a hand is waved, but no finger pointed. (With the advent of iCloud, by the way, this "out there" must have undergone a shift of meaning.)

Most mathematicians, when pressed, sidestep the issue: for them existence means being free from contradiction.

Figure 15.1. Rudolf Carnap (1891–1970) warns of pseudo-problems.

Carnap was adamant: to claim that mathematical objects such as numbers or triangles *exist* is simply meaningless. We have no way of deciding the issue one way or the other. The statement is without content. Conversely, Carnap was just as adamant in claiming that the existence of an outer world (stones, clouds, all that stuff) cannot be decided either. Again, the issue is meaningless—a mere pseudo-problem, so he says. With mathematics and the real world therefore on almost equal footing, this view seems to rehabilitate a Platonism of sorts. Carnap probably did not understand it that way.

A non-sensory reality of mathematics seems hard to reconcile with the life-forms of us humans, housed as we are in space-time and relying on our senses. It makes it even harder to account for the "unreasonable effectiveness of mathematics." While some mathematics is never applied to anything at all, that part that is does an outstanding job. Physics is full of examples where mathematical theories have been proved right to within one part in ten million. More strikingly even, purely mathematical ideas on space or symmetry have prompted physical theories, almost out of the blue, leading to predictions that eventually proved right (some after a time delay of a hundred years). This is much more than can be expected from an analytical "juice extractor" of empirical data.

How come empiricism and pragmatism alike confirm mathematical reasoning? What kind of preestablished harmony links the ideal world of mathematics with the wetware of our brain? Realists have a lot to explain.

The opposition argues that we can well do without the much-touted fairyland "out there." Mathematics is not out there, but "in there"—namely in our minds, which have evolved mindlessly. Mathematics is an altogether human enterprise; and if it were anything more than that, it would not make any difference to us. As Wittgenstein wrote, "Mathematics is after all an anthropological phenomenon."

If realists are wrong, mathematical entities are merely abstract fictions and useful conventions. The world out there knows no problems—they all are man-made, and so are their solutions. They were invented. The square root of -1 is a brainchild, a device so that $x^2 + 1 = 0$ has a solution. In a similar sense, a vector is a contraption. It is not "out there." It serves as a placeholder for an element of a vector space. That vector space is not "out there" either: it serves as a convenient concept to organize a bundle of properties, and as such it emerged only in the late nineteenth century, not out of the blue, but from well-prepared minds.

At this point, Platonists reply that mathematical "inventions" have an uncanny way of surprising their creators. It was certainly not planned that the solution of $x^2 + 1 = 0$ would be the key for solving *all* polynomial equations, *and* moreover satisfy $e^{i\pi} = -1$, so to speak as a bonus; or that the square of π would be related to the sum of the inverses of all square numbers. Nobody can dream up such a thing. If mathematics is a product of the mind, how can it be so mind-blowing?

Would it help to know whether the imaginary unit i, or the vector-space structure, or proof by induction were invented, rather than discovered? For an anthropologist, the bottom line is that mathematicians got used to it.

In mathematics, you do not understand things, says John von Neumann. You just get used to them.

Mathematics: Is It Surreal?

"The opposite of play is not seriousness," wrote Sigmund Freud, "but reality." Does it follow that the opposite of mathematics-as-reality is mathematics-as-play? In any case, the ludic aspect of mathematics cannot be overlooked.

Mathematical reasoning may have started with playfully arranging stones in squares and triangles. Indeed, the tetractys—the arrangement of the natural numbers 1, 2, 3, and 4 in a triangle—was considered by the Pythagoreans to be "the spring of all our wisdom, the perennial root of Nature's fount" (that is what their secret oath said). The tetractys embodied perfection and was venerated as the key to harmony. Any child playing with pebbles can spontaneously discover it (or invent it?).

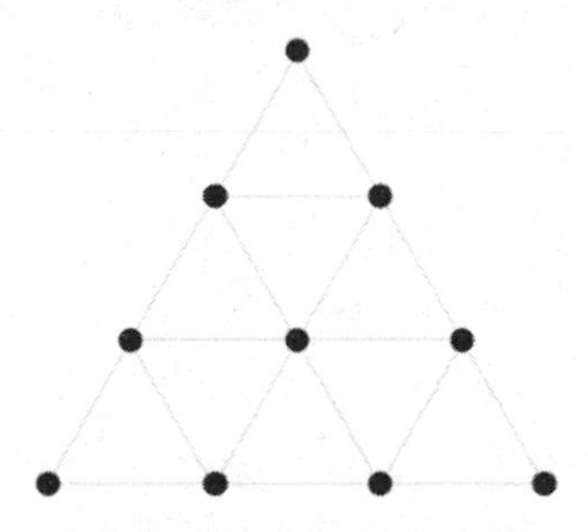

Figure 15.2. The tetractys.

Formalism, the main rival of Platonism, is often characterized as a "meaningless game with markers." This view is a caricature. Not even formalists are likely to ever overlook the uncanny applicability of mathematics to science; and indeed, mathematics is indispensable, in a way that no game can ever be. However, the purely manipulative aspect of handling strings of symbols with replacement rules stood at the core of Hilbert's program for proving the consistency of mathematics. It meant acting as if the mathematical symbols were like chess pieces, *as if* they had no other content than that assigned by the rules of the game. (Acting "as if" is itself a hallmark of playing, of course.)

Hilbert's program turned out to be overambitious, but it gave a seminal impetus for the development of computers, and subsequently launched the vast enterprise of checking all known mathematics by means of computer proofs. This is a pretty good record for what started out as a game of pretense, an acting "as if."

Mathematics-as-game is more than a useful technique for looking at the foundations of mathematics. Play is a prime mover of the cultural phenomenon "mathematics."

There is hardly a better example than the surreal numbers. They were discovered, or invented, some fifty years ago, at a time when everyone thought that the story of numbers had run its course (to be more precise, the story of those numbers, finite or other, that can be lined up, somehow or other).

Not so, as it turned out: a young Cambridge mathematician named John Horton Conway started playing around and found lots of new numbers with very odd properties. He displayed them to a colleague from the United States named Donald Knuth, who coined the only fitting name: *surreal* numbers! Knuth promptly wrote a book on them while spending a euphoric week in a hotel room in Oslo.

Entitled *Surreal Numbers* (and subtitled *How Two Ex-students Turned on to Pure Mathematics and Found Total Happiness*), the book is about Alice and Bill, two hippies who have run off from college. Now they are trying to "find themselves" on a shore of the Indian Ocean. What they find instead is a piece of rock half-buried in the sand, with an inscription in Hebrew relating how "in the beginning, everything was void, and J. H. W. H. Conway began to create numbers." This Knuth-testament version of genesis goes on for a few more lines, and then breaks off. The rest of the story is about Alice and Bill making sense of the text.

Conway created all numbers large and small out of two rules:

> Every number corresponds to two sets of previously created numbers, such that no member of the left set is greater than or equal to any member of the right set.

So, numbers are pairs of sets of numbers: they can be written $(L|R)$. For the rule to make sense, the numbers must be ordered:

> One number is less than or equal to another number if and only if no member of the first number's left set is greater than or equal to the second number, and no member of the second number's right set is less than or equal to the first number.

This will eventually produce a profusion of numbers out of numbers, but how does J. H. W. H. start? This is how: let L and R both be the empty set $\varnothing$. The command "no number of the left set is greater than or like..." is surely satisfied because there *is* no member of the left set. This first number $(\varnothing \mid \varnothing)$ will be named 0 (zero). Naming is denoted by the symbol =: (the dots indicating that it is a definition). Thus, the first number comes about. It is

$$(\varnothing \mid \varnothing) =: 0.$$

Like John von Neumann with his construction of the natural numbers, J. H. Conway starts literally from scratch. But he goes much further. His next steps are almost obvious:

$$(0 \mid \varnothing) =: 1,$$
$$(\varnothing \mid 0) =: -1.$$

The names on the right are just signs. That they actually do what we associate with 1 and −1 is left for later. Again, the empty set makes it rather easy to verify the "greater than or equal to" command. Alice and Bill are conscientious enough to check that each of the three numbers obtained so far, namely 0, 1, and −1, is less than or equal to itself; that they are different from each other; and even that $-1 < 0 < 1$. So far, so good. Next comes

$$(1 \mid \varnothing) =: 2, (2 \mid \varnothing) =: 3, (\varnothing \mid -1) =: -2, \text{etc.}$$

Hence, all these "integers" are surreal numbers. Up to now, however, these numbers form an ordered structure, and nothing more. There is no algebra within sight. By a stroke of good fortune, Alice and Bill discover the missing part of the rock, where J. H. W. H., in Jehovah-like lingo, gives the rules for adding numbers and negating them, and enjoins the numbers "to be fruitful and multiply." It is only at this point that the surreal integers show their integrity, or rather their integer-ness. This is where the going gets harder. But what would a game be if it was only smooth sailing?

Next comes a move that is surprising (but only for the unprepared mind): the surreal number $(0|1)$ is named $\frac{1}{2}$, and $\left(0|\frac{1}{2}\right)$ is named $\frac{1}{4}$, and so on. By proceeding in this way, and keeping the multiplication rule in mind, all dyadic fractions $\frac{m}{2^n}$ are obtained (with m, n being integers). This gives the go-ahead for the next big leap forward, which yields all the real numbers. They are pairs of surreal numbers $(L|R)$, where L and R are sets of dyadic fractions. At this point mathematicians feel back in safe waters: the construction is very similar to the time-hallowed method of Dedekind cuts. Doubtlessly, Dedekind cuts had suggested the whole game of introducing surreal numbers as pairs of sets of numbers fulfilling some ordering condition. Addition, subtraction, multiplication, and presently also division all work as they should. So far, however, what we have seen is just a playful way of introducing the reals. They could have been acquired at a cheaper price in any mathematics department.

When Bill and Alice start to wonder about that, their banter is interrupted by a voice from above, sounding like thunder: "Rubbish! Wait until you get to infinite sets."

Lo and behold, the number $(0, 1, 2, \ldots|\varnothing)$ turns out to be larger than all integers. It is a transfinite number that will be named ω, and opens the way to all ordinals. Conversely, the number $\left(0|1, \frac{1}{2}, \frac{1}{4}, \ldots\right)$, nicknamed ϵ, is a true infinitesimal, being positive but smaller than any positive real. Moreover, $\frac{1}{\omega} = \epsilon$. The dream of Leibniz comes true!

The surreal numbers include the hyperreals and therefore Abraham Robinson's nonstandard analysis. At this point the surreal fun really starts. You may think that the continuum of numbers is densely packed; but every real number x is surrounded by a cloud of surreal numbers that are closer to x than any other real number. Arithmetic with transfinite numbers unfolds as if by magic. Like the reals, the surreals form an ordered field (well, almost).

In fact, they contain all other ordered fields. However, they are so numerous that they form no longer a set (and hence they are no field).

The novelette by Donald Knuth has not aged these fifty years, ever since the hippies returned to college. The translations and reprintings go into the dozens. The little book still marks, as science writer Martin Gardner way back then recorded,

> The only time that a major mathematical discovery has been published first as a work fiction. An empty hat rests on a table made of a few axioms of standard set theory. Conway waves two simple rules in the air, then reaches into almost nothing and pulls out an infinitely rich tapestry of numbers.

With surreal numbers being pulled out of the hat, it seems nearly pedantic to ask whether mathematical objects are real or not.

Mathematical Life

Conway's wizardry offers a new approach to old fields. The real numbers—the "real" real numbers, that is—have been around for ages. Indeed, they have been here forever, as any decent Platonist would vouchsafe. It seems difficult to imagine mathematics without real numbers. Yet, this says, maybe, more about *us* than about *them*.

Could other civilizations have produced utterly different mathematics? Is ours a fluke? It is like asking whether there could exist, on some exoplanet or in an impenetrable cloud of stellar matter, a form of life that is not based on the same genetic code as the life we know, with its triplets of nucleotides translated into an alphabet of proteins. We know only one genetic code, and we know only one form of mathematics, but this does not necessarily mean that either of them is universal. They reign unchallenged on our planet, so far. However, their monopoly on planet Earth may simply be due to the fact that they were here first and gobbled up whatever came later. At present, we just have no way to know whether there are not any rivals "out there."

But we may well imagine that a Conway-like intellect will discover some other forms of life, or of mathematics. In a certain sense, J. H. Conway did it already (did both, in fact). Let us digress and see.

Many mathematicians are struck by the odd (not to say freakish) twists and turns of the history of their science. Is mathematics contingent, an intellectual random walk? Are there utterly different mathematics? It is admittedly not easy to imagine a mathematics without natural numbers. However, it is not too hard to imagine a mathematics without continuum, for example. The theory of cellular automata provides an example.

Cellular automata were hailed as "a new kind of science" (in the modest words of Stephen Wolfram, the mastermind behind the software platform Mathematica). They do, in fact, belong to contemporary mathematics, as a subfield of computation theory, well-integrated within their neighborhood. Yet, they offer a glimpse of what another mathematics could look like—much in the same way as some strange, translucid organisms from the deep, while belonging to the tree of life on Earth, may suggest visions of extragalactic evolution.

The best way to introduce cellular automata is to follow J. H. Conway again, that charismatic prince of the weird, and to describe his Game of Life.

Figure 15.3. John Horton Conway (1937–2020).

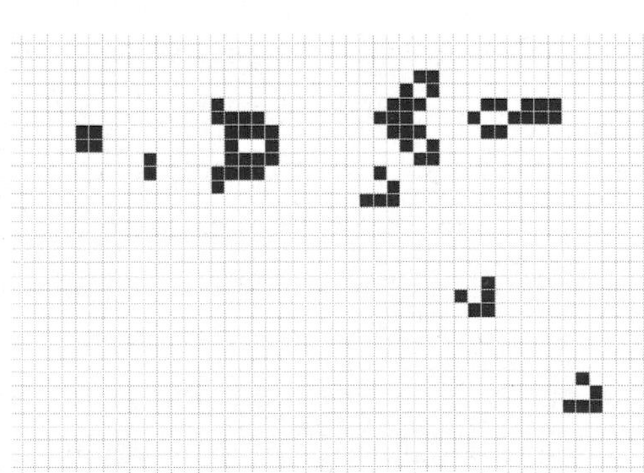

Figure 15.4. A glider gun firing a stream of gliders toward the bottom right.

The biotope is a square grid, two-dimensional and infinite. Each of the square cells touches eight neighbors. Cells can be "on" or "off." They are updated from one second to the next. An "off" cell turns "on" if and only if three of its neighbors are "on." An "on" cell remains "on" if and only if two or three of its neighbors are "on"; otherwise, it turns "off." Any given initial configuration on the grid will evolve step by step, second by second, its future preordained until eternity. Time proceeds stepwise; cells can be "on" or "off." Nothing could be less continuous. This world is discrete to the core.

An isolated "on" cell will turn "off" in the next step. A single two-by-two block of "on" cells never changes. Three "on" cells in a row keep company forever, switching from horizontal to vertical and back in a mad dance. Some small figures of five or seven "on" cells grow prodigiously, only to collapse in the end. An odd-shaped figure of five "on" cells moves crab-wise across the grid, resuming its shape (but not its place) at every fourth step. This is the glider. Another very odd-shaped figure resumes its shape and place periodically, but fires off a ceaseless stream of gliders. This is the glider gun. Glider streams can cross each other harmlessly, or annihilate each other, or deflect each other by a right angle. And so on and so on. If thirteen gliders collide in just the right way, they can build a glider gun. If two gliders collide in the right way, they can build an eater. An eater is a stationary figure that can devour gliders without leaving a trace.

The diagonals of the grid can act as neurons. Gliders can move along them like electric pulses. The pulses can transmit information. Out of this magic (based, as it will be remembered, on a few very simple rules), Conway was able to construct a universal Turing automaton: a huge figure of "on" and "off" cells on the grid. It acts on streams of gliders, just like a computer acts on the strings of bits of a program written in machine code.

It is unlikely, to say the least, that the Conway automaton will ever replace current computers; the point is rather that its behavior is equally complex. Therefore, it is fundamentally unpredictable. Indeed, we have known since Turing that such automata are unpredictable, and thus give rise to an inexhaustible wealth of challenging problems.

Moreover, Conway's automaton can be programmed to self-reproduce. It is alive, by any fair standard.

Today, cellular automata are a well-established model for computation, belonging to one of a vast number of recent fields in mathematics—a small bud on a huge tree. But what if cellular automata had arisen earlier? They require, in principle, no mathematical sophistication at all. Practitioners need only to be able to count up to eight, in order to understand the rules of transitions between "on" and "off." Suppose that the ancient Greek, instead of their infatuation with regular solids, had fallen for cellular automata. How would the alternative history of mathematics have looked?

The Game of Life is just one cellular automaton among many. The number of states could be different; so could the transition rules, and the geometry of the grid. The amount of experimentation that Conway needed before he found, with his Game of Life, a cellular automaton displaying such an interesting behavior, must have been staggering. Without computers, it could never have been done—which shows that the ancient Greeks were wise to stick to regular solids, all things considered. The development of mathematics depends on the available technology. But the underlying playfulness is surely a primeval human trait. It is evolutionarily grounded and reaches far deeper than any veneer of civilization.

While formalists often tend to compare mathematics to a game, formal proof utterly ignores the ludic aspect—the joy of playing around, of exploring, of trying out: a pleasure that is as much present in discovering as in inventing.

16

Understanding

The Proof of the Pudding—And a Flair of Its Flavor

Armchair Hedonism

Many mathematicians are Platonists; even more of them are hedonists. Most outsiders find this hard to believe. They deem it incomprehensible that one can derive pleasure from a subject matter that has dulled their years at school.

What joy can mathematics provide? First and foremost, no doubt, comes the pleasure of insight. Such moments of mental brightening are difficult to describe to those who are immune to it. They often arrive in a flash, more rarely in a slow dawning, and usually after a delay that can be irritating, frustrating, and even agonizing. Sometimes, one has to give up and try again another day. Mathematics teaches patience. It also teaches modesty. So many minds are so much smarter!

An attempt to convey the flavor of that mysterious process of understanding can only rely on examples. The choice of such examples is arbitrary—hundreds of other examples would do just as well.

To begin with, reconsider a classic result, the theorem of Pythagoras. Nothing more outworn, it may seem; it seems a mere loss of time to return to it. Yet, there exist several hundred proofs of that theorem. Would not one proof suffice? It certainly does from the logical point of view. Yet, any extra proof can provide some extra understanding. The following one has a special feel. It proves not only that $a^2 + b^2 = c^2$ holds whenever a, b, and c are the lengths of the sides of a right triangle (c being that of the hypotenuse). It also explains why.

The perpendicular through the right angle divides the triangle into two smaller right triangles (Figure 16.1). All three triangles are similar because their angles are equal. The lengths of their hypotenuses are a, b, and c. We denote their areas with S_a, S_b, and S_c. Since the two smaller triangles make up the larger triangle, we have $S_a + S_b = S_c$. Let us construct for each of the three hypotenuses the square having that hypotenuse as its side. Their areas are a^2, b^2, and c^2. Since the triangles are similar, so are the hut-like shapes (each a square with a triangle roof). All three of them have the same ratio of roof area to square area, which we denote by m. Thus, $S_a = ma^2$, $S_b = mb^2$, $S_c = mc^2$, and hence $ma^2 + mb^2 = mc^2$. Dividing by m, we are done.

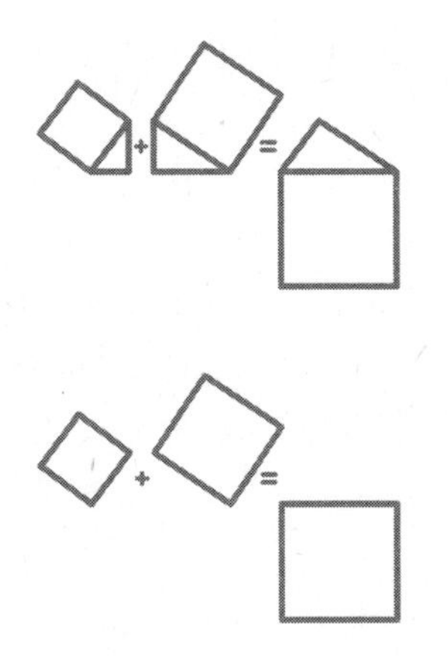

Figure 16.1. A figure can speak for itself.

According to scholars, this proof was first published in 1929 only. According to legend, Albert Einstein had discovered it as a schoolboy, too. (If so, this must have been the first time he met mc^2). Many other proofs of the theorem of Pythagoras make use of the fact that the triangles are similar, but apparently the crucial insight ("since the triangles add up, the squares must too") flashed up only thousands of years after Pythagoras. It seems hardly credible (and I am not sure that I believe it either).

Be that as it may, this proof can be looked over in one glance. So, why was it overlooked for so long?

Pick Any Polygon

In the eyes of most mathematicians, the countless unexpected connections between apparently unrelated fields are an inexhaustible source of delight. Applications of mathematics to mathematics do not really count as "applied mathematics." Yet, such applications can be, in their way, as surprising as applications of mathematics to the physical world.

Needless to say, the number of connections within mathematical fields increases with the amount of mathematics at hand. But some of them require only very little previous knowledge. To give an example, let us consider the celebrated theorem of Pick.

Only one picture of Georg Pick seems to have survived. It shows a young man with a huge tie and a conquering look. This was the fashion in the Vienna of his time. Pick was only a few years younger than Sigmund Freud, a few years older than Arthur Schnitzler. After his studies, he moved to the University of Prague, the oldest in the Habsburg empire, and spent his entire academic life there, with colleagues such as Ernst Mach and, for a brief time, Albert Einstein.

After retiring as a professor, Pick returned to Vienna in 1929. It was no longer the belle époque Vienna of his youth. The country was torn. When the Nazis took it over in 1938, Pick had to flee back to Prague. This, however, was not far enough. In 1942, he was transported to Theresienstadt, where he did not survive for long.

To this day, Pick remains known for his contributions in complex analysis and differential geometry—weighty stuff. Yet, what he published in 1899 in an obscure Bohemian journal named *Lotos* did not require any sophistication whatsoever. It turned out to be a gem.

Figure 16.2. Georg Pick (1859–1942).

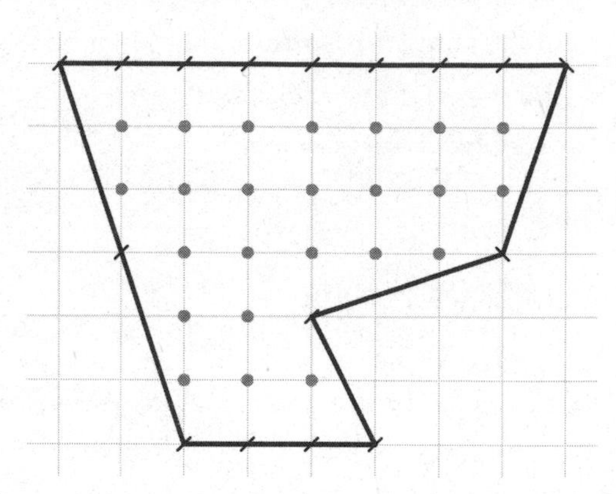

Figure 16.3. A grid polygon with Pick number $24 + \frac{18}{2} - 1 = 31$.

Let us consider a planar grid, as with the Game of Life, and let us draw a simple polygon: a closed path consisting of line segments that do not intersect. We will always assume that its vertices (the corners where two line segments join) are points on the grid (Figure 16.3). The 1×1 squares of this grid provide a natural unit measure for the area in the plane. This being said, the theorem of Pick states that the area of the polygon is given by its Pick number (of course he didn't use this term), which is defined as

$$\text{Pick(polygon)} =: P_I + \frac{P_B}{2} - 1.$$

Here, P_I is the number of grid points in the *interior* of the polygon, and P_B the number of grid points on the *boundary* (which comprise all the vertices, plus any other grid points on the edges).

Since ancient times, computing the area of a polygon has been a simple exercise, a no-brainer. Pick's formula, however, allows us to do it without measurements of lengths or angles: one simply has to count.

Let us look for a proof. The simplest case is surely the unit square itself. Four grid points on the boundary (the vertices) and none in the interior—it checks out.

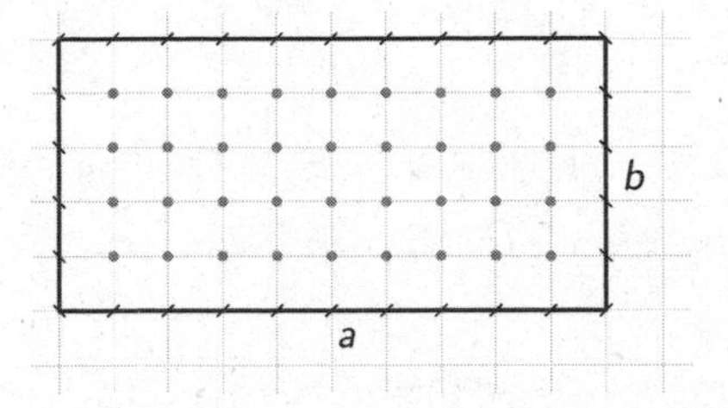

Figure 16.4. Verifying Pick's theorem for a rectangle with horizonzal and vertical sides.

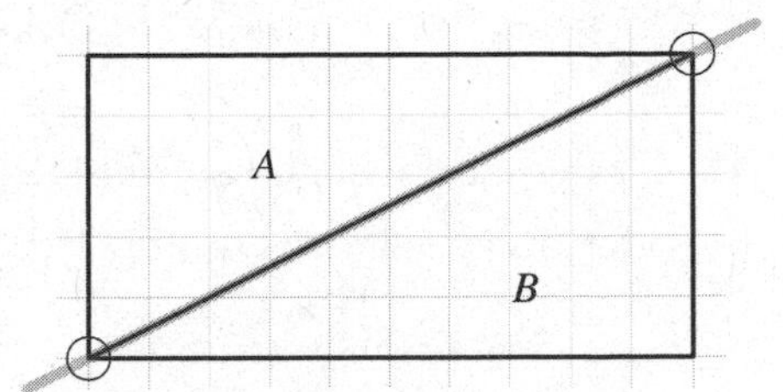

Figure 16.5. Verifying Pick's theorem for the two triangular halves of such a rectangle.

Next, we look at a rectangle whose sides are lines of the grid (Figure 16.4). If it has length a and width b (natural numbers, of course), then the number P_I of interior grid points is $(a - 1)(b - 1)$. The number P_B of grid points on the boundary is $2(a - 1) + 2(b - 1)$. The Pick number is ab, which is the area. It also checks out.

Next, let us divide the rectangle by means of a diagonal into two right triangles A and B (Figure 16.5). We show that

$$\text{Pick(rectangle)} = \text{Pick}(A) + \text{Pick}(B).$$

All grid points that are *not* on the diagonal contribute equally to the left- and right-hand sides. The same holds for the grid points (if any) that are on the diagonal, but not on a vertex. Indeed, on the left-hand side they count as interior points of the rectangle, and on the right-hand side as a boundary point of each of the two triangles A and B (twice one-half). These contributions cancel each other out. There remain the two vertices on the diagonal. They contribute $\frac{1}{2} + \frac{1}{2}$ on the left-hand side, and twice $\frac{1}{2} + \frac{1}{2}$ on the right-hand side (as there are two triangles). This is one too many, but thanks to the term -1 that belongs to each Pick number, things work out in the end. Thus, the Pick number of A is half of the Pick number of the rectangle, and hence half of its area. Therefore, it is equal to the area of A.

The same argument works for any polygon C that can be divided by a diagonal into two parts A and B:

$$\text{Pick}(C) = \text{Pick}(A) + \text{Pick}(B).$$

The Pick number is additive, just like the area. Hence, if Pick number = area holds for any two of these three polygons, it holds for the third one, too.

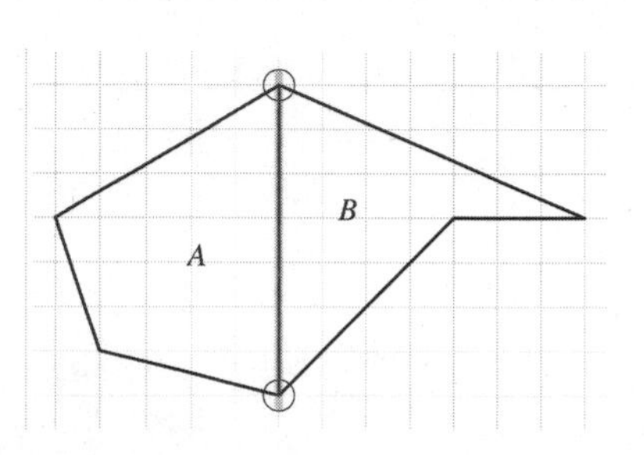

Figure 16.6. Dissecting a polygon C into two polygons by means of a diagonal.

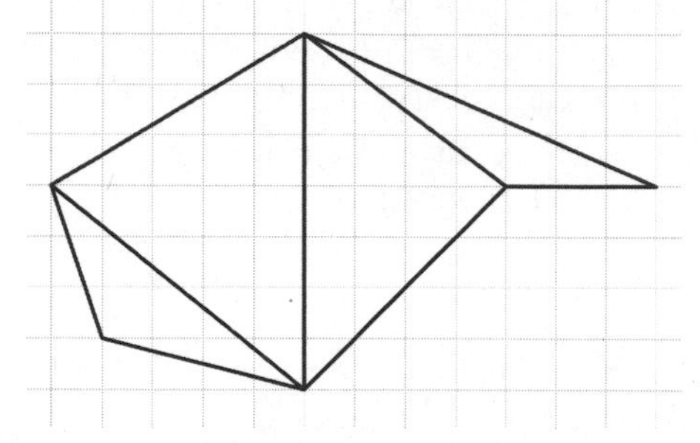

Figure 16.7. Dissecting a polygon into triangles.

Since every polygon can be divided by diagonals into triangles (Figure 16.7), we are done, right? Not quite, actually. So far, we know that Pick number = area only for those triangles that have a vertical and a horizontal edge. But here comes a trick (Figure 16.8): we can enclose any triangle in a rectangle whose sides are lines of the grid. That rectangle satisfies Pick number = area, as we know. We then snip off the surplus triangles, which all have a horizontal and a vertical edge. Since they, too, satisfy Pick number = area, so must the remaining polygon—which, after a few snips, is our triangle.

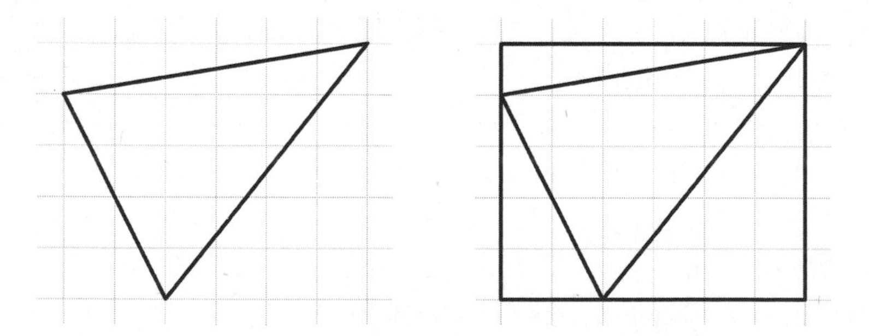

Figure 16.8. A triangle is enclosed in a rectangle with horizontal and vertical sides. The surplus triangles are then snipped off.

So, we are done: we proved Pick's theorem. Or have we? You may be worried about the division of every polygon into triangles. It seems intuitively obvious, and indeed, it is true—but it requires proof. And another objection is even more serious (and it seems even more far-fetched, at least at first glance). We have blithely spoken of *interior* grid points. Does every polygon have an interior? A diagram like Figure 16.9 shows that this is not so obvious. Toward the end of the nineteenth century, such an assertion could no longer go unchallenged. It had to be proved. It was, and is known since then as Jordan's curve theorem. The proof was not easy: it ran over many pages, and presented, a century later, quite a challenge to automated theorem provers.

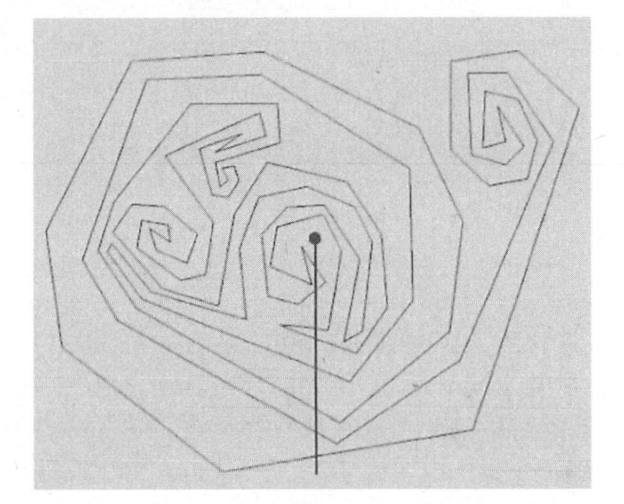

Figure 16.9. Is this point in the interior or the exterior of the polygon?

It is often claimed that one cannot understand a theorem if one does not understand its proof. Thus, we may ask: Can we understand Pick's theorem as long as we are unaware of the concern behind Jordan's curve theorem? Did we fool ourselves when we experienced this nebulous but irrefutable "understanding" that came with the proof? Is a proof worthless if it is incomplete?

George Polya, the Hungarian grand master of mathematical heuristics, gave the obvious answer: "For a strict logician, an incomplete proof is no proof at all." He added immediately: "Incomplete proofs may be useful when they are used in their proper place and in good taste."

Used in good taste! Clearly, much depends on whether you are a strict logician or a refined Epicurean. Utopians may foresee a golden age when humans relish the role of mathematics connoisseurs, like in wine-tasting parties. Arbiters of elegance will sample ingenious insights in "proof-tasting events," comparing impressions about the backbone and bouquet of a proof with their like-minded. Meanwhile, hidden from sight in chilly cellars, automated proof checkers toil to make sure that every proof that is served can be completely formalized.

Now for something entirely different: Farey series (named somewhat accidentally after a British geologist, John Farey by name, who lived from 1766 to 1826). It is not a series in the usual sense, but a sequence of ordered lists $F_1, F_2, F_3, \ldots$. Here, F_5, for instance, is the list of all fractions between 0 and 1, ordered by size, and having denominator at most 5. (Only reduced fractions are allowed: hence $\frac{2}{4}$ is out, but $\frac{1}{2}$ is in.) Thus, F_5 is the list

$$\frac{0}{1}, \frac{1}{5}, \frac{1}{4}, \frac{1}{3}, \frac{2}{5}, \frac{1}{2}, \frac{3}{5}, \frac{2}{3}, \frac{3}{4}, \frac{4}{5}, \frac{1}{1}.$$

The main theorem on Farey series is that if $\frac{a}{b}$ is any fraction from such a list, and $\frac{c}{d}$ comes next, then $bc = ad + 1$. For F_5 this is easy to check, but it also holds for F_{100} and F_{28741}. At first glance, it is a very curious result. It turns out, however, that it is an immediate consequence of Pick's theorem. This strikes one as odd. Indeed, Pick's result is about grid points and polygons, while Farey's concerns some lists of simple fractions. What does one have to do with the other?

Once you know the trick, it is obvious. The fraction $\frac{a}{b}$ corresponds to the pair of integers (a, b), which is a grid point (on the grid of mesh size 1 in the plane). In fact, the corresponding rational number is defined as the set of all grid points on the ray from the origin $(0, 0)$ through (a, b); see Figure 16.10. The slope of that ray is just $\frac{b}{a}$. As the fractions in the Farey series increase, the slopes decrease. Since $\frac{c}{d}$ is the fraction next after $\frac{a}{b}$ on Farey's list, there is no other grid point in the triangle spanned by (a, b), (c, d), and $(0, 0)$—not in the interior and not on the boundary. Hence, the area of that triangle is $\frac{1}{2}$, by Pick. Moreover, elementary geometry shows that this area is $\frac{bc - ad}{2}$ (Figure 16.11). This implies the main theorem on Farey series.

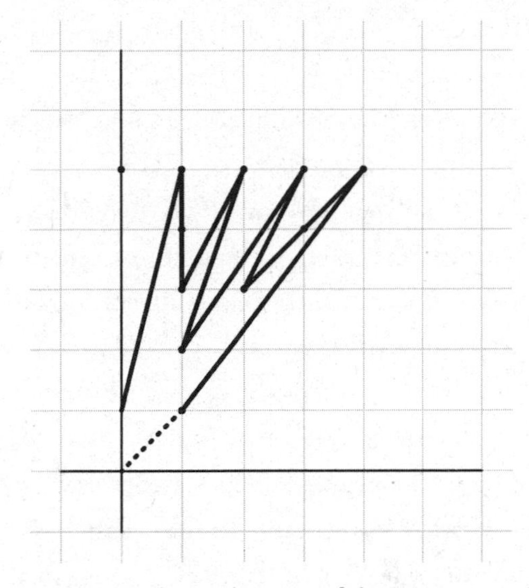

Figure 16.10. The grid points of the Farey series F_5.

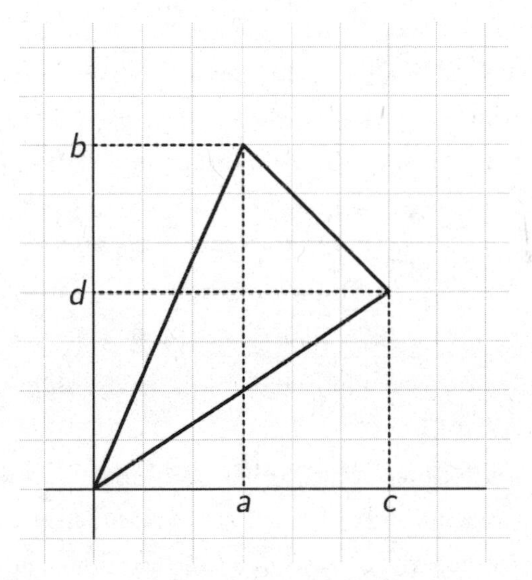

Figure 16.11. The area of the triangle spanned by (a, b), (c, d), and $(0, 0)$ is equal to the area of the triangle spanned by $(0, 0)$, $(a, 0)$, (a, b) *plus* the area of the trapezoid spanned by $(a, 0)$, (a, b), (c, d), $(c, 0)$ *minus* the area of the triangle spanned by (c, d), $(c, 0)$, and $(0, 0)$, and thus equal to $\frac{ab}{2} + \frac{(c - a)(d + b)}{2} - \frac{cd}{2} = \frac{bc - ad}{2}$.

Solid Numbers

Euler's celebrated polyhedron formula is another application of Pick's theorem.

Anyone looking up "polyhedron" on *Wikipedia* will read that "many definitions of polyhedrons have been given within particular contexts, some more rigorous than others, and there is no universal agreement over which of these to choose." This carries a whiff of scandal with it. Isn't universal agreement a hallmark of mathematics? Assuredly. But nobody is perfect.

A polyhedron is a bounded solid, that much is clear (or is it?). It has flat faces that are polygons; these faces join along edges, which are straight line segments; and the edges join at vertices. To make things simple, we shall also assume our polyhedron to be convex: every two points of it can be joined by a segment entirely contained in the polyhedron. This implies that the solid body can have neither holes nor hollow cheeks.

The Greek geometers were inordinately fond of polyhedrons, but they remained ignorant of some of the best news about them.

One fact discovered by Euler (and which Descartes, much earlier, had at his fingertips already) concerns a curious relationship between the number of vertices V, the number of edges E, and the number of faces F of any polyhedron: namely

$$V + F = E + 2.$$

This can easily be checked for the Platonic solids, for example: for cubes, F is 6, V is 8, and E is 12.

The polyhedron formula seems to have nothing to do with Pick's theorem: no grid, no area. However, it is a direct consequence, as we will see.

The first step in the proof consists in forgetting about the "solid": after all, Euler's formula concerns faces, edges, and vertices. They all belong to the two-dimensional surface of the polyhedron. Let us imagine that this surface is a rubber sheet. Let us pick any one face, name it the back face, and stretch it on a grid without tearing it. If we do enough stretching, all the other faces—the front faces—will also be stretched flat. We marked the vertices and edges, so they now form a network stretched across the back face. Neither V nor E have changed. By properly adjusting the flat rubber network, we can make sure that all the vertices end up on grid points. The stage is therefore all set for Pick.

Now that the stretching is done, the area of the back face is equal to the sum of the areas of the other faces (see Figure 16.12). Hence, the Pick number of the back face is equal to the sum of the Pick numbers of the front faces.

$$\text{Pick(back face)} = \text{sum of Pick(front faces)}.$$

From this, the Euler formula follows, by carefully counting grid points (an exercise for the ambitious, with a solution box on the next page).

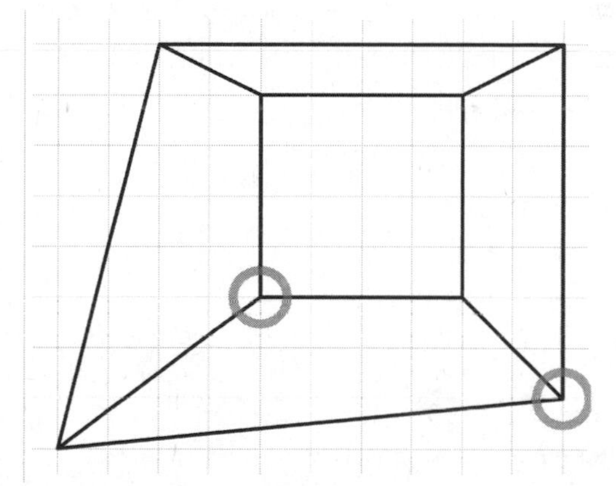

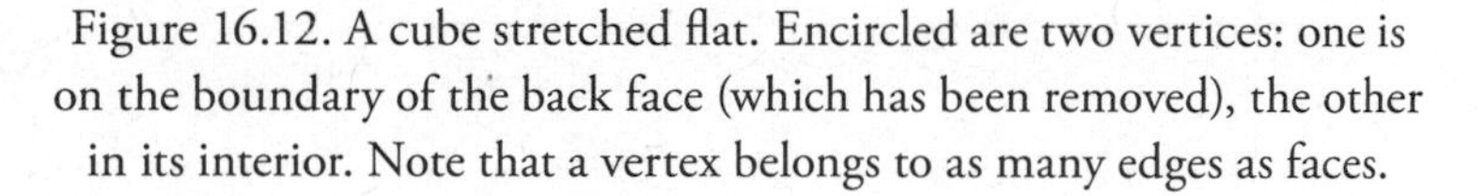
Figure 16.12. A cube stretched flat. Encircled are two vertices: one is on the boundary of the back face (which has been removed), the other in its interior. Note that a vertex belongs to as many edges as faces.

The polyhedron formula is not only one of the most widely treasured mathematical results. It plays a prominent role in the philosophy of mathematics, too. Indeed, the formula holds center stage in an iconoclastic book by Imre Lakatos, entitled *Proofs and Refutations*. The book appeared posthumously in 1976, years after the untimely death of Lakatos, but its chapters had been circulating long before, from hand to hand, almost like underground pamphlets.

We have to count grid points; but most of these occur on both sides of the equation

$$\text{Pick(back face)} = \text{sum of Pick(front faces)},$$

on the left as well as on the right, and thus will cancel. Let us neglect the vertices, for the moment. All grid points in the interior of one of the front faces are also in the interior of the back face; hence, their contributions will cancel. Away with them! Each grid point on an edge of the back polygon is also on the edge of a front polygon; away with it. Any further grid point on an edge of a front polygon belongs to two such front polygons, and therefore contributes twice weight $\frac{1}{2}$ to the sum on the right-hand side; but it is also an interior grid point of the back polygon, where it contributes weight 1. Again, we can remove these points from further consideration.

Only the grid points on the vertices are still left, and to them we now turn. There are V vertices altogether. Let us assume that v of them belong to the back face, and therefore $V - v$ to its interior. What remains on the left-hand side of

$$\text{Pick(back face)} = \text{sum of Pick(front faces)},$$

after all the canceling we have done, is

$$V - v + \frac{1}{2}v - 1.$$

On the right-hand side, we have the sum of $F - 1$ polygons, hence we will have to subtract $F - 1$. And then there are the $\frac{1}{2}$ contributions of the vertices (everything else has been canceled). To account for the number of these contributions, we have to sum, for one face after another, the number of their vertices. But to sum for each face the number of its vertices yields the same as to sum for each vertex the number of faces to which it belongs. Now a vertex belongs to as many faces as edges. The total number of edges is E, and each belongs to two vertices, hence our sum will yield $2E$. This must be corrected by the fact that we have to omit, for each of the v vertices of the back face, that back face itself. The right-hand side of

$$\text{Pick(back face)} = \text{sum of Pick(front faces)}$$

is therefore $\frac{1}{2}(2E - v) - (F - 1)$. Comparing with the left-hand side, we see that the terms with v obligingly cancel, and thus the equation yields $V - 1 = E - F + 1$, which is Euler's formula.

Lakatos had some firsthand knowledge of underground work. He hailed from Hungary, and fought with the Marxist resistance against the Nazis. After the war, he studied in Moscow, during the worst years of Stalin's rule. Some ten years later, he fled from his communist country, eventually landing at the London School of Economics, where he became a colleague of Karl Popper in the philosophy of science department.

The rebellious ideas of Lakatos fitted well with the youthful spirit of the 1960s, and acted like a whiff of fresh air at a time when the philosophy of mathematics had almost fossilized into dogmatic formalism. The playful style of his work made philosophers remember that metamathematics cannot account for what mathematicians are really doing.

Proofs and Refutations is a lively discussion between a teacher and some extremely clever students. They analyze a proof of the polyhedron formula, find loopholes and counterexamples, try to salvage the result, etc. Most of the action goes into finding what a polyhedron actually is, or rather (since "actually is" carries Platonic overfreight), how a polyhedron should be defined to make the proof work and the theorem come true. In this sense, it is language analysis. As Ian Hacking later summed it up: "The meanings of the words of the theorem have been so refined that indeed the theorem is true in virtue of what the words mean (when properly understood)."

Many of the arguments from *Proofs and Refutations* retrace, often verbatim, the opinions of eminent mathematicians. (As Lakatos says: "Philosophy

Figure 16.13. Imre Lakatos (1922–1974), playwright-philosopher.

of science without history of science is empty; history of science without philosophy of science is blind.")

By assuming our polyhedrons to be convex, we avoid most of the issues that the students from the book by Lakatos debate, but even so, our informal proof of how to deduce Euler's result from Pick's is full of holes. For instance, expanding the back face upon the grid, so that all edges are stretched taut, is at best a heuristic device. It is easy to imagine how to stretch a polygon on a rack; we need only think of rubber sheets. However, Greek geometers, who would hardly ever have encountered elastic materials, would have been outraged by the argument. To us it seems supported by intuition, but not of the sort that Kant would have recognized as a priori basis of geometry. It is clear that a formal proof must be vastly more cumbersome. Mathematicians know that they have to come up with it, sooner or later. Their duty demands it.

Such formal proofs exist, assuredly. The formula for polyhedrons was one of the earlier triumphs of automated computer proving. Yet, how dry and unenlightening must a formal proof be, how different from the spirited banter in the classroom of Lakatos! A computer has no notion of handling things in space, be they solid bodies or elastic nets.

Between Aha! and Oops!

As almost all mathematicians agree, proof is the sine qua non of mathematics. Yet, Ian Hacking has pointed out that there are two very different ideals of proof: he calls them Cartesian and Leibnizian.

On the one hand, "there are proofs that, after some reflection and study, one totally understands, and can get in one's mind 'all at once'. That's Descartes." And on the other hand, "there are proofs in which every step is meticulously laid out, and can be checked line by line, in a mechanical way. That's Leibniz."

Anyone on the lookout for pleasure must turn to the Cartesian side. Cartesian proofs are often associated with a flash of understanding, an aha!

experience, sudden enlightenment. "A new light flashed upon the mind of the first man (be he Thales or some other) who demonstrated the properties of the isosceles triangle." Thus spoke Immanuel Kant.

It may well happen that such a flash of understanding is eventually succeeded by other and more disagreeable insights, whether in a flash or in a slow dawning. One has overlooked something; there are counterexamples; the proof is incomplete, even fallacious. From "aha!" to "oops!" is only one small step. This "oops!" in turn can be followed by an insight about how to repair the proof, or adjust the conjecture, or refute it, and so on. In the end, of course, one only feels safe with a Leibnizian proof. By then the fun is gone.

Whoever wants to experience the thrill of mathematical understanding need not deeply dig into elaborate theories. So-called recreational mathematics—logical puzzles and brain-teasers—work on the same principle. They are not to be despised. Just as one can conceive of a philosophical book based entirely on jokes, one can imagine a textbook on mathematics grounded only on riddles. As George Polya, the mathematical mentor of Imre Lakatos, wrote: "Elementary mathematical problems present all the desirable variety, and the study of their solution is particularly accessible and interesting."

Here are two such pedestrian examples, well-used and homely.

Number one: Two trains leave at the same time, a slow one from A to B, a faster one from B to A. Eventually, they rush past each other. The fast train arrives at B exactly one hour later, three hours before the slow train reaches its destination A. Question: How much faster is the fast train?

Some readers may wish to close the book and figure it out by themselves. This is the type of reader one can only pray and hope for.

Well, reader, let's go on. The answer is: the faster train runs twice as fast. Indeed, suppose its speed is x times as fast as that of the slow train. After they cross, the slow train has x times more miles to cover than the fast train, and will take x times more time for each mile. Thus, the slow train will need x^2 times more time to reach its destination. We know that the fast train needs one hour, and the slow train four hours. Thus $x^2 = 4$, and so $x = 2$.

The problem can be posed in any physics class in college. It can cause some confusion, but it is not really hard. A famous Russian mathematician of the last century used this example to describe the typical level of mental effort researchers have to apply on average, in their daily work. What differs is, of course, where this mental effort is applied. College students know that they are facing an exercise that countless other students have gone through, a task that requires a few hours at most—maybe only minutes, if a lucky inspiration perks up. Researchers, by contrast, do not know beforehand whether they are confronting a stumbling block of no more than average difficulty or an insurmountable obstacle.

Example number two: We have two barrels, one filled with wine, the other with an equal volume of water. In a first step, a spoonful of wine is taken from the first barrel and added to the second. Then, a spoonful from the second barrel is added to the first. Question: is there more wine in the water barrel or more water in the wine barrel?

The first impulse is to say: there is equally much. Then comes the doubt: the first spoonful was pure wine; the second, however, consisted of water mixed with a little wine.

The first impulse was right, although some people take a while to be convinced. The understanding comes with the "insight" that both barrels, in the end, contain the same volume of liquid, just as they did before. Hence, whatever wine is now missing from the first barrel must have been made up by water from the second. Aha!

What is this "understanding," which is so dear to mathematicians and yet neglected by most philosophers of mathematics? Most philosophers, by the way, but not all. Once again, Wittgenstein walks out of step. He was fascinated by the conviction carried by a mathematical demonstration. We can understand the proof, but cannot explain why we understand it. There is no second-order understanding, so to speak.

As to Wittgenstein, *he* did not even try to explain. Explaining was *not* his business, as he insisted. But he did his best to *describe* mathematical reasoning. He stressed time and again that a proof must be "*übersehbar*" (meaning perspicuous, or surveyable, or to be taken in with one glance—Cartesian, to use Ian Hacking's term). The proof of Pythagoras at the start of this chapter would have given him satisfaction.

Unfortunately, only a few theorems can be reached by such proofs. Usually, the best one can hope for is a series of moves, one following the other, and each one perspicuous. The famous mathematician Paul Erdős, one of the most prolific mathematicians of the last century, claimed that God (in his words, "the Supreme Fascist") kept these proofs jealously locked away in a book. The highest accolade a mathematical demonstration can earn is therefore: "It is from the book."

What happens when we experience proof? Neurobiologists have scanned mathematicians' brains, and duly found that some bits of lobes light up, and others not. It does not explain why understanding proof provides pleasure. Can evolutionary biology provide the key? So far it has not. Mathematical hedonism remains mysterious.

It is well-known that mathematical discovery precedes proof—often by ages. Insight relies on induction and analogy.

Let us consider another example. Once more, let us turn to Pick's theorem. What follows is no proof, merely an argument. However, it *explains* the formula of Pick.

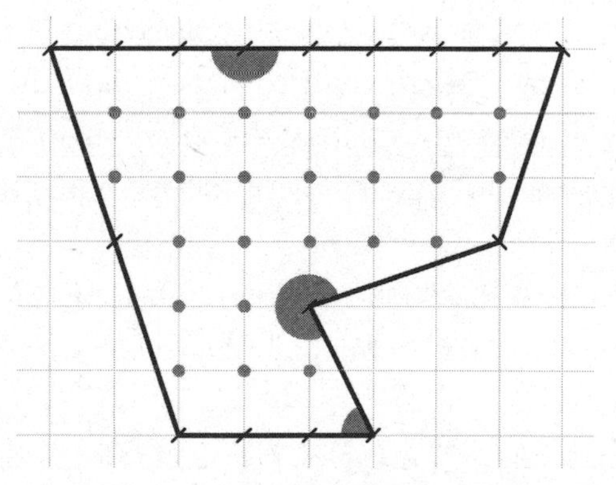

Figure 16.14. Imagine that each grid point spawns one drop of water. For a grid point on the boundary, the contribution to the interior of the polygon is proportional to the interior angle (as shown by the three shaded examples).

Suppose that each grid point, in the plane, spawns one drop of water, which then spreads evenly on the plane: one drop for every unit square of the grid. By spreading, the water will eventually cover every edge of the polygon. As much will flow from one side of the edge to the other as vice versa, simply by symmetry. Indeed, if we rotate the edge by 180 degrees around its midpoint, we get precisely the same configuration of edge and grid points. Since the flow-in–flow-out principle holds for each edge, there will be no net flow in or out of our polygon.

Now, let us compute how much each grid point contributes to the amount of water within the polygon. We can admit that grid points in the interior contribute all of their water; indeed, if some of it would flow out, as much would flow back in. As to the grid points on the boundary, each contributes in proportion to the interior angle it shares with the polygon. If this angle is 180 degrees, it contributes $\frac{1}{2} = \frac{180}{360}$; if it is 60 degrees, then $\frac{1}{6} = \frac{60}{360}$; and so on.

To compute the sum of all these interior angles, imagine a beetle running counterclockwise around the edges of the polygon. Altogether, the beetle makes a full turn of 360 degrees. At each of the grid points on the boundary, it turns by the amount (180 degrees *minus* interior angle). This holds even for the grid points that lie on an edge, but not on a vertex: at such a point, the beetle keeps going straight ahead. It also holds if the interior angle is larger than 180 degrees, as will happen at concavities of the polygon. The beetle, there, turns by a negative angle. Hence, the total turn of 360 degrees is the sum of the terms (180 degrees *minus* interior angle), summed over all P_B grid points on the boundary. Thus, the sum of the interior angles is $(180P_B - 360)$ degrees. Divided by 360, this yields $\frac{P_B}{2} - 1$. This, then, is what the grid points on the boundary contribute to the amount of water in the polygon. We obtain Pick's formula, and (as a bonus) we understand why the term -1 crops up.

This argument is certainly no proof. It is not even an informal proof. The physical intuition about the water drops oozing out from the grid points is far too slippery. Yet, physical thought experiments of the same kind—with water flowing, electrical currents circulating, weights balancing each other—have a distinguished past as heuristic tools. They have motivated some of the greatest theorems of Archimedes and Riemann.

At the same time, the physical thought experiments are notorious for playing tricks on us. One of the best-known examples for such a cognitive lure is the spring paradox.

Consider a weight hanging from a spring that hangs on a string that hangs on a spring that hangs on a hook. (Okay, better take a look at the left part of Figure 16.15.) There are two more cords of equal length, which are slack because they carry no weight. One connects the weight to the lower end of the upper spring; the other connects the upper end of the lower spring to the hook.

What will happen if the string between the two springs is cut? It is obvious, say the innocent: the weight will sag down until the two slack strings become tight. Or so one would think if one were not warned that a paradox is in the waiting. In fact, the weight will *lift up*. The two springs are still under tension, but less than before (as shown by the right part of Figure 16.15). The springs support the weight in parallel, instead of supporting it in series, as they did previously.

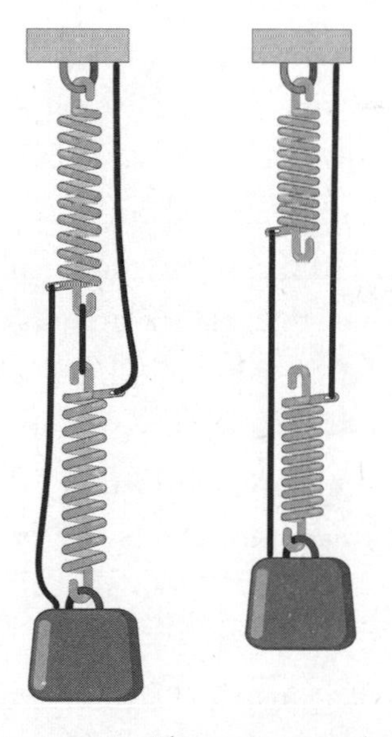

Figure 16.15. The spring paradox.

The little experiment has reappeared in a slightly changed, and even more puzzling context, as Braess's paradox. Replace the springs and strings by roads. Just as the weight can lift when a string is cut, so the traffic can flow faster when a road is blocked! In other words, if a new road is built, it can slow down the drivers. Again, this is a counterintuitive outcome; and it is related to a social dilemma. Drivers choose whatever route happens to be faster for them. This can result in a traffic jam.

Intuition, thus, can be treacherous, and understanding premature. The oscillation between aha! and oops! must have been one of the earliest motives for doing mathematics, from Thales (or some other) onward.

A Most Extraordinary Law

An insight leading to an understanding is a pleasure. But sometimes, an insight leads not to understanding, but to bewilderment. It reveals a mystery. Such an experience can be both wonderful and unsettling. It can only be conveyed by example. But which example to choose? Mathematics is teeming with them.

Let us take our cue from Leonhard Euler, that master mathematician, or rather mathemagician, whose virtuosity seems beyond compare. Portraits show him usually with a squint. His left eye turned blind when he was not yet thirty, and in his last years, he was almost completely blind in both eyes. Yet, his mathematical vision remained unimpaired, and his productivity kept its staggering pace. His completed works are still not all published. They will eventually fill more than eighty large volumes.

Here is a short sketch of what Euler termed "a most extraordinary law" ("*une loi très extraordinaire*" in his words—this was one of the rare occasions when he wrote in French, rather than in Latin).

For some reason or another, Euler computed the product

$$(1 - x)(1 - x^2)(1 - x^3)(1 - x^4)(1 - x^5)(1 - x^6)(1 - x^7)\ldots.$$

It is a product of infinitely many terms, understood as the limit of a sequence of finite products, starting with $(1 - x)$ as the first term, followed

Figure 16.16. Leonhard Euler (1707–1783), master mathemagician.

by $(1 - x)(1 - x^2)$, then $(1 - x)(1 - x^2)(1 - x^3)$, and so on. Each of these finite products is easily worked out:

$$1 - x,$$

$$1 - x - x^2 + x^3,$$

$$1 - x - x^2 + x^4 + x^5 - x^6,$$

$$1 - x - x^2 + 2x^5 - x^8 - x^9 + x^{10}.$$

And so on. It is child's play. After some time, one sees that the first terms of these polynomials "freeze." They are no longer affected by multiplication with $(1 - x^n)$, once n is larger than their degree. In this sense, it turns out that the infinite product of all $(1 - x^n)$ is the infinite sum

$$1 - x - x^2 + x^5 + x^7 - x^{12} - x^{15} + x^{22} + x^{26} - x^{35} - x^{40} + \dots.$$

Never mind, by the way, whether the series converges or not. We are just interested in the transformation of the infinite product to an infinite sum, by using the rules of algebra.

Something strange seems to go on.

In the *product*, all the powers x^n appear on equal footing.

In the *sum*, x^n occurs only for $n = 1, 2, 5, 7, 12, 15, 22, 26, 35, 40,\ldots$. The first two powers of x are preceded by a minus sign, the next two by a plus, the next two by a minus again, the next two by a plus sign, and so on. This periodic alternation of two minus signs and two plus signs is regular enough. But what about the sequence of integers? Let us inspect them again:

$$1, 2, 5, 7, 12, 15, 22, 26, 35, 40, \ldots.$$

These integers are named, for reasons that need not concern us here, the generalized pentagonal numbers—let us call them the genpen numbers. (In *The On-Line Encyclopedia of Integer Sequences*, the sequence is prosaically named A001318.)

Is there any rule behind the increase of the genpen numbers? Yes, and Euler had it certainly figured out in a flash. Look at the differences between consecutive numbers: these differences are 1, 3, 2, 5, 3, 7, 4, 9, 5, ... The odd positions in this sequence (the first, third, fifth, . . .) show simply the natural numbers 1, 2, 3, 4, . . . ; and the even positions (the second, fourth, sixth, . . .) show the odd numbers, namely 3, 5, 7, 9, ... With this rule, it is easy to extend the sequence of genpen numbers.

Now for what seems to be a complete change of topic.

Let us consider, for any natural number, the sum of its divisors. The divisors of 12, for instance, are 1, 2, 3, 4, 6, and 12, and their sum is 28. The divisors of 13 are 1 and 13, and their sum is 14. In general, the sum of divisors of a prime number p is $p + 1$. In fact, prime numbers are exactly the numbers whose sum of divisors exceeds them by 1. (For this reason, notes Euler, the number 1 should not count as a prime number: the sum of its divisors is 1, and not 1 + 1. Of course, 1 is not a composite number either: it stands on its own, the one and only unit.)

When Euler studied sums of divisors of natural numbers, he was struck by their irregularity. The first ten terms are 1, 3, 4, 7, 6, 12, 8, 15, 13, and 17. This start is not auspicious, and it does not improve. The sums of divisors for the numbers from 40 to 50, for instance, are 90, 42, 96, 44, 84, 78, 72, 48, 124, 57, and 93.

"If we examine a little the sequence of these numbers," wrote Euler, "we are almost driven to despair. We cannot hope to discover the least order."

Nevertheless, since this is Euler—

"Nevertheless," Euler continued, "I observed that this sequence is subject to a completely definite law and could even be regarded as a recurring sequence: each term can be computed from the foregoing terms, according to an invariable rule."

This rule yields the sum of divisors of the number n, if the sums of divisors of all smaller numbers are known. The rule says:

Take the sum of divisors of $n - 1$.
Add the sum of divisors of $n - 2$.
Subtract the sum of divisors of $n - 5$.
Subtract the sum of divisors of $n - 7$.
Add the sum of divisors of $n - 12$.
Add the sum of divisors of $n - 15$.
And so on.

How such a rule can be discovered seems nothing short of miraculous: but Euler had a well-prepared mind. The numbers 1, 2, 5, 7, 12, and 15 are the first genpen numbers. The alternation of signs—twice plus, twice minus, twice plus again, and so on—mirrors exactly what happens in the transformation of the infinite product into an infinite sum. So, in the rule above, you add the first two differences of the type (n - genpen number), subtract the next two, then add the next two, and so on. Of course, you have to stop when the difference (n - genpen number) becomes negative, i.e., when we reach a genpen number that is larger than n. Oh, and one last thing: if such a difference is 0 (that is, if n happens to be a genpen number itself), then replace "sum of divisors of (n - genpen number)" by the number n itself. So, that's the "invariable rule," and it is mind-boggling indeed.

Leonhard Euler was not yet able to give a proof, but he clearly enjoyed describing how he convinced himself of this "most extraordinary law." To start with, he checked it for the first twenty numbers, and then more. For instance, for the sum of divisors of 12:

Start with the sum of divisors of 12 – 1 (i.e., of 11).
Add the sum of divisors of 12 – 2 (i.e., 10).
Subtract the sum of divisors of 12 – 5 (i.e., 7).
Subtract the sum of divisors of 12 – 7 (i.e., 5).
Add the sum of divisors of 12 – 12 (i.e., 0).

This last term, as we were told, has to be replaced by the number we started with, namely 12. Altogether, the looked-for sum of divisors of 12 is, according to this rule, 12 + 18 – 8 – 6 + 12, and lo and behold, we obtain the right answer, namely 28.

Euler verified his rule for many numbers—for instance, the numbers 101 and 301—and then concluded sardonically: "I think these examples are sufficient to discourage anyone from imagining that it is by mere chance that my rule is in agreement with the truth."

Thus, the sequence of the sum of divisors follows a hard and fast law! This is all the more surprising as the task of finding the divisors of a large number, a number with twenty or sixty digits, say, is an extremely arduous and time-consuming job, the prototype of a "hard" problem.

Moreover, at first glance, the sequence of the sum of divisors should be even more mysterious than the sequence of primes. Indeed, it singles out the primes (as those numbers n for which the sum of divisors is $n + 1$). Yet, the sequence follows a clean recursive rule, whereas the sequence of primes, according to Euler, "seems not to display the least trace of order." It has resisted all efforts by mathematicians over the ages. As Euler said, "We have every reason to believe that there is some mystery which the human mind shall never penetrate." The primes seem to show up without rhyme or reason.

Euler wrote this in 1751. Some ten years later, he came up with a proof of his "most extraordinary law." Mathematics has progressed tremendously since then; yet the mystery of the primes is still with us. It has become, to quote the title of a book by John Derbyshire, a "prime obsession." It is the subject matter of the celebrated Riemannian hypothesis, which was problem number eight in Hilbert's famous list from the year 1900, and survived the twentieth century to make it into the equally famous list of the Clay Institute's Millennium Problems. In a very specific and precise sense, the conjecture says that the distribution of prime numbers cannot be distinguished from a random sequence. Nothing seems more regular than the natural numbers; and nothing more irregular than the arrangement of their building blocks, the primes.

We see only a small part of the iceberg. When thinking of all that is hidden, one cannot help being a Platonist.

The experience of mysteries that appear to be unfathomable to the human mind is not specific to mathematics. Pondering about quantum entanglement, the Big Bang, the complexities of molecular biology, or the emergence of consciousness can be very humbling. Yet, these questions concern the so-called real world, which can be viewed as an accident. It is no great scandal that the universe appears impenetrable, cold, and altogether oversized. What else would you expect? Yet, mathematicians and philosophers can experience the same sense of utter alienation without ever looking at the outside world—without leaving home, so to speak.

The earliest Platonic dialogues usually ended in puzzlement—*aporia*, as the ancients called it. As Wittgenstein wrote: "A philosophical problem has the form: I don't know my way around." Should philosophers ever lack problems (which is unlikely), all they need do is turn to mathematics.

Acknowledgments

This book has been shaped, in large part, by my former students. For several dozen years, I have been privileged to witness their first immersion into the world of mathematics, cohort after cohort. Their questions, anxieties, suggestions, doubts, and criticisms have been essential, not least because they kept me from forgetting my own youthful moments of puzzlement and thrill.

For the artwork, I have to thank Bea Laufersweiler and Alexander Helbok, as well as Karl-Heinz Gröchenig and Hannelore De Silva. Martin Mayerhofer and Markus Slawitschek were a great support.

For helping me along through catalogs and archives, I thank Marcia Tucker from the Institute for Advanced Study, Michael Nedo from the Wittgenstein Archives of Cambridge University, Stefan Sienell from the Austrian Academy of Sciences, and Thomas Maisel from the University of Vienna. Thanks to Bent Sofus Tranoy and Stephen Burch for the permission to use images of Wittgenstein and Ramsey.

I learned much from my colleagues, in particular Herlinde Pauer-Studer, Jakob Kellner, Friedrich Stadler, Christoph Limbeck, Harald Rindler, Peter Michor, Walter Schachermayer, Christoph Hauert, Klaus Schmidt, Michael Eichmair, Immanuel Bomze, Josef Hofbauer, Christian Hilbe, Georg Pflug, and Jean-Robert Tyran. I am also very grateful to Brian Skyrms, Jean-Pierre Aubin, Leopold Schmetterer, Robert Leonard, Glen Shafer, Reinhard Siegmund-Schultze, Don Saari, Douglas R. Hofstadter, and Cheryl Misak for their kind help and expertise. I owe particularly much to the good-humored probing by Martin Nowak (an incorrigible Platonist). My editor, T. J. Kelleher, helped a lot in putting the book into shape.

I hardly need to say that without the support and the patience of my son, Willi, and my wife, Anna Maria, this book would still not be finished.

That being said, I must point out that my cat Monty has proved utterly resilient to my demands for undisturbed working time. She is even apt to walk purposely across my keyboard. All remaining mistakes are therefore hers.

References

CHAPTER 1

A wonderful introduction to geometry is offered by Coxeter (1989). Some classic texts on the history of geometry are Bell (1940), van der Waerden (1954), Kline (1972), Struik (1987), and Cooke (2005).

For more on ancient Greek mathematics, see Heath (1981), and for more on *Meno* and anamnesis, see Klein (1965) and Day (1994). We will return to the Delian problem in the next chapter.

For some 360 more proofs of the theorem of Pythagoras, see Loomis (1968). One of them appears in Chapter 16. For the amazingly short proof of Morley's theorem by John H. Conway, see Conway (2001), Roberts (2015), or Cardil (2015). (The latter can serve as an example of the splendid opportunities offered by computer graphics.)

The quote by H. S. M. Coxeter is from Coxeter (1989).

The fallacious proof that all triangles are isosceles rests on the erroneous assumption that the point P (the intersection of the bisector of angle A and the perpendicular bisector of BC) is in the interior of the triangle. This is not the case. The next fallacy (rearranging pieces leads to a loss of area) is based on the fact that the two large "triangles" are not triangles at all. The line that appears to be a hypotenuse is bent. This example is from Du Sautoy (2010). For more paradoxes and fallacies, see Bunch (1997).

Fowler (1999) offers a highly readable guide through Euclid's mathematical biotope. For a biography of Hilbert, see Reid (1996); for a shorter account, see Stewart (2017b). For some more accounts of Euclid and co., see Reid (2004) and Dodgson (1879). The German geometer insisting on total obscurity is W. A. Diesterweg.

For Lobachevsky, see Stewart (2017b). The dramatic developments of non-Euclidean geometry and their impact are retold in Greenberg (2007). For more on Kant's view of geometry, see Friedman (2012). For more on Schopenhauer's caustic views of mathematics, see Lemanski (2020). The quote by Schopenhauer is from *The World as Will and Representation*, Vol. II, Ch. 13 (1891).

For the biological foundations of geometric cognition, see, for example, Hohol (2019).

Darwin's quote is from Darwin (1987).

For what mathematics has done to some (and only some) philosophers, see Hacking (2000), to whom we return in Chapter 16.

The quote by Newton is from *Philosophia Naturalis Principia Mathematica* (1687).

CHAPTER 2

The development of the number concept is a main course in every history of mathematics; see, for example, Bell (1940), Kline (1972), or Hodgkin (2005). For the legendary Kronecker quote on integers, see Weber (1893).

For mathematical introductions to number systems, see Stewart and Tall (2015), Ebbinghaus et al. (1996), and Cory (2015). A philosophical introduction from 1937 that is still highly readable today is Waismann (2003)—his quote is from there.

The quote by Leibniz is from *Gesammelte Werke*, Vol. 5.2.

On the reason why $0 \times 0 = 0$ etc., see Gowers (2003). The Gauss quote on "One should never forget..." can be found in Kline (1972). On indirect proof as irony, see Polya (1945) and Ording (2019). For the classic texts introducing real numbers via cuts, see Dedekind (1963) and Grattan-Guinness (2000). Such cuts will reappear in Chapter 15.

For constructible numbers, see Coxeter (1989), Stewart (1989a), Courant and Robbins (1996), and Kazarinoff (2003). The Gauss quotes are from his *Theoria residuorum biquadraticorum* (1831). For higher number systems, see Ebbinghaus et al. (1996).

CHAPTER 3

For der Kluge Hans, see Gross (2014). For the Ishango bone, see Huylebrouck (2019) and Stewart et al. (1996). A splendid artistic celebration of chalkboards in mathematics departments is Wynne (2021).

For more on the number sense, see Dehaene (2011). For numbers in the history of cultures, see Everett (2017) and Rudman (2007). The conflicting views of Mill and Frege are discussed in Shapiro (2000).

The letter by Lord Byron dates from December 1812.

Proof by induction is described in every introductory textbook of mathematics. For two particularly appealing presentations, see Polya (1954) and Knuth (1997). The term "passage to the next integer" was proposed by Polya (1945).

The Wittgenstein view on jokes is mentioned in Malcolm (1962). The Frege paraphrase is from Frege (1884); see van Heijenoort (1967).

For the history of mathematical induction, see Freudenthal (1953). For Poincaré's philosophical views and the quotations in the text, see Poincaré (1982) and Folina (1992). On Peano's axioms in historical context, see Segre (1994) and Grattan-Guinness (2000). The skeptical take on the analytic-synthetic dichotomy is from Quine (1953). A technical point: Dedekind's proof that Peano's axioms define natural numbers uniquely presupposes second-order logic; for first-order logic, the theorem of Skolem shows that this is not valid.

For some books on infinity, see Gamow (1946), Stewart (2017a), Rucker (1982), and Barrow (2006). For biographies of Cantor, see Purkert and Ilgauds (2013) and Dauben (2014). The Descartes quote can be found in Barrow (2006), the Locke quote is in "An Essay Concerning Human Understanding" (1690), and the Gauss quote is from his letter to H. C. Schumacher from July 12, 1831. The Wittgenstein quote "your head whirls" is from his *Lectures on the Foundations of Mathematics* (Diamond 1989); the searchlight quote is from *Philosophical Remarks*, Section 142.

For a clear introduction to set theory, see Stewart and Tall (2015). For more on cardinals, see Conway and Guy (1996). For more on space-filling curves, see Sagan (1994). The quotes by Kronecker and Cantor are from Purkert and Ilgauds (2013). See Hilbert (1925) for Cantor's paradise.

Frege's *The Foundation of Arithmetic* and Russell's *Introduction to Mathematical Philosophy* are still a pleasure to read. Translations of Frege's *Begriffsschrift* and the paper on transfinite ordinals by John von Neumann (then still Janos) can be found in van Heijenoort (1967).

CHAPTER 4

For biographies on Kurt Gödel, see Budiansky (2012), Goldstein (2005), Dawson (1997), Dawson and Sigmund (2006), Sigmund et al. (2006), and Regis (1987). For a sample of books on Gödel's work, see Baaz et al. (2011), Fefermann (1986, 1996), Franzen (2005), Hintikka (1999), and van Atten and Kennedy (2003).

The Schlick letter is quoted in Sigmund (2017). The Kant quote about logic is from his *Critique of Pure Reason* (1781). For Russell's contrary opinion, see Russell (1946). A mathematician's take of Aristotelian logic can be found in Hilbert and Ackermann (1928). For a superbly commented edition of Lewis Carroll's *Alice in Wonderland* and *Through the Looking Glass* by Martin Gardner, see Carroll (1990).

For a biography of Boole, see Nahin (2012). The quote by Russell about Boole's book is from Russell (1918). The interplay of logic and mathematics in the nineteenth century is well covered by Grattan-Guinness (2000) and Kline (1980). For Cantor's view of a set as an abyss, see Dedekind (1932). For his definition as "a collection . . . into a whole," see Cantor (1895) and Dauben (1990).

For biographies of Russell, see Russell (1967–1969) and Monk (1997, 2001). The comics-style biography by Doxiadis and Papadimitriou (2009) is a delight. The Musil quote is from Musil (1913). For introductions to set theory, see Devlin (1993) and Tiles (2004). For more on Zermelo, see Ebbinghaus (2007) and O'Connor and Robertson (1999).

For a biography of Hilbert, see Reid (1996). On Hilbert's program, see Mancosu (1998, 2010), Sieg (2013), and Zach (2019). An elegant introduction to mathematical logic is due to Quine (1981).

Gödel's proof was superbly explained in Doug Hofstadter's bestselling *Gödel, Escher, Bach* (1979). For short sketches of the main proof idea, see Gödel himself (1931) and a monologue in Hugh Whitemore's 1986 play *Breaking the Code*. The sentence-by-sentence description of Gödel's incompleteness proofs is in Hoffmann (2017). See also the books Smullyan (1992), Nagel and Newman (2005), Franzen (2005), and Chaitin (2007). The "in a sense fortuitously" quip is in footnote 15 of Gödel (1931).

Hilbert's radio address (including the final guffaw) is preserved; see Smith (2014b).

For more on Wittgenstein's take of Gödel's and Hilbert's work, see Floyd (2019). The Wittgenstein quote "Gödel confronts us . . ." is from Wittgenstein MS 121 (MS for manuscript) in his *Nachlass*; "the mathematician's superstitious fear..." is from Wittgenstein (1956), Appendix III, 17; "there is always time..." is from Diamond (1989), 209; and "hint from the gods..." is from Wittgenstein (1956), IV, 56.

CHAPTER 5

The classic biography of Turing is Hodges (2012). For two other highly readable biographies, see Leavitt (2006) and Copeland (2013).

On the Mechanical Turk, see Standage (2002) and Levitt (2000). For a more recent chess-playing automaton, see Hsu (2004). The path-breaking paper on AI is Turing (1950).

The *Entscheidungsproblem* is from Hilbert and Ackermann (1928). The seminal Turing paper is Turing (1937), in which appears the "child's arithmetic book" quote. For its history, see Copeland and Fan (2023). For the quote "It is difficult today...," see Newman (1955). The visiting card example is by Alvy Ray Smith; see Smith (2014a).

On undecidability, see Davis (2001). Davis (1965) contains many of the early contributions to this issue. The Gödel quote on the right perspective is reported in Wang (1974).

The life and death of Turing has attracted a great amount of attention, including a play, *Breaking the Code* by Hugh Whitemore, and a film, *The Imitation Game* (2014) with Benedict Cumberbatch. For the early history of computers, see Ifrah (2001), Copeland (2012), and Dyson (2012). The coroner's quote can be found in Copeland (2012).

For a double biography of Boole and Shannon, see Nahin (2012). The skeptical (and outdated) Bourbaki quote is from Bourbaki (1968). The December 2008 issue of *Notices of the AMS* is devoted to formal proofs, and this chapter draws heavily on the papers Hales (2008), Harrison (2008), and Wiedjik (2008), all from that issue. The quotes by Bacon and Bourbaki are gleaned from Hales (2007). See also MacKenzie (2001) and Davis (2001). For errors by great mathematicians, see Lecat (1935).

The Turing quotes about buzzing in his head and the child-machine can be found in Copeland (2013). Copeland (2004) is a collection of the basic papers by Turing. The Turing quotes about "men born in the usual manner" and "give a good showing in the game" are from Turing (1950).

On the *Monadology* of Leibniz, see Savile (2000).

For early calculating machines, see Ifrah (2001) and Kidwell and Williams (1992). The German Wilhelm Schickhard constructed a calculating machine even before Pascal and Leibniz, and wrote letters to Johannes Kepler about it. The Chinese thought experiment has attracted a huge amount of attention; see Searle (1980), also for the related quote. For a thoughtful discussion of consciousness, see Dennett (1991). The quote from Turing about cold porridge is from Turing (1950).

On ELIZA, see Weizenbaum (1966). An early overview on artificial intelligence is Hofstadter (1979). For recent books on what some call the golden age of AI, see Christian (2020) and Roose (2021). On the myth of the blank slate, see Pinker (2002). The Hofstadter quote and the Q and A with GPT are from a personal communication, July 2022. On AlphaZero, see Silver et al. (2018).

For the "scientific breakthrough of the year 2020," see Service et al. (2020). The GPT rhyme on primes is from Bubeck et al. (2023).

CHAPTER 6

For the history of infinitesimals, see Kline (1972), Bell (1940), Bell (2005), Edwards (1979), Barrow (2006), or Huemer (2016). For a many-perspective approach to infinitesimals and continuity, see Shapiro and Hellman (2020).

On Archimedes, see, for example, Stewart (2017b), Bell (1937), or Stein (1999). For more on the number π, see Eymard and Lafon (1999), Delahaye (1997), Posamentier and Lehmann (2004), Bailey et al. (1997), and Remmert (1996). For Spengler's views, see Spengler (1991).

The quotes by Leibniz, Cavalieri, Bernoulli, and Newton are taken from Kline (1972). For Leibniz's views of the continuum, see Leibniz (2001). A biography on d'Alembert is Hankins (1990); a novel based on him is by Denis Diderot (*D'Alembert's Dream*, 1830). For more on d'Alembert's origins, see Launay (2010).

For Zeno, see Cajori (1915), Grünbaum (1967), Barnes (1982), and Ushenko (1946). A famous treatment of the paradox of Achilles and the tortoise can be found in Hofstadter (1979). The harmonic series is used in Derbyshire (2004) as an introduction to the Riemann hypothesis (arguably the most famous conjecture in mathematics). For an elegant solution of the Basel problem, see Hofbauer (2000).

Heidegger's motto is from *Gesammelte Werke*, Vol. 65 (1989, Klostermann, Frankfurt). The quote by Hegel is from *Science of Logic* (1816), and the Jarmusch quip can be found in his film *Paterson* (2016).

For important books on the origins of calculus, see Mancosu (1996) and Baron (1969).

Berkeley (1734, 1710) is still a pleasure to read. The standard introduction to nonstandard analysis is Robinson (1996). On the life and work of Abraham Robinson, see Dauben (2014). Cantor's "cholera bacillus" comment is cited in Dauben (1990), and Russell's grumble is from Russell (1903).

CHAPTER 7

Excellent introductions to the theory of probability can be found in Billingsley (1995), Rényi (1970), and Ross (2013). Diaconis and Skyrms (2018) includes a minimalistic tutorial on probability, very much to the point. For general reading on the philosophical aspects of probability theory, see Diaconis and Skyrms (2018), von Plato (1998), Hacking (2006), Daston (1979), and Ekeland (1988). The philosophical aspects are covered in Galavotti (2005) and Hájek and Hitchcock (2016a, b).

Biographies of Voltaire are too numerous to list. For Condamine, see Condorcet's obituary in volume 2 of *Oeuvres Complètes de Condorcet* (1804) and Condamine's biography in the history of mathematics by St. Andrews University: O'Connor and Robertson (2014). On the role of roguery in the history of probability, see Bellhouse (1993). For "the first bomb thrown at the ancien régime," see Gay (1965).

For the role of Pascal, see Oystein (1960). Good historical introductions can be found in Hacking (2006), von Plato (1998), Stewart (1987), and Devlin (2008). For the bouquet of views on probability, the Maxwell quote is from Campbell and Garrett (1881); the Laplace quote from *Essai Philosophique sur les Probabilités* (1814); the Einstein quote from Isaacson (2007) (and countless other sources); the Schrödinger quote from Schrödinger (1929); the Monod quote from Monod (1971); and the Russell quote from a 1929 lecture, see Bell (1940).

Despite previous texts by De Moivre and Leibniz, the official birth of probability as a mathematical discipline dates from Jacob Bernoulli's book *Ars Conjectandi* (1713). A discussion of the "great Bernoulli swindle" can be found in Diaconis and Skyrms (2018).

The quote due to Shakespeare is from his play *Henry V.* The Musil quote is from Musil (1906).

For an English translation of Kolmogorov's classic, see Kolmogorov (1950). See also Billingsley (1995) for an introduction to measure theory.

The random walk results are a small sampler from Feller (1975). For more on the St. Petersburg paradox, see Diaconis and Skyrms (2018) and Martin (2013).

The two-envelope paradox is treated in Broome (1995). For many other paradoxes in probability, see Wikipedia (2023b). Another rich collection of probabilistic paradoxes is Szekely (1986).

CHAPTER 8

This chapter, just like the preceding one, owes a lot to Diaconis and Skyrms (2018) as well as to von Plato (1998). Some of the topics are very well covered in Delahaye (1999) and Stigler (1999, 2016). A thoughtful discussion is in Bunge (1988).

For the many facets of Richard von Mises, we refer to his biographer Reinhard Siegmund-Schultze; see, for example, Siegmund-Schultze (2009). In particular, Siegmund-Schultze (2006, 2010) shine a light on the mathematical problems with von Mises's approach. English translations of some of von Mises's works are von Mises (1951, 1981). His quote about being "outdated" is from the preface to the third German edition; see von Mises (1981).

For the early papers on *Kollektivs*, see Wald (1936), Ville (1936), and Church (1940). More on Ville's fascinating biography is in Shafer (2009). For the revival of random sequences, see Kolmogorov (1963), Solomonoff (1964), Chaitin (1967), Martin-Löf (1966), Schnorr (1971), and Li and Vitányi (1997). Good surveys on random sequences are Keller (1986) and Volchan (2002). For much more on randomness, see Calude (1994) and Chaitin (2001).

Von Neumann's joke about the state of sin is from Bhattacharya (2021), and Laplace's all-knowing intellect comment is from *Essai Philosophique sur les Probabilités* (1814).

On the butterfly effect, see Gleick (1987), Stewart (1989a), Ekeland (1988, 1993), or Peterson (1997). For biographies on von Neumann, see Macrae (1999) and Bhattacharaya (2021). For propensities, see Popper (1957, 1992).

Misak (2020) is an accomplished biography of Ramsey. Ramsey (1931) and de Finetti (1937) are the main sources for subjective probability; see also Fishburn (1970) and de Finetti (1999). The quote about "a cross between a light-house and a balloon" is in Misak (2020), and the quotes from the Moral Science Club are in Ramsey (1931).

The black sheep joke is from Cathcart and Klein (2008). The classic foundations of Bayesianism are Bayes (1763) and Laplace's 1774 article "Memoire sur la probabilité des causes par les évenements." The sunrise quote is from *An Enquiry Concerning Human Understanding* by Hume (1748). "Probability does not exist" is the motto of de Finetti (1937).

CHAPTER 9

For a first introduction to voting, see Saari (2008b). For more, see Saari (2008a, 2003, 1995), Aizerman and Alekserov (1995), and Nurmi (1999). For interesting philosophical aspects on the "will of the people," see Saari (2003).

Badinter and Badinter (1988) wrote a splendid intellectual biography of Condorcet.

More on bizarre outcomes of voting procedures can be found in Schofield (1983) and Saari (1989). On the likelihood of paradoxes, see Saari and Tataru (1999).

For some of the modern classics of voting theory, we refer to Arrow (1963), Gibbard (1973), and Sen (1970). An interesting voting scheme based not on ranking, but on grading, is presented in Balinski and Laraki (2007).

CHAPTER 10

A good general reference for decision theory is provided by Chapters 2 and 3 of Diaconis and Skyrms (2018). For a biography of Bentham, see Schofield (2009). For an introduction to utilitarianism, see Rosen (2003). For the panopticon as germ of the modern surveillance state, see Foucault (1977). Bentham's "greatest good" quote is from Bentham (1776).

A historical account of the St. Petersburg paradox is in Samuelson (1977). For the boundedness of the utility function, see Menger (1934) and Arrow (1974). The quote by Hume is from *A Treatise of Human Nature* (1739).

The first part of von Neumann and Morgenstern (1944) is devoted to the utility function. It is now a standard part of every economics curriculum. It later became known that cardinal utilities had been anticipated by Ramsey (1931). Another forerunner was Alt (1936).

For the Allais paradox, see Heukelom (2015). The main rival of expected utility is prospect theory; see Kahneman and Tversky (1979). A bestselling account of the paradoxes is given in Kahneman (2011). For good overviews, see also Szpiro (2011) and Poundstone (2010).

The views of Mill are described in Mill (2011). The US politician with the "unknown unknowns" was Donald Rumsfeld, Secretary of Defense (stated in response to a question at a news briefing on February 12, 2002). For risk and ambiguity, see Ellsberg (1961) and Raiffa (1961). For an insider report on the Pentagon Papers, see Ellsberg (2002).

For more on Pascal's wager, see Adamson (1995) and Jordan (2006). All the quotes by Pascal are from his *Pensées* (1670). The quotes by Voltaire are from the seventeenth of his *Lettres Philosophiques* (1734; also known as *Lettres Anglaises*), and the quote by Diderot is from *Pensées Philosophiques* (1746).

CHAPTER 11

In 1954, Richard Braithwaite gave his inaugural lecture in Cambridge on "The Theory of Games as a Tool for the Moral Philosopher"; see Braithwaite (1963). For excellent introductions to game theory, see Binmore (2008), Osborne and Rubinstein (1994), and Myerson (1997). Much of the material in this chapter is also in Sigmund (2010).

For the quote by Haidt, see Haidt (2007). For more on social insects, see Trivers (1985) and Dawkins (2016). The selfishness quote by Pascal is from *Pensées* (1670), and the quotes by Adam Smith from his *Wealth of Nations* (1776). The quip by Stiglitz can be found in the *Guardian* (December 20, 2002). For experiments on the donation game, see Camerer (2003) and Yamagishi (2010).

For a biography of John von Neumann, see Bhattacharya (2021). On the early years of game theory, see Leonard (2010). Von Neumann's pathbreaking paper on parlor games is von Neumann (1928).

The biography of John Nash is by Nasar (1998), and includes the re-discovering quote by John Milnor and Nash's comment on von Neumann's brush-off.

For more on the Prisoner's Dilemma, see Poundstone (1992). The first mention of the Prisoner's Dilemma in a textbook is in Luce and Raiffa (1957).

For sequential versus simultaneous donation games, see Yamagishi (2010). For more on the partners/rivals dichotomy, see Hilbe et al. (2018).

For more in-depth discussions on the Golden Rule, see Sober and Wilson (1998) and Sugden (1986). Rousseau's "natural goodness" quote is from his *Discourse on the Origin of Inequality* (1755), and Hume's corn harvesting parable is from *A Treatise of Human Nature* (1739).

The classic text on repeated games is Axelrod (1984). See also Sigmund (1995, 2010) as well as Nowak (2006) and Nowak and Highfield (2011). For more on evolutionary game theory, see Weibull (1995) and Gintis (2000).

On the role of gossip and reputation, see Dunbar (1996). For more on indirect reciprocity, see Sigmund (2010) and Nowak and Highfield (2011). On the role of forgetfulness in indirect reciprocation, see Hilbe et al. (2018). The surprising science of reputation is described in Whitfield (2012). The quote by Binmore is from Binmore (1994), and Haig's sentence is cited in Nowak (2011). For eBay's founder's letter from February 26, 1996, see Omidyar (1996). "The inner voice that warns us..." is from Mencken (1982). The Darwin quote on "man's motive to give aid" is from Darwin (1871) and the one on blushing from Darwin (1872).

CHAPTER 12

Important game theoretic approaches to the social contract are Binmore (1994, 1998) and Skyrms (1996, 2004) as well as Gauthier (1986). Being impartial, we will mostly follow Sigmund (2010).

For various aspects of Hobbes's life and work, see for example Malcolm (2002). The quote by Aubrey is from his *Brief Lives* (1669–1696). For the Hobbes–Wallis feud, Hobbes's comments on geometry, and the related quotes, see Hellmann (1998) and Jesseph (1999, 2018).

For a short biography of Rousseau, see Wokler (2001). For hunter-gatherer societies, see Boehm (2012).

For a game theoretic analysis of the stag hunt game, see Skyrms (2004).

The mutual aid game is described in Sugden (1986), and the common good game in Camerer (2003). The experiment repeated in Copenhagen, Minsk, Samara, etc. is from Hermann et al. (2008).

The common goods experiment with and without peer punishment is in Fehr and Gächter (2002). The pool punishment experiment is in Yamagishi (1986). The "nature of man" and "brutish and short" quotes by Hobbes are from his *Leviathan* (1651). The quote by Hardin is from Hardin (1968).

For field research on small-scale societies, see Ostrom (1990, 2005). On the importance of volunteering, see Hauert et al. (2007) and Sigmund (2010).

CHAPTER 13

The twentieth-century classic on fairness is Rawls (1971). For the example with interest rates, which follows a very different approach, see Gauthier (1986).

The Nash bargaining solution was proposed in Nash (1950), that of Kalai and Smorodinsky in Kalai and Smorodinsky (1975). For more on bargaining, see Binmore et al. (1994), Osborne and Rubinstein (1994), and Binmore (1998).

For more on Divide the Dollar, see Skyrms (2002, 2004) and Sugden (1986). For the ultimatum game, see Nowak et al. (2000), Sigmund et al. (2002), Camerer (2003), and Sigmund (2010).

The classic text on evolutionary games and escalation versus convention is Maynard Smith (1982); for an up-to-date treatment of game theory in animal societies, see McNamara and Leimar (2020). Lorenz wrote on "moral-like behavior" in Lorenz (1966). Rousseau's quote is from his *Discourse on Inequality* (1755). "Expropriators are expropriated" is in Karl Marx's *Capital*, Chapter 24, Section 7 (1867). For more on the endowment effect, see Thaler (2015). Canettis's view on grabbing are in his *Crowds and Power* (1960). "Property theft" is from Proudhon's 1840 book, *What Is Property.*

CHAPTER 14

The literature on Wittgenstein and Turing is enormous. For Wittgenstein, we refer to Bartley (1988), Janik and Toulmin (1973), McGuiness (2005), Monk (2001), and Nedo (2012). On Turing, see Hodges (2012), Leavitt (2006), Copeland (2013), and Dyson (2012). Diamond (1989) has collected the lecture notes of several of Wittgenstein's students. Floyd (2019) has analyzed in-depth the Wittgenstein–Turing–Gödel connection. The huge Wittgenstein *Nachlass* is electronically accessible through Alois Pichler's Wittgenstein Archives at the University of Bergen.

"In other circles" is a quote from Nagel (1936). The exchange between Wittgenstein and Turing is from Diamond (1989). Wittgenstein's "My task is not" quote is from Wittgenstein (1956), Section VII.19.

For Wittgenstein on language games and life-forms, see Wittgenstein (1953), §23. Wittgenstein's "colorful medley" is from Wittgenstein (1956), Section III.46.

On mathematics as a language, see Gowers (2003) and Hersh (1997a). The latter article cites the fallacious computation of the minimum. For the Manin quote, see Manin (2000). The quote by Connes is in Changeux and Connes (1995), the Galilei quote from *Il Saggiatore* (1623), and the Goethe quote from *Maximen und Reflexionen* (1833). Wittgenstein's commentary on writing digits is from his *Philosophical Investigations* (1953), §145. Von Foerster's anecdote is from personal communication (Frankfurt, September 1995). For more on mathese, see also Pollard (2023).

The example of the automatic theorem prover with human-style output is from Ganesalingam and Gowers (2017).

The Hilbert space example for semantic descent is taken from Gowers (2003).

More on *Deutsche Mathematik* can be found in Segal (2003) and Siegmund-Schultze (1984).

For the Arecibo message, see Wikipedia (2023a); and for a serious discussion on how to speak with non-Earthlings, see Oberhaus (2019).

Hardy (1940) is poignant and beautiful. The quote "mathematics as the ultimate of technology transfer" is from Stewart (1987). Leibniz wrote on the *characteristica universalis* to his friend Philipp Spener in July 1987. The quote "impossible to be imprecise" is from Ekeland (1988).

CHAPTER 15

For recent books on the philosophy of mathematics, see Mancosu (2010), Hacking (2014), Shapiro (2000), Corfield (2003), Aspray and Kitcher (1988), and Brown (2008). For some noteworthy (and very well-written) contributions by mathematicians, see Weyl (2009), Kline (1980), Hardy (1929, 1940), Hersh (1997b), Thurston (1994), and Rota (1991). The latter prophesizes the eventual extinction of "our latter-day philosophers" with the same aplomb as Bell prophesied the extinction of Platonists.

For the contest of the "big three," see Shapiro (2000) and Mancosu (2010). For histories of the Vienna Circle, see Menger (1994), Stadler (2015), Sigmund (2015), and Edmonds (2020). The various quotes of Vienna Circle members can be found in Hahn (1980), Schlick (1986), and Carnap (1963). Good accounts of the Vienna Circle's views on mathematics can be found in Hempel (1945) and Waismann (2003). For the eventual prevalence of Platonic views, see also Davis and Hersh (1981).

The Gödel quote is from his 1951 Gibbs Lecture; see Vol. III of his *Collected Works* (1995). Russell's amazement at finding Gödel an "unadulterated Platonist" is from his memoirs, Russell (1967–1969). The Bell quote is from Bell (1940). For the Langlands quote, see Langlands (2010).

For more on Platonism, see Balaguer (1998). The definition of realism is from Putnam (1975). Maddy (1997) defends naturalism (the view that mathematics needs no extra-scientific ratification). An articulate presentation of a Platonic view is by Penrose (1997). For two important views on why there is a philosophy of mathematics, see Hacking (2014) and Gowers (2006).

The "pure mathematics" quote by Kant is from his *Prolegomena to Any Future Metaphysics* (1783). Hacking's "AT ALL" quote is from Hacking (2014). The Plato quote is from book 7 of his *Republic*. For more on Plato on mathematics, see O'Connor and Robertson (2007). Hersh's weekday versus weekend quip is from Hersh (1997b). The "unreasonable effectiveness" alludes to Wigner (1960). The "analytical juice extractor" quote is from Hempel (1945). The Wittgenstein quote on mathematics as an anthropological phenomenon is from Wittgenstein (1956), Section VII.33.

The quote by Freud can be found in his essay "Creative Writing and Day-Dreaming" (1908). On the tetractys, see van der Waerden (1960). For a biography of Conway, see Roberts (2015). For surreal numbers, see Conway (2001) and Knuth (1974). The quote about "the only time that a major mathematical discovery..." is from Gardner (1977). Another showpiece of mathematical playfulness is Ording (2019).

For introductions to the Game of Life, see Berlekamp et al. (1982), Poundstone (1987), and Sigmund (1995). For more on cellular automata, see Wolfram (2002).

CHAPTER 16

Loomis (1968) has some 370 proofs of the Pythagorean theorem.

For the biography of Pick, see O'Connor and Robertson (2005).

Pick's theorem first appeared in Pick (1899). For more on its intriguing facets, see Steinhaus (1950) or Aigner and Ziegler (2018). More on the Jordan curve theorem can be found in Hales (2007). For the Polya quote "an incomplete proof is no proof at all," see Polya (1945).

Lakatos (1976) cannot be recommended too highly. The quote on "the meaning of the words..." is from Hacking (2014). David Eppstein (2022) presents twenty-one proofs of Euler's formula, and he is on the hunt for more.

The distinction between Cartesian and Leibnizian proofs is from Hacking (2014); see also Hacking (2000). The quote by Kant is from his *Critique of Pure Reason* (1781). The Polya quote about "elementary mathematical problems" is from Polya (1945). The two simple brainteasers are folklore among mathematicians. For wonderful examples of recreational mathematics and the aha! experience, see Gardner (1977, 1978). The remark on the average mental effort is from personal communication with V. Arnold (Berlin, August 1998). The water flow argument for Pick's theorem can be found in Spacematt (2021). For more on Braess's paradox, see Cohen and Horowitz (1991).

The section on "a most extraordinary law" is based on a paper by Euler, which was translated and commented by Polya (1954). The Wittgenstein quote is from Wittgenstein (1953), §123.

Bibliography

Adamson, D. (1995) Blaise Pascal: Mathematician, Physicist, and Thinker About God. Palgrave Macmillan, Houndmills, UK.

Aigner, M., Ziegler, G. M. (2018) Proofs from THE BOOK. (6th ed.) Springer, Berlin.

Aizerman, M., Aleskerov, F. (1995) Theory of Choice. North-Holland, Amsterdam.

Alt, F. (1936) Über die Messbarkeit des Nutzens. Zeitschrift für Nationalökonomie 7, 161–169.

Arnold, V., Atiyah, M., Lax, P., Mazur, B. (eds.) (2000) Mathematics: Frontiers and Perspectives. American Mathematical Society, Providence, RI.

Arrow, K. (1963) Social Choice and Individual Values. (2nd ed.) Wiley, New York.

Arrow, K. (1974) The use of unbounded utility functions in expected-utility maximization: Response. Quarterly Journal of Economics 88, 136–138.

Aspray, W., Kitcher, P. (eds.) (1988) History and Philosophy of Modern Mathematics. University of Minnesota Press, Minneapolis.

Axelrod, R. D. (1984) The Evolution of Cooperation. Basic Books, New York.

Baaz, M., Papadimitriou, C. H., Putnam, H. W., Scott, D. S., Harper, C. L. Jr. (2011) Kurt Gödel and the Foundations of Mathematics: Horizons of Truth. Cambridge University Press, New York.

Badinter, E., Badinter, R. (1988) Condorcet: Un intellectuel en politique. Fayard, Paris.

Bailey, D. H., Plouffe, S. M., Borwein, P. B., Borwein, J. M. (1997) The quest for pi. Mathematical Intelligencer 19, 50–56.

Balaguer, M. (1998) Platonism and Anti-Platonism in Mathematics. Oxford University Press, Oxford.

Balinski, M., Laraki, R. (2007) A theory of measuring, electing, and ranking. Proceedings of the National Academy of Sciences 104, 8720–8725.

Barnes, J. (1982) The Presocratic Philosophers. Routledge, London.

Baron, M. E. (1969) The Origins of the Infinitesimal Calculus. Pergamon Press, Oxford.

Barrow, J. D. (2006) The Infinite Book: A Short Guide to the Boundless, Timeless and Endless. Vintage, New York.

Bartley, W. W. III (1988) Wittgenstein. Cresset, London.

Bayes, T. (1763) An essay towards solving a problem in the doctrine of chances. Philosophical Transactions 53, 370–418.

Bell, E. T. (1937) Men of Mathematics: The Lives and Achievements of the Great Mathematicians from Zeno to Poincaré. Simon and Schuster, New York.

Bell, E. T. (1940) The Development of Mathematics. McGraw-Hill, New York. (Reprinted by Dover Publications in 1992.)

Bell, J. L. (2005) The Continuous and the Infinitesimal in Mathematics and Philosophy. Polimetrica, Milan.

Bellhouse, D. (1993) The role of roguery in the history of probability. Statistical Science 8, 410–420.

Bentham, J. (1776) A Fragment on Government. E. Wilson, London.

Berkeley, G. (1710) A Treatise Concerning the Principles of Human Knowledge. Aaron Rhames, Dublin. (Reprinted as Principles of Human Knowledge by Doubleday in 1960.)

Berkeley, G. (1734) The Analyst: A Discourse Addressed to an Infidel Mathematician. J. Tonson, London.

Berlekamp, E. R., Conway, J. H., Guy, R. K. (1982) Winning Ways for Your Mathematical Plays. Academic Press, New York. (Reprinted by A K Peters in 2000.)

Bhattacharya, A. (2021) The Man from the Future: The Visionary Life of John von Neumann. Penguin Allan Lane, London.

Billingsley, P. (1995) Probability and Measure. (3rd ed.) John Wiley & Sons, New York.

Binmore, K. G. (1994) Playing Fair. Game Theory and the Social Contract, Vol. I. MIT Press, Cambridge, MA.

Binmore, K. G. (1998) Just Playing. Game Theory and the Social Contract, Vol. II. MIT Press, Cambridge, MA.

Binmore, K. (2008) Game Theory: A Very Short Introduction. Oxford University Press, Oxford.

Binmore, K., Osborne, M., Rubinstein, A. (1992) Noncooperative models of bargaining. In Aumann, R. J., Hart, S. (eds.), Handbook of Game Theory with Economic Applications, Vol. I, Elsevier, Amsterdam, 179–225.

Boehm, C. (2012) Moral Origins: Social Selection and the Evolution of Virtue, Altruism, and Shame. Basic Books, New York.

Bourbaki, N. (1968) Theory of Sets. Addison-Wesley, Reading, MA.

Braithwaite, R. (1963) Theory of Games as a Tool for the Moral Philosopher. Cambridge University Press, Cambridge, UK.

Broome, J. (1995) The two-envelope paradox. Analysis 55, 6–11.

Brown, J. R. (2008) Philosophy of Mathematics. Routledge, London.

Bubeck, S., et al. (2023) Sparks of artificial general intelligence: Early experiments with GPT-4. Preprint, https://arxiv.org/pdf/2303.12712.pdf.

Budiansky, S. (2021) Journey to the Edge of Reason: The Life of Kurt Gödel. W. W. Norton, London.

Bunch, B. (1997) Mathematical Fallacies and Paradoxes. Dover Publications, New York.

Bunge, M. (1988) Two faces and three masks of probability. In E. Agazzi (ed.), Probability in the Sciences, Kluwer Academic Publishers, Dordrecht, 27–49.

Cajori, F. (1915) The history of Zeno's arguments on motion: Phases in the development of the theory of limits. American Mathematical Monthly 22, 1–7, 39–47, 77–83, 109–115, 143–149, 179–186, 215–221, 253–258, 292–297.

Calude, C. S. (1994) Information and Randomness: An Algorithmic Perspective. Springer-Verlag, Berlin.

Camerer, C. (2003) Behavioral Game Theory: Experiments in Strategic Interactions. Princeton University Press, Princeton, NJ.

Campbell, L., Garrett, W. (1881) The Life of James Clerk Maxwell. Macmillan and Co., London.

Cantor, G. (1895) Beiträge zur Begründung der transfiniten Mengenlehre. English translation by P. E. B. Jourdain, Contributions to the Founding of the Theory of Transfinite Numbers, Open Court, Chicago, 1915.

Cardil, R. (2015) John Conway's proof of Morley's Theorem. Matematicas Visuales, matematicasvisuales.com/english/html/geometry/triangulos/morleyconway.html.

Carnap, R. (1963) Intellectual Autobiography. The Library of Living Philosophers Vol. XI. Open Court, La Salle, IL.

Carroll, L. (1990) The Annotated Alice: Alice's Adventures in Wonderland and Through the Looking Glass. (With notes by M. Gardner.) Penguin Hammond, New York.

Cathcart, D., Klein, J. (2008) Plato and a Platypus Walk into a Bar: Understanding Philosophy Through Jokes. Penguin Books, New York.

Chaitin, G. (1967) On the length of programs for computing finite binary sequences: Statistical considerations. Journal of the ACM 16, 145–159.

Chaitin, G. (2001) Exploring Randomness. Springer, New York.

Chaitin, G. (2007) Meta Math!: The Quest for Omega. Atlantic Books, London.

Changeux, J.-P., Connes, A. (1995) Conversations on Mind, Matter, and Mathematics. Princeton University Press, Princeton, NJ.

Christian, B. (2020) The Alignment Problem: Machine Learning and Human Values. W. W. Norton, New York.

Church, A. (1940) On the concept of a random sequence. Bulletin of the American Mathematical Society 46, 130–135.

Cohen, J. E., Horowitz, P. (1991) Paradoxical behavior in mechanical and electrical networks. Nature 352, 699–701.

Colman, A. (1995) Game Theory and Its Applications in the Social and Biological Sciences. Butterworth Heinemann, Oxford.

Condorcet, M. J. (1804) Oeuvres complètes de Condorcet. Chez Henrichs, Paris.

Conway, J. H. (2001) On Numbers and Games. A K Peters, Natick, MA.

Conway, J. H., Guy, R. K. (1996) Cardinal numbers. In The Book of Numbers, Springer, New York, 277–282.

Cooke, R. (2005) The History of Mathematics. Wiley-Interscience, New York.

Copeland, B. J. (ed.) (2004) The Essential Turing. Oxford University Press, Oxford.

Copeland, B. J. (2012) Alan Turing's Electronic Brain: The Struggle to Build the ACE, the World's Fastest Computer. Oxford University Press, Oxford.

Copeland, B. J. (2013) Turing: Pioneer of the Information Age. Oxford University Press, Oxford.

Copeland, B. J., Fan, Z. (2023) Did Turing stand on Gödel's shoulders? The Mathematical Intelligencer 44, 308–319.

Corfield, D. (2003) Towards a Philosophy of Real Mathematics. Cambridge University Press, Cambridge, UK.

Cory, L. (2015) A Brief History of Numbers. Oxford University Press, Oxford.

Courant, R., Robbins, H. (1996) What Is Mathematics?: An Elementary Approach to Ideas and Methods. (2nd ed.) Oxford University Press, New York.

Coxeter, H. S. M. (1989) Introduction to Geometry. (2nd ed.) John Wiley & Sons, London.

Darwin, C. (1871) The Descent of Man, and Selection in Relation to Sex. John Murray, London.

Darwin, C. (1872) The Expression of the Emotions in Man and Animals. John Murray, London.

Darwin, C. (1987) Charles Darwin's Notebooks, 1836–1844: Geology, Transmutation of Species, Metaphysical Enquiries. (P. H. Barrett et al., eds.) Cambridge University Press, Cambridge, UK.

Daston, L. L. (1979) The Reasonable Calculus: Classical Probability Theory 1650–1840. Harvard University Press, Cambridge, MA.

Dauben, J. W. (1990) Georg Cantor: His Mathematics and Philosophy of the Infinite. Princeton University Press, Princeton, NJ.

Dauben, J. W. (2014) Abraham Robinson: The Creation of Nonstandard Analysis, a Personal and Mathematical Odyssey. Princeton University Press, Princeton, NJ.

Davis, M. (1965) The Undecidable: Basic Papers on Undecidable Propositions, Unsolvable Problems and Computable Functions. Raven Press, New York.

Davis, M. (2001) Engines of Logic: Mathematicians and the Origin of the Computer. W. W. Norton, London.

Davis, P. J., Hersh, R. (1981) The Mathematical Experience. Birkhäuser, Boston.

Dawkins, R. (2016) The Selfish Gene. (4th ed.) Oxford University Press, Oxford.

Dawson, J. W. (1997) Logical Dilemmas: The Life and Work of Kurt Gödel. A K Peters, Wellesley, MA.

Dawson, J. W., Sigmund, K. (2006) Gödel's Vienna. Mathematical Intelligencer 28, 44–55.

Day, J. M. (1994) Plato's Meno in Focus. Routledge, London.

Dedekind, R. (1932) Gesammelte mathematische Werke, Vol. III. Vieweg, Braunschweig.

Dedekind, R. (1963) The nature and meaning of numbers. In Essays on the Theory of Numbers, Dover Publications, New York, Part II.

de Finetti, B. (1937) La Prévision: Ses lois logiques, ses sources subjectives. Annales de l'Institut Henri Poincaré 7, 1–68.

de Finetti, B. (1999) Theory of Probability. Wiley, New York.

Dehaene, S. (2011) The Number Sense: How the Mind Creates Mathematics. Oxford University Press, New York.

Delahaye, J. P. (1997) Le fascinant nombre pi. Bibliothèque pour la Science, Paris.

Delahaye, J. P. (1999) Information, complexité et hasard. Editions Hermes, Paris.

Dennett, D. (1991) Consciousness Explained. Little, Brown and Company, New York.

Derbyshire, J. (2004) Prime Obsession: Bernhard Riemann and the Greatest Unsolved Problem in Mathematics. Penguin Books, London.

Devlin, K. (1993) The Joy of Sets: Fundamentals of Contemporary Set Theory. (2nd ed.) Springer, New York.

Devlin, K. (2008) The Unfinished Game: Pascal, Fermat, and the Seventeenth-Century Letter That Made the World Modern. Basic Books, New York.

Diaconis, P., Skyrms, B. (2018) Ten Great Ideas About Chance. Princeton University Press, Princeton, NJ.

Diamond, C. (ed.) (1989) Wittgenstein's Lectures on the Foundations of Mathematics, Cambridge 1939. University of Chicago Press, Chicago.

Dodgson, C. L. (1879) Euclid and His Modern Rivals. Macmillan, London.

Doxiadis, A., Papadimitriou, C. H. (2009) Logicomix: An Epic Search for Truth. Bloomsbury, London.

Dunbar, R. (1996) Grooming, Gossip, and the Evolution of Language. Harvard University Press, Cambridge, MA.

Du Sautoy, M. (2010) The Number Mysteries. HarperCollins, London.

Dyson, G. (2012) Turing's Cathedral: The Origins of the Digital Universe. Pantheon Books, New York.

Ebbinghaus, H.-D. (2007) Ernst Zermelo: An Approach to His Life and Work. Springer, Berlin.

Ebbinghaus, H.-D., et al. (1996) Numbers. Graduate Texts in Mathematics Vol. 123. Springer, New York.

Edmonds, D. (2020) The Murder of Professor Schlick: The Rise and Fall of the Vienna Circle. Princeton University Press, Princeton, NJ.

Edwards, C. (1979) The Historical Development of the Calculus. Springer, New York.

Ekeland, I. (1988) Mathematics and the Unexpected. University of Chicago Press, Chicago.

Ekeland, I. (1993) The Broken Dice and Other Mathematical Tales of Chance. (C. Volk, trans.) University of Chicago Press, Chicago.

Ellsberg, D. (1961) Risk, ambiguity, and the Savage axioms. Quarterly Journal of Economics 75, 643–669.

Ellsberg, D. (2002) Secrets: A Memoir of Vietnam and the Pentagon Papers. Viking Press, New York.

Eppstein, D. (2022) Twenty-one proofs of Euler's formula: $V - E + F = 2$. The Geometry Junkyard, https://www.ics.uci.edu/~eppstein/junkyard/euler/. Accessed December 12.

Everett, C. (2017) Numbers and the Making of Us: Counting and the Course of Human Cultures. Harvard University Press, Cambridge, MA.

Eymard, P., Lafon, J.-P. (1999) The Number π. (S. S. Wilson, trans.) American Mathematical Society, Providence, RI.

Feferman, S. (1986) Gödel's life and work. In Feferman et al. (eds.), Gödel's Collected Works, Vol. 1, Princeton University Press, Princeton, NJ, 1–36.

Feferman, S. (1996) In the Light of Logic. Oxford University Press, Oxford.

Fehr, E., Gächter, S. (2002) Altruistic punishment in humans. Nature 415, 137–140.

Feller, W. (1975) An Introduction to Probability Theory and Its Applications. John Wiley & Sons, New York.

Fishburn, P. (1970) Utility Theory for Decision Making. John Wiley & Sons, New York.

Floyd, J. (2019) Wittgenstein's Philosophy of Mathematics. Cambridge University Press, Cambridge, UK.

Folina, J. (1992) Poincaré and the Philosophy of Mathematics. St Martin's Press, New York.

Fowler, D. (1999) The Mathematics of Plato's Academy: A New Reconstruction. Clarendon Press, Oxford.

Franzen, T. (2005) Gödel's Theorem: An Incomplete Guide to Its Use and Abuse. A K Peters, Wellesley, MA.

Freudenthal, H. (1953) Zur Geschichte der vollständigen Induktion. Archives Internationales d'Histoire des Sciences 6, 17–37.

Friedman, M. (2012) Kant on geometry and spatial intuition. Synthese 186, 213–255.

Foucault, M. (1977) Discipline and Punish: The Birth of the Prison. Allen Lane, London.

Galavotti, M. C. (2005) Philosophical Introduction to Probability. CSLI Publications, Stanford.

Gamow, G. (1946) One Two Three... Infinity. Macmillan, London.

Ganesalingam, M., Gowers, W. T. (2017) A fully automatic theorem prover with human-style output. Journal of Automated Reasoning 58, 253–291.

Gardner, M. (1977) Mathematical Magic Show. Alfred A. Knopf, New York.

Gardner, M. (1978) Aha! Aha! Insight. Scientific American, New York.

Gauss, C. F. (1831) Theoria residuorum biquadraticorum II. Commentationes Societas Regiae Scientiarum Gottingensis Recentiores 7, 89–148.

Gauthier, D. (1986) Morals by Agreement. Clarendon Press, Oxford.

Gay, P. (1965) Voltaire's Politics: The Poet as Realist. Vintage Books, London.

Gibbard, A. (1973) Manipulation of voting schemes: A general result. Econometrica 41, 587–601.

Gintis, H. (2000) Game Theory Evolving: A Problem-Centered Introduction to Modeling Strategic Interaction. Princeton University Press, Princeton, NJ.

Gleick, J. (1987) Chaos: The Making of a New Science. Viking Press, New York.

Gödel, K. (1931) Über formal unentscheidbare Sätze der Principia Mathematica und verwandter Systeme I. Monatshefte für Mathematik und Physik 38, 173–198.

Gödel, K. (1995) Collected Works, Vol. III. (S. Fefermann et al., eds.) Oxford University Press, Oxford.

Goldstein, R. (2005) Incompleteness: The Proof and Paradox of Kurt Gödel. W. W. Norton, New York.

Gowers, W. T. (2003) Mathematics: A Very Short Introduction. Oxford University Press, Oxford.

Gowers, W. T. (2006) Does mathematics need a philosophy? In Hersh (2006), 182–200.

Grattan-Guinness, I. (2000) The Search for Mathematical Roots 1870–1940: Logics, Set Theories and the Foundations of Mathematics from Cantor Through Russell to Gödel. Princeton University Press, Princeton, NJ.

Greenberg, M. (2007) Euclidean and Non-Euclidean Geometries: Development and History. W. H. Freeman, San Francisco.

Gross, H. J. (2014) Eine vergessene Revolution: Die Geschichte vom klugen Pferd Hans. Biologie in unserer Zeit 44, 268–272.

Grünbaum, A. (1967) Modern Science and Zeno's Paradoxes. Allen and Unwin, London.

Hacking, I. (2000) What mathematics has done to some and only some philosophers. In Smiley, T. J. (ed.), Mathematics and Necessity: Essays in the History of Philosophy, Oxford University Press, Oxford, 83–138.

Hacking, I. (2006) The Emergence of Probability. Cambridge University Press, Cambridge, UK.

Hacking, I. (2014) Why Is There a Philosophy of Mathematics at All? Cambridge University Press, Cambridge, UK.

Hahn, H. (1980) Empiricism, Logic and Mathematics. (McGuiness, B., ed.). Reidel, Dordrecht.

Haidt, J. (2007) The new synthesis in moral psychology. Science 316, 998–1002.

Hájek, A., Hitchcock, C. (eds.) (2016a) The Oxford Handbook of Probability and Philosophy. Oxford University Press, Oxford.

Hájek, A., Hitchcock, C. (2016b) Probability for everyone—Even philosophers. In Hajek and Hitchcock (2016a), 5–30.

Hales, T. C. (2007) The Jordan curve theorem, formally and informally. American Mathematical Monthly 114, 882–894.

Hales, T. C. (2008) Formal proof. Notices of the AMS 55, 1370–1380.

Hankins, T. L. (1990) Jean d'Alembert: Science and the Enlightenment. Gordan and Breach, New York.

Hardin, G. (1968) The tragedy of the commons. Science 162, 1243–1248.

Hardy, G. H. (1929) Mathematical proof. Mind 38, 1–25.

Hardy, G. H. (1940) A Mathematician's Apology. Cambridge University Press, Cambridge, UK.

Harris, M. (2015) Mathematics Without Apologies. Princeton University Press, Princeton, NJ.

Harrison, J. (2008) Formal proof—Theory and practice. Notices of the AMS 55, 1395–1406.

Hauert, C., Traulsen, A., Brandt, H., Nowak, M. A., Sigmund, K. (2007) Via freedom to coercion: The emergence of costly punishment. Science 316, 1905–1907.

Heath, L. (1981) A History of Greek Mathematics. Dover Publications, New York.

Hellman, H. (1998) Great Feuds in Science: Ten of the Liveliest Disputes Ever. Wiley, New York.

Hempel, C. G. (1945) On the nature of mathematical truth. American Mathematical Monthly 52, 543–556.

Henrichs, J., et al. (2006) Costly punishment across human societies. Science 312, 1767–1770.

Hermann, B., Thöni, U., Gächter, S. (2008) Antisocial punishment across societies. Science 319, 1362–1367.

Hersh, R. (1997a) Math lingo vs. plain English: Double entendre. American Mathematical Monthly 104, 48–51.

Hersh, R. (1997b) What Is Mathematics, Really? Oxford University Press, New York.

Hersh, R. (ed.) (2006) 18 Unconventional Essays on the Nature of Mathematics. Springer, New York.

Heukelom, F. (2015) A history of the Allais paradox. British Journal for the History of Science 48, 147–162.

Higgins, P. (1998) Mathematics for the Curious. Oxford University Press, Oxford.

Hilbe, C., Schmid, L., Tkadlec, J., Chatterjee, K., Nowak, M. A. (2018) Indirect reciprocity with private, noisy, and incomplete information. Proceedings of the National Academy of Sciences 115, 12241–12246.

Hilbe, C., Traulsen, A., Sigmund, K. (2010) Partners or rivals? Strategies for the iterated prisoner's dilemma. Games and Economic Behavior 92, 41–52.

Hilbert, D. (1925) Über das Unendliche. Mathematische Annalen 95, 161–190.

Hilbert, D., Ackermann, W. (1928) Die Grundzüge der theoretischen Logik. Springer, Berlin. (Translated into English as Principles of Mathematical Logic, Chelsea Publishing Company, New York, 1950.)

Hintikka, J. (1999) On Gödel. Wadsworth Academic Publishers, Belmont, MA.

Hodges, A. (2012) Alan Turing: The Enigma. (Centenary ed.) Princeton University Press, Princeton, NJ.

Hodgkin, L. (2005) A History of Mathematics: From Mesopotamia to Modernity. Oxford University Press, Oxford.

Hofbauer, J. (2000) A simple proof of $1 + \frac{1}{2^2} + \frac{1}{3^2} + \ldots = \frac{\pi^2}{6}$ and related identities. American Mathematical Monthly 109, 196–200.

Hoffmann, D. (2017) Die Gödelschen Unvollständigkeitssätze: Eine geführte Reise durch Kurt Gödels historischen Beweis. (2nd ed.) Springer Spektrum, Berlin.

Hofstadter, D. R. (1979) Gödel, Escher, Bach: An Eternal Golden Braid. Basic Books, New York.

Hohol, M. (2019) Foundations of Geometric Cognition. Routledge, London.

Hsu, F.-H. (2004) Behind Deep Blue: Building the Computer That Defeated the World Chess Champion. Princeton University Press, Princeton, NJ.

Huemer, M. (2016) Approaching Infinity. Palgrave Macmillan, New York.

Huylebrouck, D. (2019) Missing link. In Africa and Mathematics: From Colonial Findings Back to the Ishango Rods. Mathematics, Culture, and the Arts, Springer, Cham, 153–166.

Ifrah, G. (2001) The Universal History of Computing: From the Abacus to the Quantum Computer. Wiley, New York.

Isaacson, W. (2007) Einstein: His Life and Universe. Simon and Schuster, New York.

Janik, A., Toulmin, S. (1973) Wittgenstein's Vienna. Weidenfeld and Nicolson, London.

Jesseph, D. M. (1999) Squaring the Circle: The War Between Hobbes and Wallis. University of Chicago Press, Chicago.

Jesseph, D. M. (2018) Geometry, religion and politics: Context and consequences of the Hobbes–Wallis debate. Notes and Records 72, 469–486.

Jordan, J. (2006) Pascal's Wager: Pragmatic Arguments and Belief in God. Clarendon Press, Oxford.

Kahneman, D. (2011) Thinking, Fast and Slow. Farrar, Straus and Giroux, New York.

Kahneman, D., Tversky, A. (1979) Prospect theory: An analysis of decision under risk. Econometrica 47, 263–292.

Kalai, E., Smorodinsky, M. (1975) Other solutions of Nash's bargaining problem. Econometrica 43, 513–518.

Kazarinoff, N. D. (2003) Ruler and the Round: Classic Problems in Geometric Constructions. Dover Publications, New York.

Keller, J. (1986) The probability of heads. American Mathematical Monthly 93, 191–197.

Kidwell, P. A., Williams, M. R. (1992) The Calculating Machines: Their History and Development. MIT Press, Cambridge, MA.

Klein, J. (1965) A Commentary on Plato's Meno. University of North Carolina Press, Chapel Hill.

Kline, M. (1972) Mathematical Thought from Ancient to Modern Times. Oxford University Press, New York.

Kline, M. (1980) Mathematics: The Loss of Certainty. Oxford University Press, New York.

Knuth, D. E. (1973) The Art of Computer Programming, Vol. III. Addison-Wesley, Reading, MA.

Knuth, D. E. (1974) Surreal Numbers: How Two Ex-students Turned on to Pure Mathematics and Found Total Happiness. Addison-Wesley, Reading, MA.

Knuth, D. E. (1997) The Art of Computer Programming, Vol I. (3rd ed.) Addison-Wesley, Upper Saddle River, NJ.

Kolmogorov, A. N. (1950) Foundations of the Theory of Probability. Springer, New York.

Kolmogorov, A. N. (1963) On tables of random numbers. Sankhya 25, 369–376.

Kreisel, G. (1980) Kurt Gödel. Biographical Memoirs of Fellows of the Royal Society 26, 148–224.

Lagarias, J. C., Zong, C. (2012) Mysteries in packing regular tetrahedra. Notices of the AMS 59, 1540–1549.

Lakatos, I. (1976) Proofs and Refutations: The Logic of Mathematical Discovery. Cambridge University Press, Cambridge, UK.

Langlands, R. P. (2010) Is there beauty in mathematical theories? Institute for Advanced Study, http://publications.ias.edu/sites/default/files/ND.pdf.

Launay, F. (2010) D'Alembert et la femme du vitrier Rousseau, Etiennette Gabrielle Ponthieux (ca 1683–1775). Recherches sur Diderot et sur l'Encyclopedie, https://journals.openedition.org/rde/4725.

Leavitt, D. (2006) The Man Who Knew Too Much: Alan Turing and the Invention of the Computer. W. W. Norton, New York.

Lecat, M. (1935) Erreurs des Mathématiciens des Origines à Nos Jours. Ancienne Librairie Castaigne, Brussels.

Leibniz, G. W., (2001) The Labyrinth of the Continuum: Writings on the Continuum Problem, 1672–1686. (R. T. W. Arthur, trans.) Yale University Press, New Haven, CT.

Lemanski, J. (ed.) (2020) Language, Logic, and Mathematics in Schopenhauer. Birkhäuser, Cham.

Leonard, R. (2010) Von Neumann, Morgenstern, and the Creation of Game Theory: From Chess to Social Science, 1900–1960. Cambridge University Press, Cambridge, UK.

Levitt, G. M. (2000) The Turk, Chess Automaton. McFarland, Jefferson, NC.

Li, M., Vitányi, P. (1997) An Introduction to Kolmogorov Complexity and Its Applications. Texts in Computer Science, Springer, New York.

Loomis, E. S. (1968) The Pythagorean Proposition: Its Demonstrations Analyzed and Classified, and Bibliography of Sources for Data of the Four Kinds of Proofs. National Council of Teachers of Mathematics, Washington, DC.

Lorenz, K. (1966) On Aggression. Methuen, London.

Luce, D., Raiffa, H. (1957) Games and Decisions. John Wiley & Sons, New York.

MacKenzie, D. (2001) Mechanizing Proof: Computing, Risk, and Trust. MIT Press, Cambridge, MA.

Macrae, N. (1999) John von Neumann: The Scientific Genius Who Pioneered the Modern Computer, Game Theory, Nuclear Deterrence, and Much More. (2nd ed.) American Mathematical Society, Providence, RI.

Maddy, P. (1997) Naturalism in Mathematics. Oxford University Press, Oxford.

Malcolm, N. (1962) Ludwig Wittgenstein: A Memoir. Oxford University Press, New York.

Malcolm, N. (2002) Aspects of Hobbes. Oxford University Press, Oxford.

Mancosu, P. (1996) Philosophy of Mathematics and Mathematical Practice in the Seventeenth Century. Oxford University Press, New York.

Mancosu, P. (ed.) (1998) From Hilbert to Brouwer: The Debate on the Foundations of Mathematics in the 1920s. Oxford University Press, Oxford.

Mancosu, P. (2010) The Adventure of Reason: Interplay Between Philosophy of Mathematics and Mathematical Logic, 1900–1940. Oxford University Press, Oxford.

Manin, Y. (2000) Mathematics as profession and vocation. In V. Arnold et al. (eds.), Mathematics: Frontiers and Perspectives, American Mathematical Society, Providence, RI, 154–159.

Martin, R. (2013) The St. Petersburg paradox. Stanford Encyclopedia of Philosophy Archive, https://plato.sydney.edu.au/archives/spr2014/entries/paradox-stpetersburg/.

Martin-Löf, P. (1966) The definition of random sequences. Information and Control 9, 602–619.

Maynard Smith, J. (1982) Evolution and the Theory of Games. Cambridge University Press, Cambridge, UK.

McGuinness, B. (2005) Wittgenstein: A Life: Young Ludwig 1889–1921. Clarendon Press, Oxford.

McKelvey, R. (1979) General conditions for global intransitivities in formal voting models. Econometrica 47, 1085–1112.

McNamara, J. M., Leimar, O. (2020) Game Theory in Biology: Concepts and Frontiers. Oxford University Press, Oxford.

Mencken, H. L. (1982) A Mencken Chrestomathy. Penguin Random House, New York.

Menger, K. (1934) Das Unsicherheitsmoment in der Wertelehre. Zeitschrift für Nationalökonomie 51, 459–485.

Menger, K. (1994) Reminiscences of the Vienna Circle and the Mathematical Colloquium. Reidel, Dordrecht.

Merlin, V., Tataru, M., Valognes, F. (2000) On the probability that all decision rules select the same winner. Journal of Mathematical Economics 33, 183–208.

Mill, J. S. (2011) A System of Logic, Ratiocinative and Inductive. (Classic Reprint.) Forgotten Books, London.

Misak, C. (2020) Frank Ramsey: A Sheer Excess of Powers. Oxford University Press, Oxford.

Monk, R. (1991) Ludwig Wittgenstein: The Duty of Genius. Penguin, New York.

Monk, R. (1997) Bertrand Russell: 1872–1920: The Spirit of Solitude. Vintage, London.

Monk, R. (2001) Bertrand Russell: 1921–1970: The Ghost of Madness. Vintage, London.

Monod, J. (1971) Chance and Necessity. Knopf, New York.

Musil, R. (1906) Die Verwirrungen des Schülers Törless. Wiener Verlag, Wien. (Translated into English as M. Mitchell [trans.], Confusions of Young Torless, Oxford World Classics, Oxford University Press, Oxford, 2014.)

Musil, R. (1913) Der mathematische Mensch. Gesammelte Werke vol 2. Rowohlt, Hamburg.

Myerson, R. (1997) Game Theory: Analysis of Conflict. Harvard University Press, Cambridge, MA.

Nagel, E. (1936) Impressions and appraisals of analytic philosophy in Europe, Journal of Philosophy 33, Part I: 5–24, Part II: 29–53.

Nagel, E., Newman, J. (2005) Gödel's Proof. Routledge, New York.

Nahin, P. (2012) The Logician and the Engineer: How George Boole and Claude Shannon Created the Information Age. Princeton University Press, Princeton, NJ.

Nasar, S. (1998) A Beautiful Mind: A Biography of John Forbes Nash, Jr., Winner of the Nobel Prize in Economics, 1994. Simon and Schuster, New York.

Nash, J. (1950) The Bargaining Problem. Econometrica 18, 155–162.

Nash, J. (1996) Essays on Game Theory. Edward Elgar, Cheltenham, UK.

Nedo, M. (ed.) (2012) Ludwig Wittgenstein: Ein biographisches Album. Beck, Munich.

Newman, M. H. A. (1955) Alan Mathison Turing 1912–1954. Biographical Memoirs of Fellows of the Royal Society 1, 253–263.

Nowak, M. A. (2006) Evolutionary Dynamics. Harvard University Press, Cambridge, MA.

Nowak, M. A., Highfield, R. (2011) Super-cooperators: Altruism, Evolution, and Why We Need Each Other to Succeed. Free Press, New York.

Nowak, M. A., Page, K. M., Sigmund, K. (2000) Fairness versus reason in the ultimatum game. Science 289, 1773–1775.

Nurmi, H. (1999) Voting Paradoxes and How to Deal with Them. Springer-Verlag, Berlin.

Oberhaus, D. (2019) Extraterrestrial Languages. MIT Press, Cambridge, UK.

O'Connor, J. J., Robertson, E. F. (1999) Ernst Friedrich Ferdinand Zermelo. MacTutor History of Mathematics Archive, https://mathshistory.st-andrews.ac.uk/Biographies/Zermelo/.

O'Connor, J. J., Robertson, E. F. (2005) Georg Alexander Pick. MacTutor History of Mathematics Archive, https://mathshistory.st-andrews.ac.uk/Biographies/Pick.

O'Connor, J. J., Robertson, E. F. (2007) Plato on mathematics. MacTutor History of Mathematics Archive, https://mathshistory.st-andrews.ac.uk/Extras/Plato_on_mathematics/.

O'Connor, J. J., Robertson, E. F. (2014) Charles Marie de La Condamine. MacTutor History of Mathematics Archive, https://mathshistory.st-andrews.ac.uk/Biographies/La_Condamine/.

Omidyar, P. (1996) Founders letter. eBay Feedback Forum, https://pages.ebay.com/services/forum/feedback-foundersnote.html.

Ording, P. (2019) 99 Variations on a Proof. Princeton University Press, Princeton, NJ.

Osborne, M., Rubinstein, A. (1994) A Course in Game Theory. MIT Press, Cambridge, MA.

Ostrom, E. (1990) Governing the Commons: The Evolution of Institutions for Collective Action. Cambridge University Press, Cambridge, UK.

Ostrom, E. (2005) Understanding Institutional Diversity. Princeton University Press, Princeton, NJ.

Oystein, O. (1960) Pascal and the invention of probability theory. American Mathematical Monthly 67, 409–419.

Penrose, R. (1997) The Large, the Small and the Human Mind. Cambridge University Press, Cambridge, UK.

Peterson, I. (1997) The Jungles of Randomness: A Mathematical Safari. John Wiley & Sons, New York.

Pichler, A., et al. (2016) Wittgenstein's Nachlass. The Wittgenstein Archives at the University of Bergen, https://wab.uib.no/wab_nachlass.page/.

Pick, G. (1899) Geometrisches zur Zahlentheorie. Sitzungsberichte des deutschen naturwissenschaftlich-medicinischen Vereines für Böhmen "Lotos" in Prag 19, 311–319.

Pinker, S. (2002) The Blank Slate: The Modern Denial of Human Nature. Viking, New York.

Poincaré, H. (1982) The Foundations of Science. (G. B. Halstead, trans.) University Press of America, Washington, DC. (Originally published in 1913.)

Pollard, C. (2023) Chapter two: Mathese. Ohio State University, https://www.asc.ohio-state.edu/pollard.4/680/chapters/mathese.pdf. Accessed January 2023.

Polya, G. (1945) How to Solve It: A New Aspect of Mathematical Method. Princeton University Press, Princeton, NJ. (Reprinted by Princeton Science Library in 2004.)

Polya, G. (1954) Induction and Analogy in Mathematics. Princeton University Press, Princeton, NJ.

Popper, K. (1957) The propensity interpretation of the calculus of probability and of the quantum theory. In S. Korner, M. Price (eds.), Observation and Interpretation: A Symposium of Philosophers and Physicists, Buttersworth, London, 65–70.

Popper, K. (1992) Unended Quest: An Intellectual Autobiography. Routledge, London.

Posamentier, A. S., Lehmann, I. (2004) π: A Biography of the World's Most Mysterious Number. Prometheus Books, Amherst, NY.

Poundstone, W. (1987) The Recursive Universe: Cosmic Complexity and the Limits of Scientific Knowledge. Oxford University Press, Oxford.

Poundstone, W. (1992) Prisoner's Dilemma. Doubleday, New York.

Poundstone, W. (2010) Priceless: The Myth of Fair Value (and How to Take Advantage of It). Farrar, Straus and Giroux, New York.

Purkert, W., Ilgauds, H. J. (2013) Georg Cantor 1845–1918. Birkhäuser, Basel.

Putnam, H. (1975) What is mathematical truth? Historia Mathematica 2, 529–533.

Quine, W. V. O. (1953) From a Logical Point of View: 9 Logico-philosophical Essays. (2nd ed.) Harvard University Press, Cambridge, MA.

Quine, W. V. O. (1981) Mathematical Logic. (Revised ed.) Harvard University Press, Cambridge, MA.

Raiffa, H. (1961) Risk, ambiguity, and the Savage axioms: A comment. Quarterly Journal of Economics 75, 690–694.

Ramsey, F. (1931) Truth and Probability. In R. B. Braithwaite (ed.), The Foundations of Mathematics and Other Logical Essays, Harcourt Brace, New York, 156–198. (Based on a talk originally given in 1926.)

Rawls, J. (1971) A Theory of Justice. Harvard University Press, Cambridge, MA. (Reprinted by Belknap Press in 1999.)

Regis, E. (1987) Who Got Einstein's Office? Eccentricity and Genius at the Institute for Advanced Study. Addison-Wesley, Boston.

Reid, C. (1996) Hilbert. Copernicus, New York.

Reid, C. (2004) A Long Way from Euclid. Dover Publications, New York.

Remmert, R. (1996) What Is π? In Ebbinghaus et al. (1996), Chapter 5.

Rényi, A. (1970) Foundations of Probability. Holden-Day, Inc., San Francisco.

Roberts, S. (2015) Genius at Play: The Curious Mind of John Horton Conway. Bloomsbury, New York.

Robinson, A. (1996) Non-Standard Analysis. (Revised ed.) Princeton University Press, Princeton, NJ.

Roose, K. (2021) Futureproof: 9 Rules for Humans in the Age of Automation. Random House, New York.

Rosen, F. (2003) Classical Utilitarianism from Hume to Mill. Routledge, London.

Ross, S. (2013) A First Course in Probability. Pearson, Upper Saddle River, NJ.

Rota, G.-C. (1991) The pernicious influence of mathematics upon philosophy. Synthese 88, 165–178.

Rucker, R. (1982) Infinity and the Mind: The Science and Philosophy of the Infinite. Birkhäuser, Basel.

Rudman, P. S. (2007) How Mathematics Happened: The First 50,000 Years. Prometheus Books, New York.

Russell, B. (1903) Principles of Mathematics. Cambridge University Press, Cambridge, UK.

Russell, B. (1918) Mysticism and Logic and Other Essays. Allen and Unwin, London.

Russell, B. (1946) History of Western Philosophy. Allen and Unwin, London. (Reprinted by Routledge Classics in 2004.)

Russell, B. (1967–1969) The Autobiography of Bertrand Russell. 3 vols. Allen and Unwin, London.

Russell, B., Whitehead, A. N. (1910) Principia Mathematica. Cambridge University Press, Cambridge, UK.

Saari, D. G. (1989) A dictionary for voting paradoxes. Journal of Economic Theory 48, 443–475.

Saari, D. G. (1995) Basic Geometry of Voting. Springer, New York.
Saari, D. G. (2003) Capturing the "will of the people." Ethics 113, 333–349.
Saari, D. G. (2008a) Disposing Dictators, Demystifying Voting Paradoxes: Social Choice Analysis. Cambridge University Press, Cambridge, UK.
Saari, D. G. (2008b) Mathematics and voting. Notices of the AMS 55, 448–455.
Saari, D. G., Tataru, M. M. (1999) The likelihood of dubious election outcomes. Economic Theory 13, 345–363.
Sagan, H. (1994) Space-Filling Curves. Universitext. Springer, New York.
Samuelson, P. (1977) St. Petersburg paradoxes: Defanged, dissected, and historically described. Journal of Economic Literature 15, 24–55.
Savile, A. (2000) Routledge Philosophy Guidebook to Leibniz and the Monadology. Routledge, London.
Schlick, M. (1986) Die Probleme der Philosophie in ihrer Zusammenfassung (Vorlesungen 1934/35). Suhrkamp, Frankfurt.
Schnorr, C. P. (1971) Zufälligkeit und Wahrscheinlichkeit. Springer, Berlin.
Schofield, N. (1983) Generic instability of majority rule. Review of Economic Studies 50, 695–705.
Schofield, P. (2009) Bentham: A Guide for the Perplexed. Continuum, London.
Schrödinger, E. (1929) Was ist ein Naturgesetz? Naturwissenschaften 17, 9–11.
Searle, J. (1980) Minds, brains, and programs. Behavioral and Brain Sciences 3, 417–424.
Segal, S. L. (2003) Mathematicians Under the Nazis. Princeton University Press, Princeton, NJ.
Segre, M. (1994) Peano's axioms in their historical context. Archive for History of Exact Sciences 48, 201–342.
Sen, A. (1970) Collective Choice and Social Welfare. Holden-Day, San Francisco.
Service, R. F., et al. (2020) Runners-up. Science 370, 1402–1407.
Shafer, G. (2009) The education of Jean André Ville. Electronic Journal for History of Probability and Statistics 5, 1–50.
Shapiro, S. (2000) Thinking About Mathematics: The Philosophy of Mathematics. Oxford University Press, Oxford.
Shapiro, S., Hellman, G. (eds.) (2020) The History of Continua: Philosophical and Mathematical Perspectives. Oxford University Press, Oxford.
Sieg, W. (2013) Hilbert's Programs and Beyond. Oxford University Press, Oxford.
Siegmund-Schultze, R. (1984) Theodor Vahlen—zum Schuldanteil eines deutschen Mathematikers am faschistischen Mißbrauch der Wissenschaft. Naturwissenschaften, Technik und Medizin 21, 17–32
Siegmund-Schultze, R. (2006) Probability in 1919/20: The von Mises-Polya-controversy. Archive for History of Exact Sciences 60, 431–515.
Siegmund-Schultze, R. (2009) Mathematicians Fleeing from Nazi Germany: Individual Fates and Global Impact. Princeton University Press, Princeton, NJ.
Siegmund-Schultze, R. (2010) Sets versus trial sequences, Hausdorff versus von Mises: "Pure" mathematics prevails in the foundations of probability around 1920. Historia Mathematica 37, 204–241.

Sigmund, K. (1993) Games of Life: Explorations in Ecology, Evolution and Behavior. Oxford University Press, Oxford. (Reprinted by Dover Publications in 2017.)

Sigmund, K. (2010) Calculus of Selfishness. Princeton University Press, Princeton, NJ.

Sigmund, K. (2017) Exact Thinking in Demented Times: The Vienna Circle and the Epic Quest for the Foundations of Science. Basic Books, New York.

Sigmund, K., Dawson, J., Mühlberger, K. (2006) Kurt Gödel: Das Album. Vieweg, Wiesbaden.

Sigmund, K., Fehr, E., Nowak, M. A. (2002) The economics of fair play. Scientific American 286, 80–85.

Silver, D., et al. (2018) A general reinforcement learning algorithm that masters chess, shogi, and Go through self-play. Science 362, 1140–1144.

Skyrms, B. (1996) Evolution of the Social Contract. Cambridge University Press, Cambridge, MA.

Skyrms, B. (2002) Stability and explanatory significance of some simple evolutionary models. Philosophy of Science 67, 94–113.

Skyrms, B. (2004) The Stag Hunt and the Evolution of Social Structure. Cambridge University Press, Cambridge, MA.

Smiley, T. J. (ed.) (2000) Mathematics and Necessity: Essays in the History of Philosophy. Oxford University Press, Oxford.

Smith, A. R. (2014a) His just deserts: A review of four books. Notices of the AMS 61, 891–895.

Smith, J. T. (2014b) David Hilbert's radio address—English translation. Mathematical Association of America, https://www.maa.org/press/periodicals/convergence/david-hilberts-radio-address-english-translation.

Smith, P. (2007) An Introduction to Gödel's Theorems. Cambridge University Press, Cambridge, UK.

Smullyan, R. M. (1992) Gödel's Incompleteness Theorems. Oxford University Press, New York.

Sober, E., Wilson, D. S. (1998) Unto Others: The Evolution and Psychology of Unselfish Behavior. Harvard University Press, Cambridge, MA.

Solomonoff, R. J. (1964) A formal theory of inductive inference. Information and Control 7, Part I: 1–22, Part II: 224–254.

Spacematt (2021) Pick's theorem: The wrong, amazing proof. YouTube, https://www.youtube.com/watch?v=uh-yRNqLpOg.

Spengler, O. (1991) The Decline of the West. Oxford University Press, Oxford. (Originally published in 1918.)

Stadler, F. (2015) The Vienna Circle: Studies in the Origins, Development, and Influence of Logical Empiricism. Springer, New York. (Originally published in 2001.)

Standage, T. (2002) The Turk: The Life and Times of the Famous Eighteenth-Century Chess-Playing Machine. Walker, New York.

Stein, S. (1999) Archimedes: What Did He Do Besides Cry Eureka? Mathematical Association of America, Washington, DC.

Steinhaus, H. (1950) Mathematical Snapshots. Oxford University Press, New York.

Stewart, I. (1987) The Problems of Mathematics. Oxford University Press, Oxford.

Stewart, I. (1989a) Does God Play Dice? The Mathematics of Chaos. Blackwell, Oxford.

Stewart, I. (1989b) Galois Theory. (2nd ed.) Chapman and Hall, London.

Stewart, I. (2013) Visions of Infinity: The Great Mathematical Problems. Basic Books, New York.

Stewart, I. (2017a) Infinity: A Very Short Introduction. Oxford University Press, Oxford.

Stewart, I. (2017b) Significant Figures: The Lives and Work of Great Mathematicians. Basic Books, New York.

Stewart, I., Huylebrouck, D., Horowitz, D., Kuo, K. H., Kullman, D. E. (1996) The mathematical tourist. The Mathematical Intelligencer 18, 56–66.

Stewart, I., Tall, D. (2015) The Foundations of Mathematics. (2nd ed.) Oxford University Press, Oxford.

Stigler, S. M. (1999) Statistics on the Table: The History of Statistical Concepts and Methods. Harvard University Press, Cambridge, MA.

Stigler, S. M. (2016) The Seven Pillars of Statistical Wisdom. Harvard University Press, Cambridge, MA.

Struik, D. J. (1987) A Concise History of Mathematics. (4th ed.) Dover Publications, New York.

Sugden, R. (1986) The Economics of Rights, Co-operation, and Welfare. Basil Blackwell, Oxford.

Szekely, I. (1986) Paradoxes in Probability Theory and Mathematical Statistics. Reidel, Dordrecht.

Szpiro, G. (2011) Pricing the Future: Finance, Physics, and the 300-Year Journey to the Black-Scholes Equation. Basic Books, New York.

Thaler, R. H. (1992) The Winner's Curse: Paradoxes and Anomalies of Economic Life. Princeton University Press, Princeton, NJ.

Thaler, R. H. (2015) Misbehaving: The Making of Behavioral Economics. W. W. Norton, New York.

Thurston, W. P. (1994) On proof and progress in mathematics. Bulletin of the American Mathematical Society 30, 161–177.

Tiles, M. (2004) The Philosophy of Set Theory: An Historical Introduction to Cantor's Paradise. Dover Publications, New York.

Trivers, R. (1985) Social Evolution. Benjamin Cummings, Menlo Park, CA.

Turing, A. M. (1937) On computable numbers, with an application to the Entscheidungsproblem. Proceedings of the London Mathematical Society 2, 230–265.

Turing, A. M. (1950) Computing machinery and intelligence. Mind 59, 433–460.

Ushenko, A. (1946) Zeno's paradoxes. Mind 55, 151–165.

van Atten, M., Kennedy, J. (2003) On the philosophical development of Kurt Gödel. Bulletin of Symbolic Logic 9, 425–476.

van der Waerden, B. L. (1954) Science Awakening. Noordhoff, Dordrecht.

van der Waerden, B. L. (1960) Die Pythagoreer: Religiöse Bruderschaft und Schule der Wissenschaft. Benziger, Zürich.

van Heijenoort, J. (1967) From Frege to Gödel: A Source Book in Mathematical Logic, 1879–1931. Harvard University Press, Cambridge, MA. (Reprinted in 2002.)

Ville, J.-A. (1936) Sur la notion de collectif. Comptes rendus de l'Académie des Sciences 203, 26–27.

Volchan, S. (2002) What is a random sequence? American Mathematical Monthly 109, 46–63.

von Mises, R. (1951) Positivism: A Study in Human Understanding. Harvard University Press, Cambridge, MA.

von Mises, R. (1981) Probability, Statistics and Truth. Dover Publications, New York.

von Neumann, J. (1928) Zur Theorie der Gesellschaftsspiele. Mathematische Annalen 100, 295–320.

von Neumann, J., Morgenstern, O. (1944) Theory of Games and Economic Behavior. Princeton University Press, Princeton, NJ. (Reprinted in 2004.)

von Plato, J. (1998) Creating Modern Probability: Its Mathematics, Physics and Philosophy in Historical Perspective. Cambridge University Press, Cambridge, UK.

von Plato, J. (2018) In search of the sources of incompleteness. Proceedings of the International Congress of Mathematicians 3, 4043–4060.

Waismann, F. (2003) Introduction to Mathematical Thinking: The Formation of Concepts in Modern Mathematics. Dover Publications, New York. (Originally published in 1936.)

Wald, A. (1936) Sur la notion des collectifs dans le calcul des probabilités. Comptes rendus de l'Académie des Sciences 202, 180–183.

Wang, H. (1974) From Mathematics to Philosophy. Routledge, London.

Weber, H. (1893) Leopold Kronecker. Jahresbericht der Deutschen Mathematiker-Vereinigung 2, 5–31.

Weibull, J. (1995) Evolutionary Game Theory. MIT Press, Cambridge, MA.

Weizenbaum, J. (1966) ELIZA—A computer program for the study of natural language communication between man and machine. Communications of the ACM 9, 36–45.

Weyl, H. (2009) Mind and Nature: Selected Writings on Philosophy, Mathematics, and Physics. Princeton University Press, Princeton, NJ.

Whitfield, J. (2012) People Will Talk: The Surprising Science of Reputation. John Wiley & Sons, Hoboken, NJ.

Wiedijk, F. (2008) Formal proof—Getting started. Notices of the AMS 55, 1408–1414.

Wigner, E. P. (1960) The unreasonable effectiveness of mathematics in the natural sciences. Communications on Pure and Applied Mathematics 13, 1–14.

Wikipedia (2022) Two envelopes problem. Wikipedia, https://en.wikipedia.org/wiki/Two_envelopes_problem. Accessed December 2022.

Wikipedia (2023a) Arecibo message. Wikipedia, https://en.wikipedia.org/wiki/Arecibo_message. Accessed January 2023.

Wikipedia (2023b) Category: Probability theory paradoxes. Wikipedia, https://en.wikipedia.org/wiki/Category:Probability_theory_paradoxes. Accessed April 5.

Wikipedia (2023c) Polyhedron. Wikipedia, https://en.wikipedia.org/wiki/Polyhedron. Accessed January 7.

Wittgenstein, L. (1922) Tractatus Logico-Philosophicus. Harcourt, Brace & Company, New York.

Wittgenstein, L. (1953) Philosophical Investigations. (G. E. M. Anscombe, trans.) Macmillan, London.

Wittgenstein, L. (1956) Remarks on the Foundations of Mathematics. (G. E. M. Anscombe, trans.) Blackwell, Oxford.

Wittgenstein, L. (1964) Philosophical Remarks. (R. Rhees, ed.; R. Hargreaves, R. White, trans.) Blackwell, Oxford.

Wokler, R. (2001) Rousseau: A Very Short Introduction. Oxford University Press, Oxford.

Wolfram, S. (2002) A New Kind of Science. Wolfram Media, Champaign, IL.

Wynne, J. (2021) Do Not Erase: Mathematicians and Their Chalkboards. Princeton University Press, Princeton, NJ.

Yamagishi, T. (1986) The provision of a sanctioning system as a public good. Journal of Personality and Social Psychology 51, 110–116.

Yamagishi, T. (2010) Trust: The Evolutionary Game of Mind and Society. Springer, Tokyo.

Zach, R. (2019) Hilbert's Program. Stanford Encyclopedia of Philosophy, https://plato.stanford.edu/entries/hilbert-program/.

Image Credits

1.1 The bust of Socrates is a marble copy of a bronze by Lysippus, and stands in the Louvre. Photo by Eric Gaba, Wikimedia Commons, https://commons.wikimedia.org/wiki/File:Socrates_Louvre.jpg.

1.2 The page of *Meno* dates from the tenth century. Scholia, Suppl. Grace. 7, fol. 418r, Austrian National Library, http://data.onb.ac.at/rec/AC14451800.

1.19 The photo of Hilbert was taken in 1912 for postcards of faculty members at the University of Göttingen, which were sold to students at the time; see Reid (1970). It is in the public domain: Wikimedia Commons, https://commons.wikimedia.org/wiki/File:Hilbert.jpg.

1.28 The five Platonic solids (tetrahedron, hexahedron, octahedron, dodecahedron, and icosahedron) as depicted by Johannes Kepler in his *Mysterium Cosmographicum*.

1.29 The engraving of Pascal is a work by Gérard Edelinck and dates from 1691. Wikimedia Commons, https://commons.wikimedia.org/wiki/File:FR-631136102_GRA_6025_Portrait_de_Blaise_Pascal_par_Edelinck.jpg.

1.31 The engraving of Descartes is by Balthasar Moncornet (1600–1668). Wikimedia Commons, https://commons.wikimedia.org/wiki/File:Descartes-moncornet.jpg.

2.1 The picture of Kant is by Adolf Neumann (1871), a drawing after a painting by Johann Gottlieb Becker, and appeared in the magazine *Die Gartenlaube*. It is in the public domain: Wikimedia Commons, https://commons.wikimedia.org/w/index.php?curid=15148126.

2.7 The picture of Gauss is by Siegfried Detlev Bendixen, and was published in *Astronomische Nachrichten* in 1828. It is in the public domain: Wikimedia Commons, https://commons.wikimedia.org/w/index.php?curid=2404149.

3.1 The picture of Clever Hans was taken in 1904, and published in Karl Krall's *Denkende Tiere* in 1921. Wikimedia Commons, https://commons.wikimedia.org/wiki/File:CleverHans.jpg.

3.2 The picture of the Ishango bone is reprinted by permission of the Royal Belgium Institute of Natural Sciences, ©RBINS.

3.7 The photo of Poincaré was taken in 1887 by Eugène Pirou. Wikimedia Commons, https://commons.wikimedia.org/wiki/File:Young_Poincare.jpg.

3.8 The photo of Peano is from 1910; the photographer is unknown. Wikimedia Commons, https://commons.wikimedia.org/wiki/File:Giuseppe_Peano.jpg.

3.15 The wood engraving of Galileo Galilei is by L. Dumont, after H. Rousseau. Public domain, Wellcome Collection 3375i, https://wellcomecollection.org/works/e3yjf7et/items.

3.16 The photo of Cantor is from 1870; the photographer is unknown. Wikimedia Commons, https://commons.wikimedia.org/wiki/File:Georg_Cantor3.jpg.

4.1 The photo of young Gödel is in the Kurt Gödel papers, Manuscript Division, Department of Rare Books and Special Collections, Princeton University Library. Reproduced with kind permission of the Institute for Advanced Study.

4.2 The marble bust of Aristotle is after an original by Lysippus, and is in the Museo Nationale Palazzo Altemps. The photo is in the public domain: Wikimedia Commons, https://commons.wikimedia.org/wiki/File:Aristotle_Altemps_Inv8575.jpg.

4.3 The photo of Frege dates from 1880; the photographer is unknown. It is in the public domain: Wikimedia Commons, https://commons.wikimedia.org/wiki/File:Young_frege.jpg.

4.4 The photo of Russell dates from 1916, is by an unknown photographer, and was published in his book *Justice in War-Time*. Wikimedia Commons, https://commons.wikimedia.org/wiki/File:Honourable_Bertrand_Russell.jpg.

4.7 The photo of Wittgenstein's military passport is from the Wittgenstein Archive at Cambridge University, with kind permission by Michael Nedo.

5.1 The picture of the Mechanical Turk is a print from 1789 by Joseph Racknitz, in *Über den Schachspieler des Herrn von Kempelen: Nebst einer Abbildung und Beschreibung seiner Sprachmaschine* by Carl Friedrich Hindenburg.

5.2 The image is by Bea Laufersweiler.

5.3 The business card–sized universal Turing machine is from Alvy Ray Smith, from alvyray.com (with kind permission of the author). For more, see Smith (2014a).

5.4 The photo of the Manchester Baby is used by permission from the Science Museum Group.

5.5 The engraving of Babbage is by R. Roffe (1833). Public domain, Wellcome Collection 657i, https://wellcomecollection.org/works/krsuk329/images?id=wrd93e2n.

5.6 The drawing of Lovelace is by an unknown artist: Wikimedia Commons, https://commons.wikimedia.org/wiki/File:Ada_Byron_aged_seventeen_(1832).jpg.

5.7 The photo of the trial piece of the analytical engine is by Sergei Magel/HNF (Heinz Nixdorf Museum).

5.12 The picture of the "sautoir" is by an unknown author, in *Oeuvres Complètes de Blaise Pascal*, Detunes, La Haye (1779).

6.4 The line engraving of d'Alembert is by P. Maleuvre, 1775, after A. Pujos. Public domain, Wellcome Collection 500i, https://wellcomecollection.org/works/waav8nqd/items.

6.5 The portrait of Weierstrass is by Peter Weierstrass (1820–1904) and in the public domain. The image is from the Karl Weierstrass Institute Berlin, https://www.wias-berlin.de/about/weierstrass/cv.jsp.

6.6 The picture of Oresme is from *Traité de la sphère* (1400–1420). Wikimedia Commons, https://commons.wikimedia.org/wiki/File:Oresme.jpg.

6.7 The drawing of a flying buttress from the Votivkirche in Vienna is by Moritz Thausing (1879). Wikimedia Commons, https://commons.wikimedia.org/wiki/File:Die_Votivkirche_in_Wien;_Denkschrift_des_Baucomit%27es_ver%C3%B6ffentlicht_zur_Feier_der_Einweihung_am_24._April_1879_(1879)_(14597612677).jpg.

6.10 The picture of Newton is a stipple engraving by S. Freeman after Sir G. Kneller (1702). Public domain, Wellcome Collection 7390i, https://wellcomecollection.org/works/db8mn7gz/items.

6.11 The line engraving of Leibniz is by an unknown artist. Public domain, Wellcome Collection 5505i, https://wellcomecollection.org/works/wd93ncu6/items.

6.12 The line engraving of George Berkeley is by W. Holl. Public domain, Wellcome Collection 1028i, https://wellcomecollection.org/works/ama2rsk7/items.

6.13 The photo of Abraham Robinson is by an unknown photographer, used with kind consent of the Yale University Office of Public Affairs and Communication.

7.1 The engraving of Voltaire (by an unknown artist) is from Voltaire's *Élémens de la philosophie de Neuton* (1738). Wikimedia Commons, https://commons.wikimedia.org/wiki/File:Voltaire_-_%C3%89l%C3%A9mens_de_la_philosophie_de_Neuton.png.

7.2 The engraving of Condamine is by P. P. Choffard (1730–1809), and is in the public domain. Wikimedia Commons, https://commons.wikimedia.org/wiki/File:Charles_Marie_de_La_Condamine.jpg.

7.7 The portrait of Jacob Bernoulli is by Nicolas Bernoulli (who died in 1716). It is displayed in the old aula of the University of Basel. The signature is UBH Portr BS Bernoulli J 1654, 6; courtesy of the University of Basel, received 29.11.2022.

7.8 The photo of Kolmogorov is by Konrad Jacobs. It belongs to the Bildarchiv des Mathematischen Forschungsinstituts Oberwolfach, copyright MFO, ID 7495, and can be used on the terms of the Creative Commons license: Wikimedia Commons, https://commons.wikimedia.org/wiki/File:Andrej_Nikolajewitsch_Kolmogorov.jpg.

7.9, 7.10 The graphs are by Hannelore De Silva (University of Economics and Business Administration, Vienna).

8.1 The photo of von Mises dates from 1930; see his *Collected Papers*. The photographer is unknown.

8.2 The passport photo of Abraham Wald is pre-1938, photographer unknown. Permission by Robert Wald.

8.3 The photo of Jean André Ville is from ca. 1930, by an unknown photographer. Permission by Glenn Shafer.

8.4 The photo of Ramsey is from 1925, by an unknown photographer. Permission by Stephen Burch.

8.5 The engraving of Laplace is by J. Belliard, ca. 1790.

9.1 The statue of Condorcet stands in the Cour Napoléon of the Louvre. The photo is by Jastrow, and it is in the public domain: Wikimedia Commons, https://commons.wikimedia.org/wiki/File:Condorcet_cour_Napoleon_Louvre.jpg.

9.2 The statue of de Borda is from the Galerie des Grands Hommes (identifiant 1561) and was taken before 1900 from a plaster model of a statue of Borda by sculptor Paul Aubé. Musée d'Orsay, https://anosgrandshommes.musee-orsay.fr/index.php/Detail/objects/1523.

10.1 The engraving of Jeremy Bentham is by J. Posselwhite, after J. Watts. Public domain, Wellcome Collection 106i, https://wellcomecollection.org/works/m8vudzxu/items.

10.3 The photo of Maurice Allais is from 1950, with the generous permission of Fondation Maurice Allais, photographer unknown.

11.1 The engraving of Adam Smith is after a painting by James Tassie. It is in the public domain: Wikimedia Commons, https://commons.wikimedia.org/wiki/File:AdamSmith.jpg.

11.2 The photo of Morgenstern is from the Archives of the University of Vienna, photographer unknown (ca. 1930).

11.3 The photo of von Neumann is from ca. 1940. Copyright by Los Alamos National Laboratory. Wikimedia Commons, https://commons.wikimedia.org/wiki/File:JohnvonNeumann-LosAlamos.jpg.

11.4 The graphic is by Bea Laufersweiler.

11.5 The photo of Nash is by kind permission of the Princeton University Library, Special Collections.

11.7 The engraving of David Hume is by W. and D. Lizars, Scottish National Gallery, public domain, https://www.nationalgalleries.org/art-and-artists/32038/david-hume-1711-1776-historian-and-philosopher.

12.1 The engraving of Hobbes is by W. Faithhorne (1668) and is in the public domain. Wellcome Trust Library 4261i, CC 4.0., Wikimedia Commons, https://commons.wikimedia.org/wiki/File:Thomas_Hobbes._Line_engraving_by_W._Faithorne,_1668._Wellcome_V0002798.jpg.

12.2 The engraving of Rousseau is from the collection of the Rousseau Museum Motiers, with kind permission of the rights owner J. B. Michel and the photographer Luc Bosson.

12.3 The picture is *The Stag-Hunt* by Lucas Cranach the Elder, 1506, and belongs to the Metropolitan Museum of Art, https://www.metmuseum.org/art/collection/search/383708. It is in the public domain.

12.5 The portrait of John Locke is from the Library of Congress, with permission of the Print and Photograph Division. The reproduction number is LC-USZ62-59655. The LOC indicates that it is a lithograph by de Fonroug after H. Garnier, date unknown.

12.7 The diagram was produced by Hannelore De Silva.

12.8 The photo of Elinor Ostrom is by kind permission of the Indiana University Archive, photographer unknown.

13.5 The photo of John Maynard Smith is courtesy of the University of Sussex Special Collections at the Keep, photographer unknown.

14.1 This photo is part of a photo of Wittgenstein and von Wright that was taken by Knut Erik Tranoy (1918–2012). It is on the homepage of the von Wright and Wittgenstein Archives, University of Helsinki, and reproduced by kind permission of Bent Sofus Tranoy.

14.2 The photo of Turing dates from ca. 1940, photographer unknown, with kind consent of the archive of King's College, Cambridge, AMT/K/7/38.

14.3 The photo of the whiteboard is by Karl Sigmund.

14.7 The example of HOL Light is from Wiedijk (2008), reprinted with the author's kind permission and that of the American Mathematical Society.

14.9 The picture of the International Congress of Mathematicians (Zürich 1932) was taken by Johannes Meiner, and belongs to the Library of ETH Zürich, record Portr_10680-FL, https://ba.e-pics.ethz.ch/catalog/ETHBIB.Bildarchiv/r/97286/viewmode=infoview. (Thanks to IMU and Birgit Seeliger.)

14.10 The Arecibo message in monochrome form is by Arne Norrman (norro), Wikimedia Commons, https://commons.wikimedia.org/wiki/File:Arecibo_message_bw.svg.

15.1 The photo of Carnap is by photographer F. Schmidt, from 1930, with kind permission of the Archives of the University of Vienna.

15.3 The photo of Conway is from Princeton University Archives, Department of Special Collections, Princeton University Library (photographer unknown).

16.2 The photo of Pick was taken in Prague around 1885, by an unknown photographer, and appears in the Weierstrass Photo Album from that year. It is in the public domain: Wikimedia Commons, https://commons.wikimedia.org/w/index.php?curid=15195006.

16.13 The photo of Lakatos dates from the 1960s, and is from the Library of the London School of Economics and Political Science. It is in the public domain: Wikimedia Commons, https://commons.wikimedia.org/w/index.php?curid=15336126.

16.15 The graphic is by Bea Laufersweiler.

16.16 The engraving of Euler is by an unknown artist and was published by William Darton, London. Public domain, Wellcome Collection 2775i, https://wellcomecollection.org/works/rkndgh8y.

Index

Credit: Bea Laufersweiler

Karl Sigmund is a professor emeritus of mathematics at the University of Vienna. One of the pioneers of evolutionary game theory, he has authored several other books on mathematics and philosophy, including *Games of Life* and *Exact Thinking in Demented Times*. He lives in Vienna, Austria.